# SEPARATING THE REAL FROM THE IMAGINED:

## Flight Research at the NACA and NASA, 1915-1998

Michael H. Gorn

National Aeronautics and Space Administration
March 2000

# TABLE OF CONTENTS

# INTRODUCTION

*Separating the Real From the Imagined* relates the history of flight research practiced from 1915 to 1998 by the National Advisory Committee for Aeronautics (NACA) and its successor, the National Aeronautics and Space Administration (NASA). While it covers many subjects, it is not a comprehensive, exhaustive, or encyclopedic treatment.. Rather, it represents a *selective overview* in which projects illustrative of an era, of pivotal technologies, or of advances in the art of flight research itself receive most of the coverage. Its overall intent is to emphasize some of the major themes, events, and accomplishments in this sometimes misunderstood field of aeronautics, to provide historical perspective about the development of the discipline, and to demonstrate the ways in which it contributed not just to the design and improvement of aircraft, but to that of spacecraft as well.

Perhaps the best way to begin is with an attempt at a definition. For the uninitiated, and even for those with some experience, the meaning of the term flight research is elusive. The title of this volume suggests at once both an attempt to define, as well as a glimpse into the contents of the narrative. "[T]o separate the real from the imagined" is a shorthand expression heard frequently at the Dryden Flight Research Center (DFRC) to describe its essential mission. Located at Edwards Air Force Base in the high desert of Southern California, Dryden is one of four NASA centers assigned flight research responsibilities (Langley Research Center, Glenn Research Center, and Ames Research Center are the others), but the only one devoted almost exclusively to this specialty. This short phrase so commonly used at Dryden originated with the center's namesake, the late Dr. Hugh L. Dryden, a leading American aerodynamicist, the last Director of the NACA, and the first Deputy Administrator of NASA. As Chairman of the

1

Research Airplane Committee, Dryden convened a meeting at the Langley Research Center in October 1956 to review the preliminary progress of the X-15 program. He presented a brief overview of the project and described its basic intentions: to "realiz[e] flights of a man-carrying aircraft at hypersonic speeds and high altitudes as soon as possible for explorations to *separate the real from the imagined problems and to make known the overlooked and the unexpected problems.*" [Author's italics]. Dryden's assumption that only a human being at the controls of the X-15, actually flying the machine, could unravel the full mysteries of hypersonics may be extended to the whole of flight research history. Only at the moment of flight do the true flying properties of any vehicle--and the research it represents--become distinct from the *anticipated* realities which presented themselves in such indispensable aeronautical tools as theoretical analysis, wind tunnel research, and computational fluid dynamics.[1]

Despite the unquestioned necessity of human hands and minds to guide research aircraft, in this volume the historic definition of flight research does not center on the role and actions of pilots. During the more than 80 years of the NACA and NASA flight research, aviators enjoyed a crucial, but not necessarily a marquee position in the pursuit of aeronautical knowledge. Rarely did the "kick the tires, light the fires" mentality prevail. Regardless of the images propagated by newspapers, by magazines, and even by celebrated books like *The Right Stuff*, caution and restraint characterized the behavior even of those flying such high performance, high profile aircraft as the X-15. The men, and later the women at the controls of the NACA and NASA aircraft--usually referred to as engineering pilots or research pilots--cultivated a sense of commitment, yet at the same time detachment toward their jobs. They possessed bachelors, and often advanced degrees in aeronautical engineering from prestigious schools. To most, flying to the edge of space held undeniable attractions, even thrills. But the research pilots recognized

[1]Hugh L. Dryden, "General Background of the X-15 Research Airplane Project," in *Research-Airplane-Committee Report on Conference on the Progress of the X-15 Project: A Compilation of the Papers Presented, Langley Aeronautical Laboratory, 25-26 October, 1956*, xvii-xix, Milt Thompson Collection, Dryden Flight Research Center Historical Reference Collection (quoted passage, xix). No publication data (editor(s), date, publisher, or place of publication) are noted on the conference proceedings.

their essential roles: to act as members of cohesive research teams; to fly and maneuver their aircraft in the precise patterns specified by their flight plans; and to do so in order to satisfy the broad objectives of the investigation. All other considerations (but safety) yielded to the mantra, "bring home the data." Consequently, at least between the covers of this book, the actual flying adventures and the specific personalities involved, while important, are not decisive. Rather, the emphasis here embraces aeronautical endeavors shared equally among the research pilots *and* their collaborators: engineers, mathematicians, computer simulation experts, instrumentation specialists, technicians, and mechanics.[2]

Yet, despite attempts at clarification, flight research remains, like Winston Churchill's famous description of Russia, "a riddle wrapped in a mystery inside an enigma," a layered phenomenon in which each definition seems to require a qualifying one. Kenneth Szalai, Dryden Director from 1991 to 1998, warned against the erroneous belief that flight research necessarily represented the concluding phase of the process of aeronautical inquiry. Actually, "research aircraft," said Szalai, "have been associated with each of the various phases of research itself, from fundamental studies to full-scale systems experiments." Bearing in mind its presence at each stage, Szalai concluded that research aircraft "serve to bring new technology to the flight environment to discover the actual performance and the actual penalties and burdens which may accompany the new technology." To put flight research into sharper relief, it should be distinguished from flight *test*, since both are used almost interchangeably. More commonly associated with the military services, flight test often concerns itself with flying prototypes or early production aircraft to determine whether they satisfy the requirements of the contract by

---

[2]See Tom Wolfe, *The Right Stuff* (New York: Farrar, Straus, Giroux, 1979), a classic about the Mercury astronauts. Wolfe's book both reflected prevailing American impressions of astronauts and test pilots, and also solidified these portraits in the public mind. His account did not lack grounding in fact; *machismo* attitudes did exist among some of the military test pilots. But, with rare exceptions, research pilots employed by the NACA and NASA (and by private industry) tended to avoid bravado. A fine historical treatment of flight testing with emphasis on the role of pilots may be found in Richard P. Hallion, *Test Pilots: Frontiersmen of Flight* (Washington, D.C. and London: Smithsonian Institution Press, 1988), as well as in other books and articles by the same author.

which they were designed and fabricated. Not confined merely to the latest aircraft, however, flight testing may also involve flying modified versions of workhorses long in the inventory. On the other hand, practitioners of flight research do not typically care whether aircraft behave in accordance with contractor's promised standards; rather, they attempt to understand the fundamental workings of the underlying science and engineering. Thus, the practice of flight research does not depend on any particular flying machine; the main objective is the acquisition of reliable in-flight data (including pilot experience) to illuminate a particular research problem. One long-time aerodynamicist at Dryden explained that flight researchers are oblivious to the actual aircraft in use, so long as the desired research can be performed on it. In this sense, flight research resembles wind tunnel experimentation more than flight testing. As a consequence, depending upon the purpose of the particular project, research pilots may fly experimental aircraft like the X-1, D-558, X-15, and the like; or they may be asked to perform maneuvers in military hand-me-downs, usually early production models, including such well-known Air Force vehicles as the F-100, the F-15, and the F-16, and the Navy F-8 and F-18; or they may employ time-honored commercial airliners like the Convair CV-990 or the Boeing 747.[3]

Nonetheless, in drawing a distinction with flight research, flight testing as undertaken by the U.S. Air Force and the other armed services should not be relegated to second-class status. Indeed, James Young shows amply in *Meeting the Challenge of Supersonic Flight* that the American military played a part every bit the equal of the NACA in the pursuit of travel at speeds faster than sound. He illuminates the influence of one extraordinarily able and determined civilian engineer named Ezra Kotcher who tried to persuade his superiors at Wright Field to launch a full-scale transonic research program *as early as 1939*. Kotcher failed but persisted. Four years later, General Franklin O. Carroll, the Chief of the Wright Field

---

[3]Bergen Evans, compiler, *Dictionary of Quotations*, (1968), s.v. "Russia" (first quoted passage); Kenneth J. Szalai, NASA Technical Memorandum 85913, "Role of Research Aircraft in Technology Development" (Edwards, California.: NASA, 1984), 11 (second and third quoted passages); J.D. Hunley, "Fifty Years of Flight Research at NASA Dryden," in *Conference Proceedings of the 1998 National Aerospace Conference: The Meaning of Flight in the 20th Century* (Dayton, Ohio: Wright State University, 1998), 197.

Engineering Division, asked the famed Hungarian-American physicist and engineer Prof

Theodore von Kármán of Caltech whether supersonic flight could be achieved. Kármán saw no

insurmountable obstacles and Kotcher and some of his Engineering Division colleagues visited

Caltech during 1943 to sketch the outline of a research program and a flight vehicle. Early the

next year Kotcher and his team approached Douglas Aircraft with plans for an aircraft capable of

speeds up to 1,500 miles per hour. Moreover, during the 1920s and 1930s the Wright Field brass

not only underwrote much of the NACA's best research, but as a consequence of flying the top

high performance aircraft of the day, the service often guided George Lewis and his staff toward

the essential aeronautical problems of the time. For example, the seminal pressure distribution

studies undertaken by the NACA during the 1920s owed their origins to a series of daring

maneuvers performed by none other than Jimmy Doolittle. Doolittle undertook the assignment

at a time when the increasing speed and power of service aircraft resulted in cases of acute

structural failure. Consequently, a few months after Doolittle's experimental flights, an official

letter arrived at the NACA from the Chief of the Army Air Service requesting an urgent research

program to investigate the threat to military pilots and aircraft posed by the unseen forces of air

pressure.[4]

---

[4]In this book, the United States Air Force and its antecedents are referred to by four designations, depending upon the chronological point in the narrative: from 28 August 1918 to 1 July 1926, the Army Air Service (AAS); from 2 July 1926 to 29 June 1941, the Army Air Corps (AAC); from 30 June 1941 to 17 September 1947, the Army Air Forces (AAF); and from 18 September 1947 to the present, the U.S. Air Force (USAF). Before August 1918, the Army's air operations had three incarnations: as the Aeronautical Division under the Army Chief Signal Officer (1 August 1907 to 17 July 1914); as the Aviation Section of the Army Signal Corps (18 July 1914 to 14 May 1918); and as the Army Division of Military Aeronautics (15 May 1918 to 27 August 1918). See Flint O. DuPre, compiler, *U.S. Air Force Biographical Dictionary* (New York: Franklin Watts, 1965), 273.

For a history of the pressure distribution program, see Chapter 2; for the supersonic story, see Chapters 4 and 5. For the Army Air Forces and USAF influence on supersonic flight, see James O. Young, *Meeting the Challenge of Supersonic Flight* (Edwards, California: Air Force Flight Test Center History Office, 1997), pp. 3-4. For additional reading about the contributions of the Army Air Service to flight testing and research see James O. Young, "Riding England's Coattails: The U.S. Army Air Forces and the Turbojet Revolution," in *Technology and the Air Force: A Retrospective Assessment*, eds. Jacob Neufeld, George Watson, and David Chenoweth (Washington, D.C.: Air Force History and Museums Program, 1997); and Anon., *Ad*

Notwithstanding military aviation's undeniable influence over the historic flight research agenda, this book still recounts an essentially civilian story. Beginning with the conditions leading to the founding of the NACA in 1915 and ending the narrative in 1998, it spans virtually the entire twentieth century. It also reviews the nineteenth century antecedents in chapter one. Yet, even though this volume adopts a broad chronology, it is not *comprehensive* in scope. The more encyclopedic approach has been adopted by two previous works. Richard P. Hallion's seminal book entitled *On the Frontier: Flight Research at Dryden 1946-1981* describes almost every major flight research program undertaken at Dryden during the years under consideration. An expanded second edition will carry the narrative into the late 1990s. Far more pictorial than Hallion's volume but still a highly worthwhile summary, Lane E. Wallace's *Flights of Discovery: 50 Years at the NASA Dryden Flight Research Center* also gives at least some coverage to every significant Dryden project. In contrast, *Separating the Real from the Imagined* chooses select subjects, explores them in greater depth, and uses them to illustrate recurring activities and themes. As a result, many programs of consequence are not mentioned, or mentioned only in passing, on the following pages.

There are several reasons for choosing a selective, rather than an all-encompassing treatment. The first is the scope of the book. It concentrates not on the Dryden Flight Research Center alone, but on flight research throughout the NACA and NASA, thus embracing a wider canvas and many more scenes of activity than previous studies. Yet, despite the broader reach, the sponsors of this book asked that it be completed on a tight schedule and that it be compact in order to conform to modern publishing practices. Both desires weighed heavily in decisions about what to cover and what to eliminate Another reason for narrowing the field of inquiry is a desire to examine the *processes* of flight research--the evolution of tools, techniques, and organization, for example--rather than the progress of each individual project. The X-15 program, for instance, changed the face of aeronautical science and engineering, but it also

---

*Inexplorata: The Evolution of Flight Testing At Edwards Air Force Base* (Edwards, California: Air Force Flight Test Center History Office, 1996).

altered radically the way in which flight research operated, exposing it to the bureaucratic imperatives demanded by large, complex organizations. This manuscript also attempts to bring the work of flight research engineers, mathematicians, technicians, and mechanics into clearer focus than past histories. Traditionally, the research pilots, who often flew hazardous missions to acquire data, won the most attention, so it seemed appropriate to offer a corrective in which those on the ground who designed, instrumented, simulated, and interpreted the experiments received due notice. Some projects illustrated these less heralded contributions better than others. Digital-fly-by-wire exemplifies one of several in which the greatest changes occurred not in the sky, but in the Dryden offices and conference rooms (as well as those of the Draper Lab). In this case, research engineers changed the very ground rules of flight by adapting the software revolution to the cockpit.

Accepting the premise that not every flight research activity could be covered, the question still remains why some appear in these pages and others do not. The reasons were both practical and historiographic. Ideally, those programs most essential to the development of flight research at the NACA and NASA deserved the strongest consideration. But factors other than sheer technical or institutional importance sometimes mitigated the decisions. For example, did sufficient historical documents and eyewitness accounts exist? If so, had any archives cataloged, or at least preserved them? Moreover, had other authors mined the existing sources extensively in their own published works? If they have, is their much left to be said? In attempting to provide an *original* portrait of the past, historians try to present subjects that have not become threadbare in the re-telling. Consequently, for reasons relating to the accessibility of sources and the originality of the narrative, several possible subjects were eliminated as candidates. But this still left a welter of possibilities. To reduce the field further, only those possessing one or more specific qualities were chosen. Did they yield pivotal technical breakthroughs? Did they span a long period of time? Did they absorb relatively large amounts of money and manpower? Did they result in long-lasting administrative adaptations? Did they attract some of the leading minds in the field? Did they win the interest of the military services or the private sector? Some clear choices emerged from this process: the X-15 program, the pressure distribution investigations of the 1920s, the flying qualities research the following decade, and the high velocity research airplane programs launched in succession during the 1940s, 1950s, and 1960s. Still, many of the selections, designed to cover the most broadly influential discoveries in the NACA's and NASA's long flight research tradition, will doubtless disappoint some. So will the many omissions. But this work is intended to open the discussion about the national role of flight research, not to close it. As such, this book serves as an invitation to those unhappy with the choices to rectify the record with their own contributions to scholarship.

Finally, since this book incorporates not just the activities of Dryden, but also of its sister flight research locations, some archival questions assumed importance. Langley, and only recently Dryden, both have established historical reference collections, rendering their achievements more accessible to scholars. As a consequence, *Separating the Real From the Imagined* gives extensive coverage to Langley flight research activities until the opening of the Muroc Flight Test Unit at Edwards (the forerunner of the Dryden Flight Research Center). At this point, the story of Hampton diminishes and that of Walt Williams' contingent comes into focus, not only because the research airplane program assumed a dominant role in the annals of flight research, but also because other published accounts (such as Lane E. Wallace's *Airborne Trailblazer: Two Decades with NASA Langley's 737 Flying Laboratory*) discuss Langley flight research in the later period. Unfortunately, the contributions of Ames and Lewis proved more difficult to assess; during the period of research for this book, neither possessed discrete historical reference collections. In the absence of coherent archival holdings, the Lewis and Ames flight research accomplishments received only partial representation. Still, there is a lengthy discussion of one major flight research program from each of these centers: the Ames Tilt-Rotor story in chapter 7, and the Lewis icing flight research project, treated in chapter 8.

Despite its limitations, then, this book traces the history of flight research during the nine decades in which it has been practiced by the NACA and NASA. The story straddles not only time and technology, but institutional evolution as well. For instance, with the creation of NASA in 1958, the NACA's reliance on in-house research yielded to the space agency's preference for partnerships with powerful contractors. Yet, despite this transformation, Dryden and the other flight research centers often succeeded in remaining faithful to the NACA tradition of employing local talent and resources in the conception, design, and construction of flying prototypes. Moreover, the eight chapters suggest that more than merely separating real from imagined problems, flight research gives rise to a bounty of dividends: it systematically discovers unexpected and overlooked aeronautical phenomena; it accounts for the human capacities and frailties of pilots under the demands of high technology; it represents the highest standards in flight safety; it insists upon *understanding fundamentally*, not just correcting, the surprises resulting from flight; and it hastens technology transfer by compelling industry and government authorities to share innovations freely during their collaborative investigations.[5]

---

[5]High speed flight research ended formally at Langley in 1958, as recounted in W. Hewitt Phillips, *Journey in Aeronautical Research: A Career at NASA Langley Research Center*, Monographs in Aerospace History Number 12 (Washington, D.C.: National Aeronautics and Space Administration, 1998), 151, 168; Szalai, "Role of Research Aircraft in Technology Development," 9-10.

The full citations for the three notable books mentioned above are: Richard P. Hallion, *On the Frontier: Flight Research at Dryden, 1946-1981* (Washington, D.C.: NASA SP-4303, 1984); Lane E. Wallace, *Flights of Discovery: 50 Years at the NASA Dryden Flight Research Center* (Washington, D.C.: NASA SP-4309, 1996); Lane E. Wallace, *Airborne Trailblazer: Two Decades with NASA Langley's Flying Laboratory* Washington, D.C.: NASA SP-4216, 1994).

# ACKNOWLEDGMENTS

Despite the common perception that history is written by solitary figures who labor by candlelight, it is actually a highly cooperative venture. During the production of this book, not just fellow historians, but librarians and archivists, photo curators, museum professionals, and editors offered counsel, suggestions, and corrections. Moreover, the engineers and scientists involved in the many projects presented on these pages read sections of the manuscript and provided much needed criticism. In all, dozens of persons contributed. Some found errors of fact, others took exception to interpretation. Some presented new sources, and more than a few recommended entirely new lines of research. To describe the manuscript as merely *improved* by their efforts would be an understatement.

The most far-reaching reviews and assistance occurred among the staff of the Dryden Flight Research Center, Edwards California. Above all, center historian Dr. J. D. Hunley-- a fine scholar in his own right--scoured three drafts of the book with meticulous attention to detail and supported the project in other ways too numerous to relate. Lee Duke, a widely experienced DFRC administrator and engineer, served as the Dryden caretaker of the project, a task he assumed with good cheer, enthusiasm, and desire for historical accuracy. In addition, Dr. Kenneth Iliff, the DFRC Chief Scientist, and Edwin Saltzman, a senior aerodynamicist, generously contributed their immense flight research experience and their extraordinary technical knowledge. A debt is owed to the following individuals who made time, in the face of many responsibilities, to review the manuscript: Ted Ayers, Roy Bryant, William Dana, Dwain Deets, Hubert Drake, Lee Duke, J.D. Hunley, Kenneth Iliff, Gary Krier, Wilt Loch, Betty Love, Tom McMurtry, John McTigue, Peter Merlin, R. Dale Reed, Edwin Saltzman, Mary Shafer, Gary Trippensee, Gerald Truszynski, and Ronald "Joe" Wilson. Peter Merlin, Curt Asher, and Betty

Love were most helpful in navigating the center's historical reference collection, while Tracy Edmondson, the Dryden Records Manager, provided similar assistance for on-site federal records. Center librarian Dennis Ragsdale and his assistant Lisa Carbaugh offered invaluable help finding published materials. Finally, thanks are due to the persons who agreed to be interviewed for this book, most of whom are or were employed at DFRC. Their participation is reflected in the footnotes.

At the Ames Research Center in Mountain View, California, a number of people shaped the narrative. In particular, the Ames historian Dr. Glenn Bugos and archivist Helen Rutt cooperated in finding many primary documents relating to the XV-15 Tilt-Rotor aircraft, even as they struggled to establish a history office and a historical reference collection. Roger Ashbaugh presented a generous and thoughtful initiation into the workings of Ames while Dr. Jack Franklin and Seth Anderson offered a synthesis of the center's past and present flight research program. Ames Chief Scientist Dr. John Howe and long-time Ames administrator Jack Boyd made general insights into the historic objectives of the center. A number of the center's information services staff were helpful to this study. Librarians Dan Pappas and Mary Walsh were flexible and accommodating and Carla Snow and Catherine Garcia gave a useful introduction to the center's records management system.

At the Glenn Research Center in Cleveland, Ohio, Kevin Coleman provided a general introduction to the facility and much needed logistical guidance and advice. Michael Ciancone represented the center's engineering perspective, kindly suggesting a number of worthwhile flight research subjects and corresponding points of contact. Among the flight research staff, engineer Porter Perkins and research pilot William Rieke freely contributed their time, their sources, and their illustrative material to the discussion of icing research. Augmenting their assistance, Margaret Shannon of Washington Writers Research found many documents from the National Archives relating to the early aircraft icing investigations at the NACA. Finally, Mary Ellen Carson of the Glenn graphics office opened the center's large photograph collection, a modern marvel of computer storage, retrieval, and reproduction.

The historical reference collection at the Langley Research Center in Hampton, Virginia, is an important documentary source for early American aeronautics. Organized and maintained by Richard Layman, now retired from the center, its holdings--and Layman's expert knowledge of the materials--were a great asset in preparing the first three chapters of the book. Engineer William Hewitt Phillips, associated with Langley flight research for almost 60 years, offered crucial insights, as well as an early version of his memoir (since published by NASA as *Journey in Aeronautical Research*, 1998). Also, librarian Garland Gouger made available the center's published holdings and photo archivist Alicia Tarrant diligently searched the extensive Langley collection for appropriate images. Keith Henry authorized the reproduction of these vintage images.

In addition to the NASA centers, NASA Headquarters made significant contributions to this book. The tireless NASA Chief Historian Roger Launius not only originated the concept of a volume on NACA/NASA flight research, but persuaded Headquarters Code R (Aeronautics) to be its sponsor. Dr. Launius also opened his large historical reference collection to the project, read the manuscript, and suggested many worthwhile changes. With his able staff, he also shepherded the final manuscript through the production labyrinth. Among those who maintain the Headquarters historical reference collection, Mark Kahn worked hard to retrieve many valuable documents and many excellent photographs.

Outside the bounds of NASA, the Air Force Flight Test Center, which co-habits Edwards Air Force Base, California, with the Dryden Flight Research Center, holds much historical material relevant to NACA and NASA supersonic flight research. The Flight Test Center's Chief Historian Dr. James Young, and his assistant Cheryl Gumm, cordially made available hundreds of documents relative to projects pursued jointly by the USAF and the NACA/NASA, as well as many fine photographs.

Finally, two academic historians reviewed the manuscript and provided wise counsel. Prof. James Hansen of Auburn University and Dr. Deborah Douglas of the Massachusetts Institute of Technology (formerly associated with the Langley Research Center) evaluated the text with great care and suggested some essential interpretive and narrative revisions.

The final acknowledgment should be paid to Annette Gorn who once again saw at close quarters the writing of a history book yet tolerated it with good humor.


Michael H. Gorn

Camarillo, California

March 2000

**Win a Free Lunch**

Primi Piatti
Italian Cuisine
202/223-3600
2013 I St. NW
Washington, DC 20006

Mon -Fri 11:30am- 2:30pm
Mon -Thu 5:30pm-10:30pm
Fri -Sat 5:30pm-10:30pm
Tel: 202/223-3600
Fax: 202/296-3725
Email: primipiatti@aol.com
American Express, Diners Club, Mastercard, Visa

"...the best Italy can offer...European flair and grace"
**Phyllis Richman, Washington Post Magazine**

"...the choice...among the pretty, the powerful...the best in authentic Italian cooking"
**Robert Shoffner, The Washingtonian Magazine**

Primi Piatt brings authentic Italian cuisine and ambiance to downtown Washington and T Corner, Virginia.

# CHAPTER 1

## Early Flight Research

### THE FIRST CENTURY

Among the technical achievements unique to the twentieth century, human flight holds a privileged place. Before 1900 no person had ever flown successfully in a powered, heavier-than-air machine. Toward the end of 1903 an American broke the thrall of gravity when he shook his brother's hand, tripped the release of their slender biplane, and flew for 12 seconds over 120 feet before his craft shuddered to a halt in the sand.[1] Less than one hundred years later, engineers and scientists conceived the X-33 lifting body, designed to be launched vertically, to race through the atmosphere at hypersonic speeds, and to re-enter the atmosphere with a glide return and a horizontal landing. Few endeavors of any kind began the century unproved and ended with such confidence.

Yet, unparalleled as the story of modern flight may be, developments during the long period preceding it have equal importance. A catalog of daring and inventive engineers, technicians, and pilots labored throughout the nineteenth century to sustain themselves aloft. Like their successors today, these early researchers usually started with a theoretical insight. To verify their speculations they designed and constructed earth-bound equipment and subjected their hypotheses to hours of repetitive testing. After extracting the data they re-examined their

---

[1]Among many histories of the Wrights' achievement, see Peter L. Jakab, *Visions of a Flying Machine: The Wright Brothers and the Process of Invention* (Washington, D.C. and London: Smithsonian Institution Press, 1990), 209-210.

initial suppositions in an effort to obtain a convergence between empirical and abstract knowledge.

Once satisfied, nineteenth century aeronauts took to the air with small gliders and full-scale flying machines to determine whether their vehicles behaved in ways hoped for and predicted. But the initiation of the airborne stage of research did not signify an end to the earlier phases. On the contrary, carefully designed flight programs epitomized the experimental process itself, augmenting the mathematical predictions, the trials on the ground, and providing entirely new evidence to complement these other forms of inquiry. Furthermore, the understanding of aeronautical behavior gleaned from systematic flying appeared in scientific journals the world over, becoming the indispensable body of literature without which routine human flight would have been delayed, or even denied.[2]

The concept of turning the open air into a flight laboratory began early in the 1800s. In fact, the father of aerial navigation actually began his explorations in the eighteenth century, during the height of the French Revolution. Sir George Cayley (1773-1857), an unassuming English baronet born to the Yorkshire gentry, not only discovered the fundamental processes of horizontal flight, but established the methodological framework for their investigation.

A self-taught polymath, Cayley packed many careers into his 84 years. He served in Parliament, studied artificial human limbs, delved into land drainage and reclamation, designed caterpillar tractors, advocated education, and participated in the founding of the British Association for the Advancement of Science. Perhaps inspired by the remarkable balloon flights of the Montgolfier brothers over Versailles in 1783 and by a mother who valued open-mindedness, at an early age he undertook studies of the physical make-up of birds, paying careful attention to the shape of their wings, their weight, and their speed in flight. In 1799 the 26 year old arrived at the theoretical groundwork which not only guided research throughout the

---

[2]Nineteenth century glider and airplane experimentation is best explained in two books by the same author: Tom D. Crouch, *The Bishop's Boys: A Life of Wilbur and Orville Wright* (New York and London: W.W. Norton, 1989); and Tom D. Crouch, *A Dream of Wings: Americans and the Airplane, 1875-1905* (Washington and London: Smithsonian Institution Press, 1989).

nineteenth, but well into the twentieth century as well. Cayley postulated four forces acting on vehicles in flight: lift, gravity, thrust, and drag. Moreover, for the purposes of aeronautical investigation he successfully proposed that researchers concentrate *either* on thrust or drag, treating them as entirely separate problems requiring independent lines of investigation. His imagination produced sketches of aerial machines not unlike the shapes familiar to this century, distinguished by long fuselages, large wings in the front, and small tail surfaces at the rear. Cayley subjected his speculations to extensive ground tests. Borrowing the whirling arm device commonly used to measure air pressure on windmill blades, he fitted a square foot wing at one end, counterbalanced it with weights at the other, and calculated the amount of weight lifted by the wing at varied velocities and pitches.

This extraordinary auto-didact then applied his results to a rigorous program of flight research which, in 1804, yielded the world's first successful model glider, a craft five feet long with wings and a tail plane made of kites. After five years of testing its qualities he successfully launched from the Yorkshire hills an unpiloted, full-sized behemoth borne aloft by 200 square feet of wing area. Finally, having theorized, tested, and flown his ideas he published his findings in a seminal three-part essay appearing in *Nicholson's Journal of Natural Philosophy, Chemistry and the Arts* in 1809 and 1810. Entitled "On Aerial Navigation," this highly influential treatise set out the principal research agenda for the next 100 years: "The whole problem," wrote Cayley, "is confined within these limits--to make a surface support a given weight by the application of power to the resistance of air."[3]

Sir George devoted much of the rest of his long life to flight research, mostly with full-sized pilotless gliders. For a time, he tried to conquer the problem of thrust, but after much searching found no engine light enough to elevate its own weight, an airframe, fuel, and an aviator. He therefore returned to aerodynamics, investigating designs which offered the least

[3]Jakab, *Visions of a Flying Machine*, 21-23 (quoted passage, 22); Crouch, *A Dream of Wings*, 27-28; William H. Longyard, *Who's Who in Aviation History: 500 Biographies* (Shrewsbury, U.K.: Airlife, 1994), s.v. "Cayley, George"; H. Guyford Stever and James J. Haggerty, *Flight* (New York: Time, 1965), 10-11.

resistance to the flow of air, conceiving of moveable tail surfaces, and considering new wing positions to increase stability. After decades of experimentation, and well into old age, Cayley pursued his ultimate flight research projects: the design, construction, and flight test of two full-sized, piloted gliders. Actually, the first human being to be transported was not an adult, but a ten year old boy. In 1849 Cayley placed the child in a two-wheeled gondola attached to a tall superstructure of wings and a tailplane. After rolling down a hill, the machine "floated off the ground for several yards," constituting the first recorded flight of a human being. During his 80th year the indomitable experimenter undertook one last flight experiment, even more daring than the last. This time he enlisted his unwilling coachman to mount his latest glider, push off from a hill, and sail across a small valley. Although the pilot flew with moderate success, upon landing he quit Cayley's service on the spot, muttering, "I was hired to drive, not to fly." Nonetheless, based on his analysis of these and the earlier flight research experiments George Cayley evolved his greatest insight of all, the concept of the airplane itself: a vehicle sustained in flight by the three separate (but coordinated) systems of lift, propulsion, and control.[4]

While many advanced airborne testing during the decades after Cayley's death, few rivaled the contributions of the German Otto Lilienthal (1848-1896). Born in Pomerania, he and his younger brother Gustav nourished their imaginations much like the Yorkshire master: by examining the flights of birds (especially storks) and reading romantic accounts of ballooning. From 1862 to 1879 Otto and Gustav constructed many ornithopters (gliders with moving, strap-on wings) while attending technical schools. After studying mechanical engineering and working as a machine shop apprentice, the elder Lilienthal opened a factory fabricating light steam engines, boilers, and mining equipment. He enjoyed enough success to concentrate his energies on his boyhood passion and by 1878 opened a aeronautics laboratory at his home in the Berlin suburbs. He decided to abandon the more complex ornithopter and delve instead into simpler fixed-wing gliders. He also left behind the intuitive methods of his youth. Like Cayley,

---

[4]Stever and Haggerty, *Flight*, 11-12 (quoted passage, 11); Crouch, *A Dream of Wings*, 28; Jakab, *Visions of a Flying Machine*, 22-23; Longyard, *Who's Who*, "George Cayley."

he relied on the whirling arm machine as an essential instrument and with it painstakingly measured the forces of air pressure. Also like his famous predecessor, he construed the riddle of flight as a set of problems, each requiring its own answer before the final, integrated objective could be achieved. Therefore, at his home workshop he devoted years of systematic and serious study to but one leg of Cayley's triumvirate of flight: to understanding the forces of lift. After thirteen years of ground experimentation he concluded what others (including Cayley) had only surmised; that a cambered, or curved wing cross-section offered the greatest aerodynamic advantage. Further, a simple, circular arch--at its highest point 1/12th the distance from the leading to the trailing edge of the wing--seemed to be the ideal shape. Lilienthal also conducted experiments to find shifts in the center of pressure as wings moved at varying angles. Of incalculable value to other researchers, his experiments resulted in an air pressure table listing the necessary wing area for gliders based on their weight and speed. Lilienthal's research program included no theoretical studies; an engineer, he took the approach of solving each problem as it arose, rather than searching for a fundamental scientific explanation for the many observed phenomena. Nonetheless, when he published the book *Birdflight as the Basis of Aviation* in 1889, it caused an international sensation.

But Otto Lilienthal had only begun to surprise the world's small aeronautical community. Once he completed his bench research and released the results, he decided to initiate a flight research program, much like George Cayley's. Rather than employing models, the engineer decided to construct full-scale gliders and to pilot them himself, thus adding the indispensable ingredient of human experience to the mass of technical evidence. While Lilienthal was not fated to enjoy decades pursuing his flying experiments, the short period open to him proved highly eventful. From 1891 to 1896 he flew nearly *2,000* times in the Rhinow Mountains near Berlin, systematically gathering data from each launch, charting the results, and modifying his vehicles slightly or significantly as each series of flights progressed. In all, he flew 16 different designs. At first his contrivances looked like the creatures which so gripped his imagination: big monoplane wings from 10 to 20 meters square, covered with cotton and opened wide like those of a soaring bird, with stabilizing surfaces at the rear. Lilienthal hung vertically in them and never perfected any mechanical controls; twists of the body gave his craft direction. But aerodynamically nothing could match the Lilienthal machines. He eventually flew as far as 1,000 feet in twelve seconds, launching himself by facing into the wind and running down the slope of a hill until, attaining sufficient speed, he jumped off of the ground, opened the wings, and became airborne. For every eight feet of forward motion his gliders averaged one foot of vertical fall. In due course he totaled more time aloft than all previous researchers combined.

The German experimented with a number of daring variations of this simple structure. He incorporated collapsible wings into the design for easier storage; he designed, constructed and flew biplanes. But both of these innovations resulted in decreased stability in the air. He even attempted to harness machine power to his gliders. In 1893 he found a novel (carbonic-acid and gasoline) two horsepower engine which flexed the craft's wings, although he never attempted to fly it with the existing glider. But two years later he increased the wing surface of the same basic design and prepared a series of flight tests. The mechanized flier failed to work and Lilienthal returned to the unpowered program. Finally, as he piloted one of his monoplanes on Sunday, August 9, 1896, it stalled in a gust of wind, pitched nose up, and plummeted from an altitude of 50 feet. Lilienthal broke his back in the accident and died in a Berlin hospital the following day at the age of 48.[5]

Yet, Otto Lilienthal proved so convincingly the air-worthiness of his inventions that others duplicated his successes and conjured ways to surpass them. His tragic death may have added to the allure. Curiously, his greatest following appeared not in Europe, but in the United States. Here the famous civil engineer Octave Chanute (1832-1910) exercised a patriarchal influence. One of the nation's most distinguished railroad and bridge designers, he specialized in the most difficult challenges and played a decisive role in the settlement of the Midwest. In his

---

[5]Crouch, *A Dream of Wings*, 162-165; Jakab, *Visions of a Flying Machine*, 32-35; Crouch, *Bishop's Boys*, 142-145; Stever and Haggerty, *Flight*, 12-13, 15; Longyard, *Who's Who*, "Otto Lilienthal."

forties he became interested in heavier-than-air flight due to its formidable technical hurdles but contented himself with reading all of the existing literature and experiencing the growing passion of an enthusiast. Many years later, when Chanute was about to retire, a friend and editor asked him to write a series of articles on the past and present state of aeronautics for *Van Nostrand's Railroad and Engineering Journal*. He began the project during the same year Lilenthal not only initiated his glider flights, but published his research in 27 installments, collected in 1894 as a book entitled *Progress in Flying Machines*. The German's writings persuaded Chanute to risk some of his own time and capital in planning, building, and flying his own gliders. At the end of 1894 he revealed his design, which would be tested aloft during summer 1896 when Chanute and several young aeronauts conducted flying experiments above the sands at Miller, Indiana, on the southern shore of Lake Michigan. They encountered many rough moments. Time and again designs needed to be altered to wring better results from the machines and to repair crash damage. Clashes occurred between Chanute and his engineer-pilot, A.M. Herring of New York. In the end, the old engineer's tall, eight-winged multiplane achieved fairly stable but short flights. But quite unexpectedly, a much plainer vehicle proved to be the summer's great success. The synergistic result of Chanute's intimate understanding of truss supports and of Herring's past flying experiences, it offered a far simpler, rigid, straight-winged biplane configuration which ultimately yielded a stable glide of 359 feet in fourteen seconds. Although embroiled over the process of creation, these two men produced in this long, light, box-like structure a great leap over all previous efforts, constructing a vehicle much more like an airplane than any of Lilienthal's bird-like machines.[6]

## TWO OHIOANS

---

[6]Crouch, *A Dream of Wings*, 21-41, 61-77, 175-202; Longyard, *Who's Who*, "Octave Chanute."

Chaotic though the Lake Michigan interlude may have been, it resulted in a body of written literature and flight research which edified and inspired the two pivotal figures in the history of powered flight. As adults, Wilbur (1867-1912) and Orville (1871-1948) Wright recounted the gift of a wonderful toy from their imposing father Milton, Bishop of the Church of the United Brethren of Christ. In 1878 he presented them with a rubber-band-powered helicopter which they copied in different sizes. The excitement of this little machine lay dormant for many years, during which time (1892) the brothers opened a shop in their native Dayton, Ohio, for the rental and repair of bicycles. These new and inexpensive modes of travel swept the country during this period and the Wrights sold them under their own brandnames: Wright Flyer, St. Clair, and Van Cleve. Still in their twenties, these two men with pleasant manners and keen mechanical skills won a loyal clientele and a successful business. But the routine failed to satisfy their inquiring dispositions. Even before starting their company they read with fascination newspaper and magazine stories about Otto Lilienthal, his research, and his flying exploits. His death in 1896 riveted their attention on the problems of flight. They scanned every local source for books and articles on ornithology and aeronautics and in so doing convinced themselves that human flight could be attained. Once they exhausted sources in Dayton, Wilbur Wright wrote to the Smithsonian Institution in May 1899 requesting further reading and the names and addresses of the leading researchers. Chanute's *Progress in Flying Machines* proved to be the museum's most important bibliographic suggestion, in addition to an 1897 edition of the *Aeronautical Annual* which featured an essay on the Chanute-Herring braced biplane. These sources embraced all of the significant aeronautical developments to date, outlined as yet uncharted avenues of research, and disclosed which lines of inquiry seemed to end in blind alleys. Finally, alert to Chanute's encyclopedic knowledge and diverse connections, the Smithsonian correspondent suggested the Ohioans open a discussion with the old master himself.[7]

With surprising speed and assurance, the Wrights blended the observations of others with their own insights and arrived at a systematic research program.. Their success reflected a keen instinct for the best work of their predecessors and a knack for integrating such knowledge into one coherent canon. At the same time, they imposed upon themselves the discipline to modify received wisdom in a deliberate and orderly fashion, resisting the temptation to jump ahead or to skip steps. They also approached perhaps the most daunting task in the history of engineering with breathtaking simplicity and confidence. "If the bird's wings," wrote Orville, "would sustain it in the air without the use of any muscular effort, we do not see why man can not be sustained by the same means." They wasted no time showing their mettle, embarking immediately on a flight research program which, although conceived quickly, nonetheless exhibited a degree of sophistication absent in all of the other experimenters. However, in order to weave the pilot's experience into the loom of technical data, the brothers emulated Cayley, Lilienthal, and Chanute in one important respect: they decided to fly their machines themselves. Until then, however, kites and models allowed them to ascertain the handling qualities and the safety of their craft. They relied on Lilienthal's air pressure tables, assuming his eminence as an aeronaut testified to the accuracy of his experimental data. Essentially, they chose one design--the stable, elegant two surface Chanute-Herring glider--as their testbed. But they endeavored to avoid the technical and the personal chaos that gripped the Chanute camp in 1896 by adhering to just one design and making incremental, calibrated changes in it; and by submerging all disputes and engineering disagreements under the amalgamated public persona of two brothers united in a single purpose.

During the initial phase of flight research the Wrights chose *control*--the least understood of Cayley's three phenomena--as their first experimental problem. *Lift*, though certainly not fully understood, had at least been well documented by Lilienthal and Chanute. *Propulsion* never

---

[7] I am indebted to Tom Crouch's *Bishop's Boys* and *A Dream of Wings* for the portion of the narrative describing the Wright Brother's contributions to flight research. Crouch, *A Dream of Wings*, 227-229; Crouch, *Bishop's Boys*, 28, 159; See also Jakab, *Visions of a Flying Machine*, 39-45; Longyard, *Who's Who*, "Wilbur Wright, Orville Wright."

worried Wilbur and Orville, who assumed that among all the lighter and more powerful engines being produced at the turn of the century one would be found to suit their purposes. While engines and aerodynamics would vex them in many ways over the next four years, *stability and control*--the interdependent forces with which they became intimately familiar as bicycle builders--posed the highest hurdles. The brothers believed they found the answer in wing warping. Orville observed that buzzards, "regain their lateral balance, when partly overturned by a gust of wind, by a torsion of the tips of the wings. If the rear edge of the right wing tip is twisted upward and the left downward the bird becomes an animated windmill and instantly begins to turn, a line from its head to its tail being the axis."[8]

With this metaphor from nature much on their minds, the brothers built and flew their first flying machines. Because of the inherent danger in mastering the mysteries of control, they began flights with a prototype kite possessing a span of five feet. Wilbur flew it at the end of July 1899, narrowly missing some boys who ducked to avoid the swooping pine and fabric creature. Using the Chanute-Herring model, the brothers braced the edges of the parallel wings with eight vertical posts, leaving the broad surfaces between the leading and trailing edges unbraced in order to test their theories of warping and lateral control. They attached four wires to the kite, where the front, outer posts joined the upper and the lower wings. Wilbur held the ends of these wires on two sticks. When he moved them in opposite directions, twisting the wings, it caused exactly the effect they expected: one wing dipped and the kite banked, and then the same on the other side. But their experiment not only achieved both control in roll and lateral balance; Wilbur also directed changes in pitch, causing the kite to ascend and descend at will. He and Orville modified the Chanute design by attaching a flat horizontal stabilizer to the front center posts. When Wilbur moved the sticks in unison he guided the wings fore and aft in relationship to each other, which in turn directed the stabilizer up or down according to the movement of the posts to which it was attached. The air flowing off of the stabilizer's surface

---

[8]Crouch, *Bishop's Boys*, 160 (first quoted passage); Crouch, *A Dream of Wings*, 227-230 (second quoted passage, 230); Jakab, *Visions of a Flying Machine*, 45-52.

pitched the kite either up or down, as Wilbur wished.

Then followed a series of full-scale piloted flying tests. They constructed their flier in the familiar box kite configuration of the Chanute-Herring machine with about 150 square feet of wing area, a 17.5-foot span, and a five-foot chord (the distance between the leading and the trailing edges of a wing at its widest point). Unlike all of their famous predecessors, however, the Wrights, acting mostly on intuition, curved their wings not in a circular arc, but rather with the top of the arch nearer the leading than the trailing edge, resulting in more predictable upwards and downwards motion in flight. Wilbur and Orville also added a forward elevator just in front of the lower wing for safety purposes; this surface helped maintain balance fore and aft and allowed instantaneous control in case of stall and nose-dive. The brothers "flew" their kite-like prototype by tethering it to a tower and guiding it from the ground by wires. Once satisfied with this machine and familiar with its control mechanisms, they looked ahead to 1900 and to strapping themselves into the glider and testing it in free flight.

In preparation, they searched for a suitable landing strip; one open and unobstructed, one private enough to be secluded from curious onlookers, and one freshened by steady breezes. After considering San Diego, California, as well as sites in Florida and Georgia, the brothers heeded Weather Bureau advice about Kitty Hawk, North Carolina, a fishing village on the northern rim of the Outer Banks. While readying themselves for these experiments, Wilbur contacted Octave Chanute. The senior engineer welcomed the correspondence and soon realized these young men possessed a diligence and a seriousness most others lacked. Chanute offered financial assistance, but the brothers declined, desiring to be their own masters. Nevertheless, they gained greatly from his engineering experience, his encouragement, and his moral support. The first encounter with Kitty Hawk in September and October proved less than rewarding. The men and their assistants were consumed by mosquitoes and dismayed by the isolation and the primitive housing. Perhaps still timid about their powers, the Wrights flew some piloted tethered flights but continued to operate their machine like a kite. While its wing-warping qualities seemed borne out, weighing scales on the wires gave some disturbing news: their machine

11

produced less lift and less drag than expected. Confusion reigned on their return to Dayton. By May 1901 they decided to increase both the surface area and the camber of the wings to remedy the problem. The changes resulted in the largest glider ever flown, with a span of 22 feet and a seven-foot chord. They returned to North Carolina in early July, determined equally to make their camp more permanent and to fly their machine successfully. After building a 16 by 25 foot hangar (also used for housing) they began the flight tests on July 27. The first attempts revealed difficulties; despite sailing up to 315 feet in 19 seconds, they found control to be erratic and the distances, disappointingly short. Wilbur narrowly avoided crashing after a near stall. Having considered every other possibility, they began to think that the lift and drag tables they relied on might be faulty, noting that their craft delivered only one third of the lift predicted by Lilienthal's calculations. With Octave Chanute in attendance for the first time, the brothers tried again in early August, coaxing 335 feet from their flier only to see it crash land in a nose-dive to the ground. Discouraged, the Wrights went home.[9]

But Chanute dispelled their gloom. He invited Wilbur to speak to the prestigious Western Society of Engineers in Chicago, acted as his host, and took the opportunity to confer with him at length about Lilienthal's airfoils. The Wrights decided to conduct their own laboratory tests using a homemade wind tunnel. Only sixteen inches square inside and just six feet long, it attained wind speeds of 25 to 35 miles per hour. After two months of operation the little instrument proved the inaccuracy of Lilienthal's tables. Airfoil models suspended on balances suggested the optimal wing cambers for their own machines and also provided the wherewithal to revise and correct the German's published data. Moreover, they derived from the wind tunnel experiments important evidence about aspect ratio (the proportion of wingspan to wing thickness). Armed with such knowledge, they again let their business go slack for a summer, spending August in Dayton constructing a new glider and then traveling in September to Kitty Hawk where they patched up their housing and finished the machine. Bigger than the previous

[9]Crouch, *A Dream of Wings*, 232-244; Jakab, *Visions of a Flying Machine*, 58-60.

models, the 1902 glider measured 32 feet in wingspan with a five-foot chord and a camber of 1 to 25 (the highest point of arch in the wing being 1/25 the chord of the wing). The brothers also added a vertical tail plane five feet by fourteen inches. Yet, it still looked like a much enlarged version of the 1896 Chanute-Herring glider.

The experiments undertaken in summer and fall enjoyed smashing success after an initial period of puzzlement. The first flights covered up to 200 feet, allowing the pilots to learn the feel of the craft. But an accident and a related anomaly resulted in one last, crucial innovation. On September 23 Orville noticed one wing tip rising during a normal glide. He tried to correct, but the opposite wing raked the ground as the vehicle descended at least 25, perhaps 50 feet. The pilot emerged from the wreckage unharmed and research resumed after a few days of repair. The glides became longer and the pilots grew increasingly adept at maneuvering the machine, but the dangerous problem of the rising wing tip persisted. Orville argued that the new tail structure caused the difficulty; as one wing tip rose and the other dipped, the rudder's surface slowed the speed of the sinking side so much that it stalled. Wilbur arrived at a brilliant answer: connect the wing warping system to a *moveable* rudder so the airfoil and tail surfaces might be adjusted in tandem. Once installed, this mechanism gave the Wright Glider a superiority over all other machines known at the time. By October 23, 1902, flights as long as 622 feet had been recorded. Writing to Chanute just before Christmas, Wilbur expressed the confidence of the two brothers. "It is our intention next year," he declared, "to build a machine much larger and about twice as heavy as our present machine. [I]f we find it under satisfactory control in flight, we will proceed to mount a motor."[10]

Despite the calm words, two formidable obstacles remained: finding and adapting a suitable engine and devising an appropriate propeller. Once again, the Wrights almost made such conundrums seem simple. They soon found that no literature existed to guide them on the

---

[10]Crouch, *A Dream of Wings*, 246-254 (quoted passage, 254); Crouch, *Bishop's Boys*, 218-241; Stever and Haggerty, *Flight*, 23.

aeronautics of the propeller and that references to the nautical screw did not apply. Therefore, they relied again on their own wits, reasoning that the propeller actually operated like a rotary airfoil whose trailing and leading edges required analysis just like that of an aircraft wing. The speed at which the propeller turned allowed the camber to be fixed correctly for each part of this rotating wing. At the same time, their earlier hunch proved to be right; sufficiently powerful but light motors did exist for their purposes. However, they finally decided to design and build their own powerplant--not because none could be found, but simply to reduce costs. It weighed 140 pounds, delivered 16 horsepower from four cylinders, and was ready for installation in May 1903. Not only did the Ohioans overcome the vexing engine and propeller questions with relative ease, but the efficiency with which they resolved them allowed the pair to concentrate on the other ingredients essential for safe flight: achieving plenty of lift, attaining good control, and lowering air resistance.

The Wrights departed for North Carolina on September 23, 1903. They brought the 1902 glider along for practice while the new flying machine rose in the hangar at Kitty Hawk. By late October the airplane required only minor work. Nonetheless, some frustrating difficulties presented themselves. The propeller shaft required repeated attention. The weather deteriorated rapidly. The first launch on December 14 had to be aborted as the plane stalled just after leaving the specially made starting rail, sending Wilbur (who attempted to climb too rapidly) and the machine crashing to the ground. After repairs and delays in connection with the winds, at 10:35 a.m. on the 17th Orville ended the Wright Brothers four-year flight research program by realizing the objective of flying the first powered aircraft under pilot control.[11]

## A GOVERNMENT IMPERATIVE

[11]Crouch, *A Dream of Wings,* 293-305; Crouch, *Bishop's Boys,* 253-272.

During the first century of flight research, experimenters in Europe and America pursued a mythic desire to fly like birds on the wing   But once Wilbur and Orville Wright accomplished this feat, aeronautical inquiry lost much of its poetic quality.  As their epochal achievement slowly gained credence around the world--an event so unbelievable it required some five years to be universally appreciated and accepted--the inherent possibilities of flight for commerce, for transport, for travel, for sport, and for war dawned on people everywhere.  The Wrights' deed, once recognized, assumed heroic proportions in the public mind.  Scientists, engineers, experimenters, inventors, tinkerers, and even lay people rushed to their benches to pursue aspects of this incredible phenomenon which puzzled and thrilled them.  Some wanted to de-code the underlying scientific principles which explained the Wright's achievement; others wanted to engineer entirely new machines; some raised questions about new structural materials; others sought improvements in specific components like propellers, engines, and wires.  Among statesmen, the Europeans first grasped the implications of the airplane to national well-being.  In an age of intense nationalism, on a continent where states lay in close proximity, every advanced government sought to guide and to nurture this powerful but unknown technology.  Their tradition of state-encouraged, sponsored, and organized laboratories and institutes differed widely from the individualistic model present in the United States.  Thus, soon after Wilbur and Orville stunned and excited European audiences with aerial exhibitions in 1908, all of the major European powers initiated some form of a national aeronautical laboratory.

France rose first to the challenge, acting even before the Wright Brother's flying exhibitions.  The Central Establishment for Military Aeronautics at Chalais-Meudon near Paris worked cooperatively with Gustave Eiffel during the famous experiments conducted between 1902 and 1906 on the tower bearing his name.  Eiffel also directed wind tunnel facilities at Champs-de-Mars and in Auteuil and affiliated himself after 1912 with the privately funded Aerotechnical Institute of the University of Paris at St. Cyr, operated by a director who reported to an advisory committee of scientists drawn from government, universities, and private entities.

15

In Russia, non-governmental agencies combined to open the Aerodynamic Institute of Koutchino, connected to the University of Moscow. In Germany, the eminent professor of fluid mechanics Ludwig Prandtl opened with state, industrial, and private assistance the Aerodynamical Laboratory of the University of Göttingen in 1903, specializing in theoretical aerodynamics. Like the director of the French Aerotechnical Institute, Prandtl received advice from a board of prominent engineers and scientists. But the most coherent approach to aeronautical research occurred in the United Kingdom. Here Prime Minister Herbert Asquith announced in 1909 the creation, at significant public expense, of the British Royal Aircraft Factory at Farnborough, formed from the sinews of the National Physical Laboratory. To oversee it, Asquith recruited no less that John Strutt, 3rd Baron Rayleigh, 1904 winner of the Nobel Prize in Physics, who presided over an Advisory Committee for Aeronautics. Organized to coordinate the air research of all government institutions, under Raleigh's leadership it attracted eminent scientists and engineers from the universities, from learned societies, and from the civil service.[12]

Not only did American attempts to erect a parallel national aeronautics structure fail during the same period, but two respected and established regional centers actually closed their doors. Unlike the Europeans who acted quickly, the United States lost precious years in its aeronautical research program due to rivalries among federal agencies, to wavering political support, and to public indifference. Indeed, the year after the Wrights' conquest at Kitty Hawk the Smithsonian Board of Regents shuttered Samuel Langley's Aerodynamical Laboratory after he lost his contest with the two brothers. Further short-sighted behavior resulted in the closure of Professor Albert F. Zahm's wind tunnel at Catholic University (which he used to calculate airflow around dirigibles) because of insufficient funds. An initial effort to rectify the trans-

---

[12]For an overview of the sensation caused by the Wrights' success, see Joseph J. Corn, *The Winged Gospel: America's Romance with Aviation, 1900-1950* (New York and Oxford: Oxford University Press, 1983), 3-11; Alex Roland, *Model Research: The National Advisory Committee for Aeronautics, 1915-1958*, vol.1 (Washington, D.C.: NASA SP-4103, 1985), 3-4; James R. Hansen, *Engineer in Charge: A History of the Langley Aeronautical Laboratory, 1917-1958* (Washington, D.C.: NASA SP-4305, 1987), 3.

Atlantic imbalance occurred in April 1911 at the first annual banquet of the U.S. Aeronautical Society, which announced plans to campaign for a national laboratory devoted to flight. Not only President William Howard Taft, but such notables as the Secretary of the Smithsonian Institution, the Chancellor of New York University, and the Secretary of the Navy accepted invitations to attend the Society's gala and to lend their support to the call for a federal research institution. But all the hopes of the air enthusiasts vanished the day before the banquet when the *Washington Star* published a report that the new laboratory would be supervised by the Smithsonian and built on the grounds of the National Bureau of Standards. The story aroused the ire of the Navy Department whose admirals felt the Bureau of Construction and Repair represented the appropriate home for a federal aeronautical facility. When Navy Secretary George Meyer pressed this viewpoint on President Taft, the Army opened its own initiative for control of aerial research. Other government agencies threatened to enter the contest. Choosing prudence, Mr. Taft withdrew his endorsement of the Aeronautical Society's plans.[13]

Now the proposition faced longer odds. During 1912 the President received a report on the subject drafted by the same figures who supported the Smithsonian proposal in 1911. It envisioned an institution modeled on those of Europe: a national laboratory which folded the many existing research centers into one structure. Taft agreed to form a commission to investigate the problem, but not before he received a humiliating third-place finish in the Presidential elections in November. Still, the 19 member panel actually drafted legislation bearing a striking similarity to the British Advisory Committee for Aeronautics, establishing a research center with federal funds and an oversight panel comprised of six representatives from government institutions and ten figures from private life. But the commission stalled in its tracks when advocates of a laboratory under Smithsonian aegis again pressed forward and Congress refused to consider the proposed bill. The impasse showed signs of clearing a month after President Woodrow Wilson's inauguration. The Smithsonian Board of Regents voted to re-open

---

[13]Roland, *Model Research*, 1: 4-6; Hansen, *Engineer in Charge*, 2-3.

the Langley Laboratory and Secretary Charles D. Walcott organized a meeting for May 1913 attended by such luminaries as Orville Wright, Albert F. Zahm and many scientists in the civil service. They agreed to support an advisory entity comprised of 16 permanent subcommittees which answered to the research objectives of a central board of oversight.

This outcome placed in Walcott's hands the keys to a solution, but the approach of hostilities in Europe gave him his biggest opening. He sent two American authorities--Physicist Alfred Zahm of Catholic University and Dr. Jerome C. Hunsaker, Naval Academy graduate and founder of the aeronautical engineering program at M.I.T.--on an extensive tour of the Continent's leading aeronautical facilities. Their report, released in 1914, decried the comparative backwardness of U.S. scholarship and infrastructure. The findings, combined with the outbreak of war in Europe during summer of that year, persuaded the Secretary to launch a legislative offensive for a federal aeronautical laboratory. His labors paid off in two short paragraphs buried in the naval appropriations act of 1915 and passed on March 3, the last working day of the session. The Smithsonian removed itself from a permanent role of leadership by agreeing only to form an advisory committee which would then take into its own hands the task of coordinating air research in *existing* institutions. Gone, too, were the Institution's earlier attempts to place Langley's old laboratory at the center of the new endeavor. Indeed, the precise wording passed by Congress requested *no* national laboratory at all, but left open the possibility with sublime artifice: "In the event of a laboratory or laboratories, either in whole or in part, being placed under the direction of the committee, the committee may direct and conduct research and experiment in aeronautics in such laboratory or laboratories...." The brief statement also borrowed directly from the British experience, empowering the President to select not more than 12 members of an Advisory Committee for Aeronautics "to supervise and direct the scientific study of the problems of flight, with a view to their practical solution, and to determine the problems which should be experimentally attacked, and to discuss their solution and their application to practical questions." The board, all unpaid, included two representatives from the War Department's military aeronautics departments; two from the Navy; one each from the

Smithsonian, the Weather Bureau, and the Bureau of Standards; and a maximum of five others "acquainted with the needs of aeronautical science...or skilled in aeronautical engineering or its allied sciences...." Finally, the legislation appropriated $5,000 annually for five years "to be immediately available, for experimental work and investigations undertaken by the committee, clerical expenses and supplies, and necessary expenses of members of the committee in going to, returning from, and while attending meetings of the committee...." [14]

The Main Committee of the Advisory Committee for Aeronautics met in the offices of the Secretary of War just seven weeks after Congress voted to conceive it, under the chairmanship of General George P. Scriven, Chief Army Signal Officer. Its first action involved its own name; adding the word *National*, it henceforth became known as the National Advisory Committee for Aeronautics, or the NACA. Scriven and the others then turned to the structural questions. The Main Committee, which constituted an independent agency reporting directly to the President of the United States, fashioned from its number an Executive Committee of seven. Elected for one year, the Executive Committee members commonly lived near Washington, D.C., allowing them to meet more frequently than the Main Committee, which convened only twice yearly. The smaller group represented the true governing authority of the NACA. Under the chairmanship of Charles Walcott and his successors it wielded control over the research agenda and executed the broad directives of the Main Committee. The Executive Committee also created and appointed such technical panels as Aerodynamics, which in turn divided its labors among various subcommittees like Airships, Seaplanes, and Aeronautical Research in Universities. But erecting an organizational entity represented only half of the NACA's initial travails. Although dormant, the idea of a national aeronautical research center still stirred the imaginations of many. As a consequence, the Executive and the Main Committees met again in

---

[14]All quoted passages from Alex Roland, *Model Research: The National Advisory Committee for Aeronautics, 1915-1958*, vol. 2 (Washington, D.C.: NASA SP-4103, 1985), 394-395; Roland, *Model Research*, 1: 6-25; Hansen, *Engineer in Charge*, 3-5; Roger E. Bilstein, *Orders of Magnitude: A History of the NACA and NASA, 1915-1990* (Washington, D.C.: NASA SP-4406, 1989), 3.

mid October 1916 and voted to request $85,000 from Congress for fiscal year 1917. Over $53,000 of it would be allocated for the site preparation and construction of the new laboratory. At first, leaders like Charles Walcott assumed the NACA might continue to ride the Navy's fiscal coattails to obtain this appropriation. But Navy Secretary Josephus Daniels, thinking no doubt of his own budget, rebuffed any more handouts for the NACA. Walcott therefore took his case directly to Congress and succeeded; on August 29, 1916, it approved the entire $85,000 request.[15]

Even while engulfed in the process of locating a site, drafting plans, and constructing buildings for the new laboratory, the NACA activated itself with surprising speed and purpose and contributed to the American War effort in several ways. Secretary Walcott initiated a survey of American aeronautical programs and projects, contacting over 100 universities, 22 aero clubs, ten manufacturers, and eight government agencies. He discovered a shocking lack of the systematic and sustained research being pursued in Europe. The Committee also negotiated between the uniformed services and the nation's engine manufacturers an agreement to produce a motor suitable for military aircraft, embodied finally in the Liberty powerplant; settled a bitter patent dispute between the Curtiss Aeroplane and the Wright-Martin Company over rights to the aileron system devised by the Wright Brothers; and dispatched to Europe in 1917 Stanford University's eminent Professor of Engineering William Frederick Durand (General Scriven's successor as NACA chairman) and distinguished Johns Hopkins University physicist Joseph S. Ames to hasten technical cooperation among the Allies and the U.S. The NACA likewise succeeded in stimulating an impressive range of American engineering projects. It contracted with Durand and Stanford for extensive propeller experiments, participated with the Bureau of Standards in engine testing, underwrote research in the Washington Navy Yard's model basin, and evaluated aeronautical inventions for the War Department. Choosing of a permanent labor force also received a high priority. John F. Victory, a secretary in the Navy Aeronautical Laboratory, agreed to serve the NACA in the same capacity and so became its first paid staff

---

[15]Roland, *Model Research*, 1: 27-32; Hansen, *Engineer in Charge*, 5-10.

member. The Committee followed Victory's induction with its initial technical hire, a former Curtiss Company engineer and draftsman named John H. DeKlyn. Finally, the NACA established a few operating practices. It opened an Office of Aeronautical Intelligence to amass all literature related to the Committee's mission, it endorsed Post Office Department subsidies for airmail operations, and issued a rule that all NACA technical papers must be released first as attachments to the Annual Report before being eligible for publication elsewhere.

All of these useful activities occurred in the absence of any true research center. Although Congress appropriated funds to build a home for American aeronautical research, it materialized slowly and fitfully. It began with good fortune and nearly ended in collapse. None other than General Scriven, in charge of Army aviation, received orders from the War Department in 1915 to identify a location for an experimental airfield and facilities. Scriven assured the NACA that its laboratory would be welcome on any of the possible sites. After considering 15 separate alternatives his selection board announced the winner: a tract of 1650 acres just north of Hampton, Virginia. Planners liked its relatively good climate, its proximity to skilled labor at Newport News, and its closeness to Washington, D.C. and the institutions of national power. As early as 1916 Scriven proposed to Charles Walcott naming the field for Samuel Langley, a suggestion never challenged. Inevitable delays in construction resulted from American entry into the war in April, 1917, but other factors also made progress difficult after the ground-breaking for the first laboratory in July. Work gangs exhausted themselves turning shovels to fill the endless marshland and digging deep to uproot the swarm of tree stumps. Deadly influenza killed dozens of laborers and the mosquitoes bedeviled everyone. These travails postponed by months the pouring of concrete and the laying of runways. Moreover, under pressure to prosecute a war, the Army abandoned Langley as its experimental air station, and although it retained the facility for operational use, established McCook Field in Dayton, Ohio as its center for air research. This decision hit NACA officials hard, denying them the close technical cooperation they expected from the service. By summer of 1919, this news and slowdowns in the completion of the essential buildings prompted John Victory and construction supervisor John DeKlyn to advise Main Committee member Joseph Ames to shut down the entire project and to transfer the laboratory to Bolling Field in Washington, D.C. But Congress, aware of the heavy investment already sunk in Hampton, declined to abandon the laboratory on the Tidewater.[16]

Even in the face of miserable conditions and interminable delays the postwar research program of the NACA got underway almost immediately. To guide it, the Main and Executive Committees required a full-time technical administrator. During the war the NACA leaders tried unsuccessfully to attract one of several eminent engineers and scientists to be director of research. By 1919 the need became critical. William Frederick Durand found the answer in a young professor-turned-engineer named George W. Lewis. The two men met during the Great War when Lewis, formerly a teacher at Swarthmore College, worked as chief engineer and an engine specialist for Clarke Thomson Research, a Philadelphia aeronautical research foundation which contracted with the NACA late in 1917. The following year the 36 year old Lewis, a graduate of Cornell University with Bachelors and Masters degrees in aeronautical engineering, joined the Subcommittee on Powerplants and befriended Joseph Ames, soon to assume the role of Executive Committee chairman from Walcott. Ames saw qualities of leadership in the forceful, outgoing, yet modest Lewis and nominated him to be the Committee's executive officer. He assumed the role in November 1919 in Washington, D.C., thus positioned himself to manage the NACA's political affairs with one eye and the administration of the laboratory with the other. Becoming Director of Research five years later, Lewis proved to be the NACA's indispensable man, cultivating Congress and the services for funds and equipment while allowing his "boys" at Langley wide latitude to pursue their research interests.

THE NACA TAKES FLIGHT

---

[16]Roland, *Model Research*, 1: 30, 33-47, 80-83; Hansen, *Engineer in Charge*, 8-22.

Well before George Lewis assumed his duties in Washington, flight research already ranked high on the NACA agenda. Before the wind tunnels roared to life and the test stands held the fury of firing engines, the NACA's nascent program manifested itself in airplanes flying test patterns over fields still wet with mud. In fact, the Main Committee concurred in the selection of the Hampton area expressly because of its conduciveness to air operations. In 1916, four years before Langley opened officially, Professor Durand extolled the region's aerial advantages: the prevailing mildness of the weather, the laboratory's propinquity to the mouths of the Chesapeake Bay and the James River (affording flights over both land and water), and the highly varied surroundings which simulated most of the conditions pilots encountered under regular circumstances.

During the same year Durand declared Langley's superiority in these respects, a researcher writing in NACA Technical Report Number 12 pondered the experimental methods available to the NACA to study the as yet mysterious effects of air pressure on flying machines. Relying heavily on the French aerodynamics program at St. Cyr, the author touched on three known techniques. One involved anchoring the object of the investigation to an instrumented carriage and measuring air resistance as the apparatus moved in various directions and at various speeds. In the second method, "[i]nstead of moving the body under test, a fixed position is given to such body placed in an artificial current of air." During such wind tunnel tests, Eiffel and his associates also employed balances to measure the total air resistance, as well as the particular resistance at given points, of aerodynamic shapes in the laboratory. The third and final possibility promised "a very considerable practical value" but yielded "complex results often difficult of analysis." Here the author referred to airplanes in free flight. In contrast, he recognized the principal limitations of the laboratory. Simulated flight only approximated the real atmosphere and the process of scaling up data derived from models often distorted the true aerodynamics of the full-scale aircraft. But flight testing also imposed difficulties. The pilot

23

often found it necessary to avoid essential maneuvers due to safety, resulting in the collection of incomplete, and perhaps inferior information. Moreover, to be effective such tests needed to be "sufficiently systematic [and] numerous," a failing admitted by French test pilots as early as 1910.[17]

The NACA ventured into flight research in 1918. It began when John DeKlyn proposed to the Executive Committee in June a project to compare propeller performance in full scale flight, in models, and in theoretical calculations. He began with a review of the published literature, as well as a previous NACA analysis of experimental propellers. DeKlyn concentrated on such variables as pitch ratio, distribution of pitch, the shape and width of blade contour, and type of blade section. To conduct research he proposed testing four propellers, two with straight blades (one cambered and one non-cambered), and two with tapered blades (one cambered, the other non-cambered). A new device invented by Professor Alfred Zahm promised to speed and simplify DeKlyn's work: a "computer" designed to measure propeller characteristics, it still awaited manufacture by Langley technicians. DeKlyn hoped ultimately to compare the data from free flight, from wind tunnel studies of scale models, and from mathematical analysis and to arrive at a set of standard propeller characteristics for optimal performance.

His prospects for success seemed remote. Langley Field remained a crude jumble of mud and timber, its construction bogged down woefully. No flying tests could begin until thrust and torque meters were ready. DeKlyn could not evaluate the propeller blade sections so long as the wind tunnels remained uncompleted. Despite these serious impediments--outweighed perhaps by a desire to get *some* type of major research underway--the Executive Committee gave its unofficial assent when Professor Ames marked "OK" on the proposal and signed his initials below. Accordingly, on July 18, 1918, the Executive Committee issued Research Authorization

[17]Hansen, *Engineer in Charge*, 11; L. Marchis, National Advisory Committee for Aeronautics Technical Report (hereafter NACA TR) 12, "Experimental Researches on the Resistance of Air" (Washington, D.C.: NACA, 1917), 555-558.

(RA) Number 1, "Comparison of Mathematical Analysis and Model Tests of Air Propellers."[18]

The go-ahead for DeKlyn marked an important event in the history of the NACA. Research Authorizations evolved from this point into the process by which the Executive Committee guided the labors of the Langley staff. Requests to inaugurate research might arrive in Washington from laboratory employees, from the military services, from other federal offices, or from industry. Those proposed by the uniformed services or by government bodies went straight to the Executive Committee and met with approval provided they did not duplicate existing work. If peer reviewed, those generated from Langley stood a high probability of acceptance but first needed to be scrutinized by the appropriate technical subcommittees before arriving at the Executive Committee. Aircraft manufacturers faced the same review process. Yet, despite the outward appearance of formality, the system did not inhibit the experimenters at Langley. Joseph Ames and George Lewis both gave broad latitude to new projects and often found themselves attaching promising new work to existing RAs, or even to winking at projects conducted without any Research Authorization at all.[19]

DeKlyn's, however, did not prove to be one of the rogue RAs. By January 1919 he and his new partner, power plant engineer Marsden Ware, familiarized themselves with the "all new developments in propeller design" reported by Britain's Royal Aircraft Factory and delivered similar designs to the Langley machine shop. They also contracted the services of Professor Everett Lesley of Stanford University to test four propellers and to record absolute values at designated points. Three months later, however, the Executive Committee began to lose

---

[18]"Memorandum of Suggested Research on Propeller Sections by John DeKlyn," 29 June 1918, Research Authorization Number 1 File, Langley Aeronautical Research Center Historical Reference Collection (hereafter referred to as LaRC Historical Reference Collection); John H. DeKlyn to John F. Hayford, 19 July 1918, RA 1 File, LaRC Historical Reference Collection; National Advisory Committee for Aeronautics Research Authorization (hereafter referred to as NACA RA) No. 1: "Comparison of Mathematical Analysis and Model Tests of Air Propellers," John F. Hayford, 17/18 July 1918, RA 1 File, LaRC Historical Reference Collection; "Memorandum Regarding Status of Research Authorization No. 1," John H. DeKlyn and Marsden Ware, July 1918, RA 1 File, LaRC Historical Reference Collection.
[19]Hansen, *Engineer in Charge*, 36-38; Roland, *Model Research*, 1: 103-104.

confidence in the project. When John Victory requested a report in April, DeKlyn replied the "research is the same" as that detailed the previous year. Other evidence did not inspire optimism: due to lack of materials, neither the Zahm computer nor the wind tunnel models had yet been fabricated. As a consequence, the Subcommittee on Aerodynamics canceled the NACA's first Research Authorization on July 23, 1919. The burden of conducting this project, in addition to the exertions of supervising construction of the laboratory, proved to be too much for DeKlyn. "Neither Mr. Ware nor myself," he finally confessed to Victory, "have any time to give to Research work." So ended the first NACA foray into flight research.[20]

Prospects for experimental flying brightened during the summer of 1919. While the laboratory itself remained in crisis during this period, the appearance in Hampton of an extraordinary young Massachusetts scientist and engineer transformed the bleak situation. Edward Pearson Warner arrived at Hampton in early 1919 to be the lab's first Chief Physicist. He also took a seat on the NACA's Aerodynamics Committee. No doubt shocked by the abysmal conditions referred to by Victory and DeKlyn, he nonetheless recognized that the government's investment could not be abandoned and acclimated himself to the prevailing circumstances. Langley benefited greatly from the 25 year-old's decision to stay, for he brought qualities and background needed desperately during this formative period. Born in 1894, Warner grew up in Concord and attended a private academy in Boston.. His father Robert, an electrical engineer, supported the family comfortably. A quiet student who always looked unkempt, Edward often seemed overwhelmed by the bulk of note pads, slide rules, and pens stuffed into his pockets. At the same time, he exhibited an astonishing mathematical gift, exemplified by the

[20]"National Advisory Committee for Aeronautics, Office of Aeronautical Engineer, Work Authorization," John H. DeKlyn, 14 January 1919, RA 1 File, LaRC Historical Reference Collection; John H. DeKlyn to John F. Victory, 5 April 1919, RA 1 File, LaRC Historical Reference Collection; "Memorandum Regarding Status of Research Authorization No. 1, March 31, 1919," John H. DeKlyn, RA 1 File, LaRC Historical Reference Collection; John F. Victory to John H. DeKlyn, 23 July 1919, RA 1 File, LaRC Historical Reference Collection; John H. DeKlyn to John F. Victory, 28 July 1919, RA 1 File, LaRC Historical Reference Collection; Hansen, *Engineer in Charge*, 43, 53.

ability to multiply four digit figures in his head. He graduated from Harvard University in 1916 with honors and enrolled in the Massachusetts Institute of Technology. There he embarked on a career in aeronautical engineering under the tutelage of Professor Jerome Hunsaker, a figure who rivaled and perhaps surpassed William Frederick Durand and Alfred Zahm as founders of the discipline in America. After studying in Paris with Pierre Eiffel, Hunsaker returned to M.I.T. in 1914, constructed his own four foot wind tunnel, and began to teach the subject in fall of that year. Warner immediately established himself as Hunsaker's leading protegé, a dynamo who talked fast and solved equations even faster. He taught a class during his first year of graduate study and dazzled his students by the speed with which he solved the most complex differential equations. During World War I the young professor taught advanced courses on aeronautics to Army and Navy cadets, attended by such men of future distinction as Leroy Grumman and Theodore P. Wright. By the time he received his Masters degree in 1919 he had won a permanent place on the M.I.T. faculty.

Warner probably accepted the NACA Chief Physicist position at the prompting of Hunsaker, but saw his mentor often due to frequent absences from Hampton. When he did appear in Virginia, however, he worked at a frantic pace. Soon after his arrival he became absorbed in John DeKlyn's propeller project. More important, having conducted many experiments with the M.I.T. wind tunnel, Warner designed and built the first such instrument for the NACA. The Committee assigned him a more formal task on June 20, 1919. Research Authorization Number 7 directed him to lead a "Comparison of Various Methods of Fuselage Stress Analysis," measuring the impact of landings on representative "stick and wire" aircraft. Warner evaluated the types of landings that most stressed the fuselage, established general rules for assessing such impacts, and calculated "the amount of error to be expected from the use of the simpler and less accurate methods."[21]

---

[21]Roger E. Bilstein, "Edward Pearson Warner and the New Air Age," in *Aviation's Golden Age: Portraits from the 1920s and 1930s*, ed. William M. Leary (Iowa City: University of Iowa Press, 1989), 113-115; Hansen, *Engineer in Charge*, 29-30; Roland, *Model Research*, 1: 82; Longyard, *Who's Who*, "Jerome Clarke Hunsaker"; John H. DeKlyn to John F. Victory, 28 July 1919, RA 1

But his chief contribution--a series of intensive flight experiments representing the NACA's first wholly indigenous research--molded the laboratory's technical style as well as its early reputation. Research Authorization Number 10 entitled "Free Flight Tests," began on the same day as the fuselage loads project. "It is very important," the RA intoned, "that data on the characteristics of airplanes in flight be secured for comparison with wind tunnel results for the same machine." More to the point, the tests would determine whether the "actual characteristics" discovered during free flight "differ[ed] from those predicted in tests on models in wind tunnels...." If this chore did not quite fill his hours, the RA also instructed Warner to investigate stability and control, the complexities of which could "only be carried on in free flight." To complete these tasks required sophisticated research techniques which measured simultaneously a bewildering variety of factors: "angle of incidence, air-speed, rate of climb, r.p.m., elevator position,...force on the stick, the lift and drag coefficients and balancing characteristics." Edward Warner may have been brilliant and quick, but even he could not produce the findings for such a project in the mere three weeks allotted by Research Authorization 10. Perhaps the short deadline resulted from impatience with John DeKlyn's performance on RA 1 or from young Warner's inflated expectations of himself. In any event, the experiments actually ran through the rest of the summer.

Yet he did move rapidly. In his rapid-fire pursuit of aerodynamics knowledge he enlisted the assistance of a fellow student of Jerome Hunsaker who happened also to be Langley's first permanent employee. Frederick H. Norton arrived in Hampton in Autumn 1918, a rookie 22 year old with limited wind tunnel experience. Although only three years his senior, Warner won the lead role due to his superior education and unique mathematical talents, and Norton learned much from him. The team worked primarily with two test pilots, both military men: Lieutenants H.M. Cronk and Edmund T. "Eddie" Allen. Warner had high regard for the role of his fliers.

---

File, LaRC Historical Reference Collection; Quoted passage in NACA RA Number 7: "Comparison of Various Methods of Fuselage Stress Analysis," Charles D. Walcott, 20 June 1919, RA 7 File, LaRC Historical Reference Collection.

"Test flying," he wrote, "is a very highly specialized branch of work, the difficulties of which are not generally appreciated, and there is no type of flying in which a difference between the abilities of pilots thoroughly competent in ordinary flying becomes more quickly apparent." Allen, in particular, possessed the right qualities; one of the outstanding research fliers of his era, he flew in the United Kingdom as well as at McCook Field, attended both the University of Illinois and M.I.T., and later became chief test pilot and director of aeronautical research for Boeing Aircraft Company. Warner, Norton, and the test pilots also had the assistance of a 24 year old airplane mechanic named Robert E. Mixson, an ambitious young man who served in the Great War and with no college degree eventually worked his way onto the Langley engineering staff. Finally, to conduct their study the small team obviously needed airplanes and a wind tunnel. Since the NACA possessed not a single aircraft, they turned to the Air Service authorities on Langley Field who agreed to loan two Curtiss JN4H Jennies, the famous war-horses aboard which many of the Army's pilots and observers learned photo-reconnaissance, gunnery, and bombing skills. Since the NACA's first tunnel had yet to be completed, Warner relied on the familiar M.I.T. model. So equipped with men, flying machines, and laboratory equipment, Warner launched his project.[22]

The obstacles that often accompany flight research soon became apparent to Edward Warner. Both of the aircraft supplied by the Air Service, equipped with 150 horsepower Hispano-Suiza powerplants, experienced engine overheating during the intense Hampton summer, making it all but impossible to execute climbs with their throttles open fully. Apparently alike in all other respects, the planes actually differed both in obvious and in subtle ways. Machine Number 1 had an oil radiator suspended below the fuselage and a reserve gasoline tank attached to the center part of the upper wing, but lacked the standard aluminum

[22]First, third, and fourth quoted passages from NACA RA Number 10: "Free Flight Tests," Charles D. Walcott, June 20, 1919, RA 10 File, LaRC Historical Reference Collection; Edward P. Warner and Frederick H. Norton, NACA TR 70, "Preliminary Report on Free Flight Tests" (Washington, D.C.: NACA, 1920), 571, 581 (second and fifth quoted passages): Hansen, *Engineer in Charge*, 41-43, 61, 162-163, 419, 538.

doors forward of the wings; Machine Number 2 had no additional radiator or fuel appendages, but did include the aluminum doors. The wooden propellers of the two appeared to be the same; in fact, Number 1's was virtually identical to the drawings while the other had warped so much that its pitch became "considerably less" than desired. Most telling of all, the lifting surfaces of the airplanes exhibited "extreme divergences between the cambers at corresponding points on the different wings [which] were by no means negligible." Fortunately, the differences balanced out one another so that the *mean* sections matched very closely. Observing these distinctions, the Chief Physicist discovered a critical but obvious factor likely to taint comparisons between free flight data and wind tunnel tests. The fatal discrepancy occurred before any plane flew and before power animated any laboratory equipment. To avoid it, designers needed to take pains to fabricate

> wind tunnel models to represent the airplane as it is actually built, or to be built, not merely according to specifications which the shop may find [itself] quite unable to
follow. It is of little use to construct model aerofoils accurate to within 0.002 inch if the full-sized wing which they represent departs as much as three-eights of an inch from the section which it is supposed to follow. Secondly, these measurements should serve to remind experimenters engaged in the design of wing sections of the futility of drawing forms which it is impossible to construct by ordinary methods. For instance, no airplane wing is constructed with the upper and lower surfaces running out until they intersect in a perfectly sharp trailing edge. Indeed, it is practically impossible to construct a model aerofoil for the wind tunnel with such a trailing edge, yet aerofoils are repeatedly drawn up in such forms. The result is that the model maker exercises his own judgment as to the extent to which the trailing edge should be rounded over, the airplane builder introduces a strip of wood or [a] steel tube for a trailing edge, and the drawing, the model, and the full-sized wing are likely ultimately to be of three quite different forms.[23]

Warner also pinpointed errors likely to result from test instruments themselves. He eliminated as significant culprits the altimeter (whose readings below 4,000 feet could be in error without much affecting the overall data) and the tachometer (which either recorded accurately or failed to work at all). But he found serious difficulties with the air-speed meters which often required re-calibration in wind tunnels. Even more telling, because the air-speed pitot tubes

---

[23]Warner and Norton, NACA TR 70, "Preliminary Report on Free Flight Tests," 571-575 (small quoted passages on 574 and 575, block quote on 575).

measured *turbulent* air passing in proximity to the wings, not undisturbed flow, false readings might result. In some reckonings, an error of half a mile per hour in air speed yielded as much experimental deviation as a 400 foot miscalculation in altitude. Warner discovered a satisfactory yet simple method of correction by laying out a measured (5,600 foot) speed course on the emerging Langley grounds and positioning observers at the ends of the course to time the aircraft flying overhead, assisted by telephone communication. Comparing figures derived from the air-speed meter to data gathered from the speed course resulted in significant improvements in the on-board equipment and a more exact understanding of its deficiencies.

After gaining an understanding of the planes and equipment, Warner's small cadre began their tests. "It is very desirable," they all agreed, "that data be obtained on the lift and drag in free flight of full-sized airplanes and parts thereof, in order that the designer may gain some knowledge as to the corrections to be applied to wind-tunnel results and as to the extent to which those results can be trusted. The problem is an extremely difficult one for many reasons...." Rather than measure the forces of lift just at the wings, the experimenters chose the simpler method of deriving the data for the airplane as a whole. As well as the air-speed meter, the tachometer, and the altimeter, Warner's group outfitted the Jennies with Langley-made inclinometers to measure the incidence of angle of attack (the angle between the direction of air flow and the direction of an aircraft's wings or fuselage). The flight regime involved altitudes between 1,500 and 4,000 feet depending on air conditions and required the pilots to fly perfectly level, to steer straight over the speed course, and to achieve a constant rate of speed for one to two minutes per test. To further the degree of difficulty, the aviators received instructions to raise the angle of flight to equal or exceed the maximum angle of lift by throttling the engine to the lowest velocity for level flight and then to open the throttle gradually. The aircraft thus flown stayed level but in a highly stalled condition, at the same time courting the danger of lateral instability. One of the NACA fliers became so adept at these delicate maneuvers that after considerable practice he flew the plane level, with throttle open wide, at an 18 degree angle of attack for an indefinite period. By plotting the curve of data from the inclinometers on the

31

two Jennies and comparing it to wind tunnel tests on the similar Curtiss JN2, a clear discrepancy emerged between the two sets of data. Above the six degree point, higher angles of attack and a greater lift coefficient proved possible in actual flights than were predicted by the tunnel experiments--greater by about *15* percent. Although the final results awaited JN4H wind tunnel tests, Warner and his team could be quite sure that the measurement of lift on models could not be relied on with certainty. On the other hand, plotting for lift/drag ratio showed a "reasonably good" correspondence between free flight and wind tunnel data.[24]

Warner and Norton also flight tested the Jennies to determine longitudinal balance. First they calculated the center of gravity by the standard method of weighing the machines under each wheel and under the tail skid, and accounting for the weight of the crew. It proved to be roughly 2.5 feet behind the leading edge of the upper wing. The pilots gathered data using two main instruments: a position indicator mounted on the elevator rocker-arm shaft to measure the angle setting of the elevator at any moment in flight; and an elevator force indicator consisting of a scale mounted between two springs to measure the tension applied by the pilot to the stick. The planes flew at altitudes of 1,500 to 4,000 feet during the initial tests operated with the elevator controls locked to reduce the number of variables. The results established a close correlation between the Langley flights and the M.I.T. wind tunnel experiments conducted on the JN2. With the controls free, "just as with the controls locked, the statical longitudinal stability is greatest at low speeds of flight,...the machine becomes unstable at speeds in the neighborhood of the maximum attainable, and...the stability is greater in gliding than in throttle open." One important distinction did emerge between the two flying approaches: equilibrium could be achieved at *any* speed by locking the controls in the correct position, while balance could be achieved with free controls *only at one speed* for a given elevator position. Finally, Warner noted that the experiments revealed the Jennies suffered from nose-heaviness and from some instability. He proposed the counterintuitive solution of moving the center of gravity *forward*,

[24]Warner and Norton, NACA TR 70, "Preliminary Report on Free Flight Tests," 575-588 (first quoted passage 578, second 588).

not aft, to increase stability over a wider range of speeds, and to change the angle of the stabilizer to allow greater downward force on the tail and thus improve balance.[25]

## THE WAY FORWARD

Edward Warner's investigations represented two milestones for the NACA: the committee's first attempts at systematic technical inquiries, and Langley's initial foray into flight research. Not in themselves benchmarks in aeronautical knowledge, Warner's projects nonetheless suggested the value of a vigorous program of government-sponsored flight research. Indeed, during the NACA's first five years, this discipline evolved from a makeshift practice dependent on a few tools and techniques into a field in which the roles of engineers, pilots, technicians, mechanics, the flying vehicles, instrumentation, and the corresponding laboratory equipment became better defined and integrated. During the interwar years flight research transformed itself into an indispensable ingredient of aeronautical inquiry.

---

[25]Warner and Norton, NACA TR 70, "Preliminary Report on Free Flight Tests," 589-597 (quoted passage, 594).

# CHAPTER 2

## Flight Research Takes Off

### MODEST BEGINNINGS

After three years of hard toil the day finally arrived to dedicate the Langley Memorial Aeronautical Laboratory. In contrast to the pessimism felt over the past year and a half, the dignitaries attending the event on June 11, 1920, saw an inspiring show. Brigadier General William "Billy" Mitchell put a 25-plane formation through its paces and other aerial exhibitions flew overhead. Rear Admiral David Taylor, Chief Constructor of the Navy, called the laboratory no less than "a shrine to which all visiting aeronautical engineers and scientists will be drawn." Other civilian and military speakers followed, heaping praises on the lab that emerged from the swamps. Then the guests went on tours of the new buildings, seeing the research laboratory and the engine-dynamometer facility. In the third structure they witnessed an event second only to the Mitchell fly-over: a demonstration of NACA's first wind tunnel, a five foot open-end design conservatively patterned after that in use at Farnborough in the U.K. Its roar duly impressed the attendees. The visitors also noticed a flight research fleet of only two planes; the two Jennies used for the 1919 flights.

Yet, together these random beginnings represented a long stride over the conditions under which the first investigations took place the year before. The immediate improvement involved not so much the physical plant as the staff peopling the buildings and the flight line. When Warner and Norton conducted their experiments, they and two others constituted the entire professional workforce. These four employees plus seven blue collar workers totaled a payroll

1

of eleven. But by 1920 the complement of engineers and pilots had *trebled* and the cadre of mechanics and craftsmen had doubled, yielding a workforce of 26 including administrative personnel. Many had only recently left school; the median age was only 28. Also, there emerged clearly defined staff roles and three functional divisions: *Aerodynamics* headed by a Chief Physicist in charge of the Aerodynamical Laboratory and the Wood Shop, responsible to the Aerodynamics Committee in Washington; *Powerplants* run by a Senior Staff Engineer in charge of the Dynamometer Laboratory and the Machine Shop, under the control of the Powerplant Committee in Washington; and *Administration, Maintenance and Purchasing* directed by a Chief Clerk who operated the Langley Field Station and the Drawing Room, reporting to the Personnel Committee in Washington. The two key technical positions were filled by men of ability. Edward Warner left Langley in Fall 1920 to return to M.I.T. as an associate professor and Frederick Norton, the deputy who blossomed under his tutelage, assumed the position of Chief Physicist. During the same year, William Frederck Durand persuaded Leigh Griffith, a middle aged Californian with a mechanical engineering degree from the California Institute of Technology, to join the laboratory as Senior Staff Engineer in charge of the high performance engines program. Griffith and Norton, as well as the Chief Clerk, reported not to a local director, but directly to George Lewis in Washington and through him, to the NACA Main Committee.[1]

From its inception, the Langley Memorial Aeronautical Laboratory centered its research agenda on aerodynamics, and on the main instrument of this discipline, the wind tunnel. Since this facility represented a *national* center designed to rival those of the Europeans, all of the instruments of research deemed essential to the field needed to be provided. The wind tunnel not only represented the latest in research equipment; it both suited and formed the research style of the laboratory's engineers and scientists. Much like the technique of discovery employed with

[1]Hansen, *Engineer in Charge*, 21, 29-30, 41-43, 69, (quoted passage, 21); Roland, *Model Research*, 1: 83.

2

such success by the Wright Brothers, Langley's engineers and scientists practiced a careful, systematic, collegial approach to their investigative work. "The Langley way," observed a commemorative book on the 75th anniversary of the lab, "was one of systematic parameter variation: that is, meticulous, exacting variation of one component at a time to identify configurations that would produce the best results. At Langley, no researcher ever really worked alone. Successful application of aeronautical research demanded collaboration." This approach to technical inquiry lent itself perfectly to the demands of the wind tunnel. Vast, man-made environments subject to exacting control and manipulation, one historian called them "complicated mechanized marvels, national resources, great and powerful monuments to the modern age." The first one, Edward Warner's five foot Atmospheric Wind Tunnel, was neither advanced, big, nor powerful, and perhaps even obsolete by its completion in June 1920. The second one, however, formed the backbone of Langley's subsequent distinction for advanced research. It sprang from the immensely fertile, yet haughty and irritable mind of Dr. Max Munk, a German aerodynamicist who signed on with the NACA in 1920 as a technical assistant. (See below in this chapter for more about Munk). Soon after his arrival he began a campaign for the construction of a wind tunnel with a pressurized air stream. Known afterward as the Variable-Density Tunnel (or VDT), it started operation in late 1922 at cost of $262,000, roughly *seven times* that of the Atmospheric Tunnel. But it paid handsomely both in terms of its research applications, as well as the notoriety it bestowed on the NACA due to its advanced capabilities. Its accuracy, derived from the higher Reynolds Numbers possible under denser pressures, far surpassed that of any other tunnel of its day. Henceforth, Langley found itself with a reputation to maintain, and lived up to it with a string of increasingly costly and complex machines: the Propeller Research Tunnel (operational 1927); the 11-inch High Speed Tunnel (1928); the 5-Foot Vertical Wind Tunnel (1929); the 7 X 10 Foot Atmospheric Wind Tunnel (1930); and the Full Scale (*30 by 60 foot*) Tunnel (1931).[2]

---

[2]James Schultz, *Winds of Change: Expanding the Frontiers of Flight, Langley Research Center's 75 Years of Accomplishment, 1917-1992* (Washington, D.C.: NASA NP-130, 1992), 10 (first

At least in its early incarnation at Hampton, flight research began at the point where the wind tunnels could not provide meaningful data. While the contours of full-sized aircraft could be duplicated exactly in small scale, duplicating the complexity of the movements of piloted flight often eluded wind tunnel technicians. Only gradually did the aerodynamicists realize that the data accumulated by flying carefully instrumented aircraft not only *corroborated* the tunnel findings, but often yielded data not even conceived under laboratory conditions. On the other hand, flight researchers came to recognize the crucial role of wind tunnels in preparing for the rigors of actual flying and to appreciate their capacity to perform experiments too impractical or dangerous for a real airplane and pilot. This process of drawing the boundaries between formal laboratory research and flight research took some time and many projects, a relationship which matured as Langley matured. Meantime, flight testing itself needed to shed its old persona. Even as early as 1920 (in no small part because of Edward Warner's research in 1919) it became increasingly clear that for progress to be made in aeronautics, the cocksure attitude of "give me the stick and I'll fly it" needed to be supplanted by a systematic, engineering approach to the problems of flight. The transformation occurred as soon as the NACA staff began to delve into these conundrums and to sense the actual dangers and difficulties. Once the experimenters realized how little they knew about the fundamental mechanisms at work, modesty replaced whatever egotism may have prevailed. The professionalization of flight research followed quickly. Government agencies and private organization involved in aviation published handbooks and guides for the crews and the fliers. *A Manual of Flight Test Procedure*, an early example of this growing literature written by an Army Air Service practitioner, added structure and process to the serious business of flying the unknown. Few men better represented this sober approach to a field heretofore (and often subsequently) dominated by colorful characters than the NACA's Chief Test Pilot, Thomas Carroll. Like Eddie Allen, Carroll arrived at Langley Field with university credentials. In fact, he started at the NACA in 1920 upon completion of a law degree from Georgetown University in Washington, D.C. Born in 1890, Carroll learned the

quoted passage); Hansen, *Engineer in Charge*, 23 (second quoted passage), 65, 74-75, 442-447.

4

pilot's art during World War I and later taught air tactics to fliers in France. Bright and thorough, he brought the perfect blend of experience and education to the role.[3]

But Carroll and his cohort Eddie Allen--soon joined by pilots William McAvoy and Paul King--quickly found themselves besieged by the demands of the NACA Executive Committee. The ceremonies of June 21 hardly ended when the NACA's official flight research program began. Indeed, Joseph Ames signed *four* Research Authorizations that very day. Together, they represented the classic lines of flight research inquiry followed during the long history of the NACA: the stability and control of aircraft, the influence of aerodynamics on flight loads and other factors, and innovations in powerplants.

"Controllability Testing" (Research Authorization 2) directed Carroll and the other pilots to obtain simultaneous measurements of acceleration, attitude, air-speed, and force using three controls during normal flights, during stunt flying, and during landings. The Aerodynamics Committee hoped to retrieve concrete data about response to controls and to derive quantitative standards from the test results. Unfortunately, like the three other flight research projects initiated by Washington, the investigations conducted under RA 2 offered no clear and final answers, and what could be gleaned only suggested the need for more thorough and intensive research. Researchers Frederick Norton and his assistant William G. Brown--who directed two of the four initial flight research RAs assigned by Ames--confessed that after more than a year of labor,

> [t]he study of controllability and maneuverability has been particularly difficult, first because the subject is so intangible and second because there is so little previous work to follow. It is felt that the present investigation leaves much to be desired in the way of completeness, but it at least places the subject on a much more scientific footing than before, and will serve as a basis for further investigation.[4]

---

[3]Richard P. Hallion, "Flight Testing and Flight Research: From the Age of the Tower Jumper to the Age of the Astronaut," in *Flight Test Techniques: AGARD Conference Proceedings No. 452* (Neuilly sur Seine, France: NATO Advisory Group for Aerospace Research and Development, 1989), 24-2; Hansen, *Engineer in Charge*, 42, 164, 166, 415, 419.
[4]Four NACA RAs were issued on the same day in June 1920: numbers 2, 3, 11, and 35. The reasons for the gaps in their numerical sequence are mysterious. See RA No. 2: "Controllability Testing," Joseph S. Ames, 11/28 June 1920, RA 2 File, LaRC Historical Reference Collection;

Norton also designed and oversaw flight tests related to aerodynamic loading under the guidance of Research Authorization 3, entitled "Tail Pressure Distribution." It empowered the Langley researchers to evaluate flight stresses on the rear surfaces by compiling "a continuous record of the variation of pressure at a large number of points while maneuvering" during accelerations. The findings again made an important contribution to a subject lacking in published work, but Norton once more conceded that since "the value of a research is not only in answering questions but also in finding questions to answer, ...a short discussion of the difficulties encountered in this investigation and the problems for which a satisfactory solution has not been arrived at will be of value in guiding future work...." The answer not given involved the profile of the horizontal rear surfaces, "[o]ne of the most important problems, and one on which there has been only a little light shed...." Norton called for new studies on tail plane cross-sections to determine which cambers and shapes offered the greatest stability with the most even distribution of loads. Another of Ames' four initial flight research projects involved Research Authorization 11, like RA 3 but targeted not on tail surfaces but on technically similar wing aerodynamics to assess the influence of air pressure on thick airfoils, especially near the critical angle, and to learn how airflow around tapered wings affected cantilevered bracings.[5]

Frederick H. Norton and William G. Brown, NACA TR 153, "Controllability and Maneuverability of Airplanes," (Washington, D.C.: NACA, 1923), (block quote, 552).
[5] RA No. 3: "Tail Pressure Distribution," Joseph S. Ames, 11/28 June 1920, RA 3 File, LaRC Historical Reference Collection; Frederick H. Norton, NACA TR 118, "The Pressure Distribution Over the Horizontal Tail Surfaces of an Airplane" (Washington, D.C.: NACA, 1923), 255 (quoted passages, 255); RA No. 11: "Wing Pressure Distribution," Joseph S. Ames, 11/28 June 1920, RA 11 File, LaRC Historical Reference Collection.

# A BIG PROJECT

During the 1920s, many flight research projects vied for the limited resources available to the NACA. Among the early undertakings, one in particular assumed a special importance. Its complexity, its high technical value, its power to attract the attention of military sponsors, and its interest to the NACA leadership separated it from the others. Begun as the fourth and final flight test RA issued by Joseph Ames on June 11, 1920, it bore the personal imprimatur of George Lewis. When Lewis worked for Clarke Thompson Research he designed an engine supercharger at the request of the Executive Committee. While not the first of its kind, the Roots Experimental Supercharger apparently offered many advantages over competing engine enhancers, including "efficient, simple, and durable" operation.. The NACA leadership decided to let the powerplant laboratory determine its feasibility. If perfected, it represented an important advance in aeronautics which promised to "prevent or reduce the diminution of power output which is experienced with engines of the conventional type as altitude is gained and the air pressure and air density are correspondingly reduced. This is effected by compressing the air charge before it enters the engine cylinders." Research Authorization 35 instructed the Langley staff to first test the device by itself, then to run experiments by fitting it onto a Liberty engine in the Dynamometer Laboratory. "If the results...prove the desirability of further development, it is proposed to continue the tests in free flight under service conditions." Because this research marked the first major project for the engine lab, inevitable delays occurred as equipment and materials were begged and borrowed. Indeed, once the initial ground experiments on the supercharger were finished, George Lewis himself cast about for a spare engine on which to mount his machine. He found one on the other side of Langley Field in a new DeHavilland DH-

7

4B and persuaded the Air Service Engineering Office to loan the aircraft for the testing period.

Because of its obvious potential to boost aircraft performance, the NACA engine investigation drew notice from several quarters. Not only did the Air Service lend an airplane; engineers at McCook Field had already undertaken a similar research program of their own and Lewis instructed Leigh Griffith to visit the Ohio facilities "before any plans are made for the equipping of the DH4B [with the] ...Root[s] type supercharger...." Meanwhile, the Navy Department offered to design and build a propeller suited to the specific requirements of the Roots Supercharger. Even industrialists took an interest. Just one month after the issuance of the Research Authorization from Washington, Leigh Griffith received a letter from a young friend in Santa Monica, California. Donald Douglas wrote with salutations from Griffith's "old town," announced the recent opening of his aircraft plant, and mentioned plans for a new commercial airplane powered by a Liberty engine. Douglas then inquired about the Roots Supercharger, a project he discovered through Griffith's father, the proprietor of a Los Angeles machine shop who held the contract to fabricate the Navy's custom propeller. (The senior Griffith also supplied parts to Douglas Aircraft). Douglas, seeking to increase the speed of his innovative *Davis-Douglas* aircraft, asked the younger Griffith whether he could see the drawings for the supercharger. Griffith obliged but warned that the device required far more testing before being placed in operational use.[6]

Indeed, years of testing lay ahead. After a half year of static tests on the Liberty powerplant, the groundwork for flight research on the supercharger began in summer 1921. But these preparations proved to be time consuming as the transfer of crucial laboratory equipment,

---

[6]RA No. 35: "Roots Type Positive Driven Supercharger," Joseph Ames, 11/28 June 1920, RA 35 File, LaRC Historical Reference Collection (first and fourth quoted passages); Arthur W. Gardiner and Elliott G. Reid, NACA TR 263, "Preliminary Flight Tests of the N.A.C.A. Roots Type Engine," (Washington, D.C.: NACA, 1928), 207, 217 (second and third quoted passages); George W. Lewis to Chief Clerk and Property Officer, Langley Memorial Aeronautical Laboratory, 18 February 1921, RA 35 File, LaRC Historical Reference Collection; George W. Lewis to Leigh Griffith, 1 August 1921, RA 35 File, LaRC Historical Reference Collection (fifth quoted passage); Hansen, *Engineer in Charge* (reprint of a letter from Donald W. Douglas to Leigh Griffith, 8 July 1920), 570-571 (sixth quoted passage).

engine parts, and fittings from the McCook Field Engineering Division took some time. Even with the Air Service's generosity, many one-of-a-kind items still needed to be machined by hand in the Langley Lab's metal and wood shops, further delaying the flight program. Writing on the last day of 1921, George Lewis expressed impatience with the slow pace of bringing his machine to fruition. He sent the powerplants staff at Langley a recent paper on air vibration in intake pipes, a recent cause for concern among participants in the project. Further, he reminded Griffith of their conversation in Washington in which Lewis "deem[ed] it advisable that you complete the first part of the...supercharger report as soon as possible...." Yet, only when the DeHavilland finally arrived at the NACA workshop in September 1922 did its mechanics finally comprehend the complexities of mating the Roots enhancer--complete with strange wind-driven fuel pumps (or blowers)--with the unfamiliar French aircraft before them. Engineers Arthur Gardiner and Elliott Reid encountered a device powered by the crankshaft through a flexible coupling. Below it were intake ducts which opened outside the engine cowling and above it, a cylindrical receiver with two outlets. The two outlets consisted of a short open-end pipe on the top of the receiver with a butterfly valve to control the supercharger; and a duct which extended along the top of the engine and connected to the intake passages of the carburetors. The pressure in this duct varied with the amount of air allowed to escape into the atmosphere through the by-pass valve.[7]

The flight test program uncovered a persistent flaw: radiator heating became a nagging problem associated with the Roots Supercharger. In order to put the system through the most grueling conditions, the flight test maneuvers placed the engine under the most severe stresses. The pilots launched the plane into continuous--not the more leisurely and commonly flown "saw-tooth"--climbs, both to duplicate the military environment and to subject the motor, and its cooling system, to the maximum duration of uninterrupted output. Under such trials the supercharger raised water temperatures at all altitudes to the boiling point. The DH-4 attained

---

[7]George Lewis to Leigh Griffith, 20 September 1921, RA 35 File, LaRC Historical Reference Collection; George Lewis to Leigh Griffith, 31 December 1921, RA 35 File, LaRC Historical Reference Collection (first quoted passage); Gardiner and Reid, NACA TR 263, "N.A.C.A. Roots Type Engine," 208; Hansen, *Engineer in Charge*, 480.

19,500 feet but the engine failed to supercharge over the 10,000 foot level. Small air leaks and wider rotor clearances than used in the dynamometer apparently contributed to the poor performance. The overheating difficulties only receded with the acquisition by the NACA of costly, French-made Lamblin radiators. Their price--$700 to $800 apiece with shipping--drove George Lewis to look for an angel to pay for them and left Griffith wondering whether the manufacturer "want[ed] to bring enough [American] money to their country to pay the interest on their war indebtedness." Admiral William Moffett, Chief of the Bureau of Aeronautics, saved the day for the NACA by loaning as many of these French units as needed--but not before Navy inspectors added to the frustrations by rejecting the first shipment due to extensive damage.[8]

Once the radiator went into service and the cooling problems abated, the Navy became all the more interested in the Roots Supercharger. A Bureau of Aeronautics inquiry routed to Griffith in February 1923 proposed a much expanded program. Rather than the Liberty engine, the Navy preferred adapting the Roots to *three* different powerplants: a Lawrance D-1, a Wright E-2, and an Aeromarine U-S-D, all mounted on airframes other than the DeHavilland. Griffith submitted to Lewis an intensive eight month program based on the Navy overture and requested the authority to hire eight new employees, including engineers, draftsmen, and machinists. His plan provided for continued flying tests of the Liberty-equipped DH-4; design, fabrication, dynamometer and flight research on the three Navy models; improvements in the rotary type of supercharger represented by the Roots; development of a fan supercharger; investigation of hand versus automatic controls; and inquiries into drive shaft coupling. Griffith estimated total costs of at least $37,500, of which all but $4,000 would be supplied by the Bureau of Aeronautics. Lewis agreed with Griffith's overall assessment, said the Navy's interest in the program "is really an excellent thing," but astounded his junior colleague in one particular: he told Griffith to hire

---

[8]George Lewis to Leigh Griffith, 13 September 1922, RA 35 File, LaRC Historical Reference Collection; Gardiner and Reid, NACA TR 263, "N.A.C.A. Roots Type Engine," 210; Leigh Griffith to George Lewis, 14 September 1922, RA 35 File, LaRC Historical Reference Collection (quoted passage); George Lewis to Langley Memorial Aeronautical Laboratory (hereafter, LMAL), 22 December 1922, RA 35 File, LaRC Historical Reference Collection; William A. Moffett to the NACA, 23 January 1923, RA 35 File, LaRC Historical Reference Collection.

only *one* new person (an engineering draftsman) until five months into the eight month endeavor. At that time Griffith could increase the staff and buy equipment with funds appropriated for the new fiscal year starting July 1, 1923. The canny Lewis really had no choice; the fledgling NACA had already run up a deficit and he thought it wise to use the Navy windfall to balance the books. Griffith, of course, had no choice but to comply, despite the implications of workload and scheduling.[9]

By Spring 1923 the Roots Supercharger had demonstrated its value in bench evaluations and in flight tests. After 120 hours in the dynamometer and 20 hours in free flight the results looked almost too good to be believed, demonstrating a new technology as useful for commercial as for military applications. The DH-4B/Liberty engine/Roots Supercharger combination proved capable of achieving an altitude of 20,000 feet in just 20 minutes. Moreover, while an unaided engine could propel an average aircraft at 100 feet per minute up to 6,300 feet, a supercharged one could maintain the same rate of climb *up to 11,500 feet*. Furthermore, the Roots device allowed aircraft to travel faster at high altitudes than similar planes with regular engines flying near ground level. Carrying a load of 1,000 pounds, a supercharged aircraft reached 8,000 feet in forty minutes compared to an hour for its unaided counterpart. Whatever mechanical difficulties the researchers encountered (such as interruptions in the smooth operation of the motor due to the discharge from the supercharger's blower) appeared to be solved. Even at 17,000 feet the Roots device consumed only 40 horsepower from an engine capable of 400 horsepower at high altitude. Moreover, those involved in the tests believed the weight burden of the experimental model (some 185 pounds) could be reduced 40 percent for a production version. The only decline in performance, and a slight one at that, occurred at low altitudes as a result of the non-variable pitch propeller demanded by the supercharger.

---

[9]Leigh Griffith to George Lewis, 3 February 1923 with enclosure: "Supercharger Development Enlarged Program to Cover U.S. Navy Request" and "Estimated Cost Enlarged Supercharger Program, Period of Six Months," RA 35 File, LaRC Historical Reference Collection; George Lewis to LMAL, 6 February 1923, RA 35 File, LaRC Historical Reference Collection; Quoted passage from George Lewis to LMAL, 17 February 1923, RA 35 File, LaRC Historical Reference Collection.

Among others, M.I.T.'s Jerome Hunsaker declared himself "greatly interested and pleased" with these findings. By now, Hunsaker had progressed far in a dual career; both a distinguished academic *and* a Navy Commander designated Chief of Design in the Bureau of Aeronautics. In the latter role, Hunsaker oversaw a Navy contract with Donald Douglas, his former pupil, to build the Davis-Douglas airplane. Hunsaker agreed with Douglas that the aircraft taking shape on the factory floor in Santa Monica--equipped with a Liberty engine and capable of high performance--might be a worthy candidate for the NACA power booster. The commander found the developments so impressive that he asked Lewis for the free flight data in order to make his own calculations about altitudes possible with the Roots system. The Langley engineers sent Hunsaker detailed sketches of the device and its installation, as well as the recent flight results.[10]

The enthusiasm of this powerful Navy friend of the NACA broadcast the importance of the research to supercharging existing naval aircraft. In November 1923 a Curtiss TS-1 airplane powered by a Lawrance J-1 motor arrived at Langley. Since the J-1 motor was cooled not by water but by air, it offered distinctly different problems from those encountered during the first three and one half years of the Roots project. Moreover, the researchers knew of no one who had yet undertaken any analyses of air-cooled supercharging, so they began without any instruction from the past. For example, it remained to be seen whether the Lawrance powerplant even radiated sufficient heat to permit supercharging. Marsden Ware and Arthur Gardiner, the two mechanical engineers leading the investigation, decided they could determine this critical factor in only one, rather dangerous way: by first flight testing the powerplant and measuring the temperature of its cylinder walls; then by inspecting the engine's physical integrity after the tests.

Their flight research program "progressively increas[ed] the amount of supercharging in successive flights...with a view to obtaining the maximum amount of data with the least

---

[10]Progress Report, "Roots Type Supercharger," 26 May 1923, RA 35 File, LaRC Historical Reference Collection; Gardiner and Reid, NACA TR 263, "N.A.C.A. Roots Type Engine," 207, 214, 215; George Lewis to LMAL, 17 February 1923, RA 35 File, LaRC Historical Reference Collection (quoted passage).

likelihood of delays due to engine failure, either from overheating or from insufficient strength." It reflected a typically cautious NACA approach. Ware and Gardiner established a baseline of ground level carburetor pressure at 15,000 feet; once achieved, the Bureau of Aeronautics' objective of "very good performance at 15,000 to 18,000 feet" would be attainable. They decided to precede the supercharged flights with a standard aircraft performing the same maneuvers, providing a basis of comparison for cylinder wall heating. To achieve the upper limits of supercharging, the engineers conducted careful inspections of the engine between flights, took temperature readings, and monitored pulsations in the air ducts. In actuality, Ware and Gardiner realized this intensive research only began the exploration of air-cooled supercharging. "The successful completion of the program," they wrote, "will establish the suitability of the engine for supercharging, but the *limit* of supercharging will not necessarily have been reached."[11]

Their modest claim proved to be prophetic. In only a few years, 20 of the aircraft in the U.S. Pacific Fleet benefited from the boost provided by the NACA Supercharger, not only in climbing to higher altitudes, but in catapulting off the decks of the Navy's great ships. In 1928 George Lewis made the flat claim that the Roots Supercharger represented one of the NACA's outstanding innovations, capable of increasing engine horsepower "at least fifty percent above that at normal sea-level operation."[12]

[11]Hansen, *Engineer in Charge*, 481; George Lewis to LMAL, 20 March 1923, RA 35 File, LaRC Historical Reference Collection; George Lewis to Marsden Ware and Thomas Carroll, 2 July 1923, RA 35 File, LaRC Historical Reference Collection; Quoted passages from Marsden Ware and Arthur W. Gardiner, "Supercharging the Lawrence J-1 Air Cooled Engine," 15 November 1923, RA 35 File, LaRC Historical Reference Collection.
[12]E.S. Land to LaRC, 8 January 1924, RA 35 File, LaRC Historical Reference Collection; Leigh Griffith to George J. Mead, 26 July 1924, RA 35 File, LaRC Historical Reference Collection; Marsden Ware and Arthur Gardiner, "Preliminary Report on Supercharging the Lawrance J-1 Engine (Air-Cooled)," 30 August 1924, RA 35 File, LaRC Historical Reference Collection; Roland, *Model Research*, 2: 651 (quoted passage).

PLANES FALLING FROM THE SKY

The worthwhile flying experiments conducted by Edward Warner in 1919 and by the supercharger investigators in the years just after the laboratory opened proved the value of the NACA's flight research program. To continue to exploit its promise, the NACA turned its attention to its flying infrastructure. For example, Warner had already demonstrated the great utility of a simple speed course to improve the accuracy of charting airplane velocity. An NACA Subcommittee on the Speed Course proposed a far more advanced system at the end of 1923. It conjured a state-of-the-art, two mile long runway on the western side of Langley Field, one capable of far more exacting measurements than the simple visual system used in 1919. The actual timing portion--framed on either side by a half mile approach for safety--occupied a mile long track bounded on its northern extremity by a hangar and by the NACA machine shop 5,600 feet to the south. Used for high speed tests over its full length, the course was halved from the midway point to the hangar to test slower aircraft. Fifty yards west of the flight line, gun cameras and precision chronometers mounted on four concrete stations measuring four by eight feet served as observation posts. The data received from these instruments, as well as from telephone communications, flowed by cable from each platform to the main laboratory.

The frugal George Lewis approved the plan but offered to support it with only $500 to $600, a token figure to clear terrain and build structures on land beset by undrained swamp, army gun butts, a field of tree stumps, a stand of trees, high weeds, and holes. As in the past, he depended upon his friends in the armed forces for the costly necessities. He pressed Leigh Griffith not only to gain the necessary approval for the new facility from Major Oscar Westover (the Langley Field Commander), but to persuade the officer to detail some enlisted troops to ready the land for runway construction. The NACA agreed to absorb the costs only of the concrete structures and the purchase and installation of the required equipment. In explaining the request, Griffith gamely reminded Westover that the "[r]esearch work now in hand for the Air Service makes it highly desirable that the speed course be completed and available for use by the first of March, 1924." Griffith conferred with Westover at the start of that year, and the Langley Commander approved the project in a "spirit of cooperation and...helpful attitude," resulting in an agreement in which the service promised to spend $6,000 to cover the cost of all improvements. In February, Westover gave Griffith the go-ahead to fabricate the platforms and designated a group of soldiers to pave the landing strip. Thus, Lewis' low budget scheme succeeded; he even persuaded Air Service authorities to donate 8700 feet of underground cable for the link-ups between the platforms and the laboratory. In fact, while the concrete was being poured for the four stations, the Optical Shop at the Washington Navy Yard gave the NACA five gun cameras and thirty rolls of film to use on the observation posts. In the end, Lewis opened a modern speed course for under $1,000--less than 1/300th of the NACA's 1924 appropriations.[13]

At the end of the year in which Westover's labor force filled and smoothed the western side of the Army reservation, a pivotal new flight research project materialized. It possessed many of the features of the supercharging project, but on a far grander scale: powerful patrons in both the military services; a research subject of the highest importance to safe military and

---

[13]RA "F", no date, RA F File, LaRC Historical Reference Collection; Everett P. Lesley to Leigh Griffith, 19 November 1923, RA F File, LaRC Historical Reference Collection; George Lewis to LMAL, 30 November 1923, RA F File, LaRC Historical Reference Collection; George Lewis to LMAL, 3 December 1923, RA F File, LaRC Historical Reference Collection; Everett Lesley and D. L. Bacon to Leigh Griffith, 20 December 1923, RA F File, LaRC Historical Reference Collection; Leigh Griffith to Commanding Officer Langley Field, 2 January 1924, RA F File, LaRC Historical Reference Collection (quoted passage); Leigh Griffith to George Lewis, 19 January 1924, RA F File, LaRC Historical Reference Collection; Oscar Westover to Leigh Griffith, 6 February 1924, RA F File, LaRC Historical Reference Collection; Leigh Griffith to George Lewis, 9 February 1924, RA F File, LaRC Historical Reference Collection; George Lewis to LMAL, 11 February 1924, RA F File, LaRC Historical Reference Collection; Leigh Griffith to George Lewis, 19 February 1924, RA F File, LaRC Historical Reference Collection; W.H. Frank to Charles Walcott, 10 March 1924, RA F File, LaRC Historical Reference Collection; George Lewis to LMAL, 14 March 1924, RA F File, LaRC Historical Reference Collection; Hansen, *Engineer in Charge*, 166, 428.

commercial flight; a project whose demands stretched the resources of the lab and required eight years to complete; and an investigation which absorbed some of Langley's best minds in the pursuit of an aeronautical mystery. In pursuing this line of inquiry--the study of the pressure loading of aircraft structures-- not only did the researchers enlist the long new speed course and its air strip, they pressed into service every other flight research asset at the laboratory. The human and material wherewithal had increased greatly during the early 1920s. Compared to 1920, the number of engineers, pilots, and office staff rose threefold to 36 while the machinists, woodworkers, and other blue collar employees now numbered 62, totaling a full-time complement of 98. The $307,000 budget at Lewis' disposal in 1924 represented about a 45 percent rise over 1920. Aircraft inventories rose accordingly. From a mere two on hand at the opening of the lab, the hangars now held ten. While it constituted a varied stock of Voughts, DeHavillands, Curtisses, Douglases, and others, all arrived on loan. Only in 1924 did the NACA order the first airplane that it purchased outright, a Boeing PW-9. Indeed, the sturdy PW-9 apparently held the distinction of being the first American vehicle since the Wright Flyer built expressly for flight research. In placing this order with the Seattle manufacturer, the Committee gambled on arousing the suspicion of a Congress convinced that plenty of surplus military planes existed in the Army and Navy reserves for NACA work. But George Lewis again won the day. First, he instructed his keen-eyed secretary John Victory to review the statutes and determine whether the purchase violated federal law. Then, convinced of its legality, he took Victory's advice and informed the House Appropriations Committee that the purchase had been made. Lewis appeared before the House Independent Offices Subcommittee and persuaded the members of the necessity of the purchase, arguing that the punishing flight loads tests planned for this particular specimen required Boeing to strengthen the tail and fuselage to exacting NACA specifications. No production model, either military or commercial, would suffice.[14]

Why did the NACA take risks to pursue this one project? Partly, it allowed Lewis to

---

[14]Hansen, *Engineer in Charge*, 161-162, 413, 428, 479-482; John Victory to George Lewis, 19 August 1924, RA 138 File, LaRC Historical Reference Collection.

establish the precedent that the NACA needed to *own* at least a small fleet of aircraft in order to accomplish its flight research mission. But far more importantly, the acquisition represented the dawn of some very important work, indeed. Informal discussions about the structural loads assignment started as early as spring 1924, well before Lewis and Victory decided to buy the PW-9. The undertaking originated in a rash of fatal Air Service crashes resulting from catastrophic failures of aircraft in flight. Four deaths and many near-disasters occurred in a variety of machines, including a Fokker PW-7, a Curtiss PW-8, and a Boeing PW-9, the latter of which suffered a total structural collapse in the air. Incidents involved the wings, rudders, struts, and stabilizers. The Engineering Division at Wright Field attempted to learn the dimensions of the mystery first-hand. It assigned a man who embodied extraordinary piloting skills, great technical sophistication, and uncommon courage to conduct flight tests to determine the forces causing military planes to fall from the sky. James H. Doolittle grew up in Alameda, California, and studied engineering at Berkeley before joining the Aviation Section of the Army Signal Corps in 1917. After the war he returned to the University of California and entered its military aeronautics program, passed flight training at Rockwell Field, and received his bachelor's degree in 1922. Over the next few years Doolittle mixed daring flight achievements with advanced scholarship. He became the first to fly coast-to-coast (from Pablo Beach, Florida, to San Diego) in less than 24 hours. Then he enrolled in Jerome Hunsaker's aeronautical engineering program at M.I.T. and eventually earned a doctorate. Looking for a way to combine his flying acumen with his academic studies, he found the perfect Master's Thesis subject: an investigation of the strange failures experienced by the Air Service planes.

The twenty-eight year old flier returned to McCook Field in March 1924 and conducted a carefully conceived yet perilous series of flights. His approach differed somewhat from the conservative, cautious flying practices followed typically by the Langley pilots (although these men also threw caution to the wind occasionally, as described later in this chapter). Indeed, because of his unusual background he acted both as flight research pilot and as chief engineer. Strapped into an Air Service PW-9--the NACA's own model was still under negotiation--he flew acceleration patterns designed to bring the aircraft to the brink of disaster. Doolittle used an accelerometer (similar to the one devised by Frederick Norton of the NACA), which he placed in a box packed with rubber sponges to dampen incidental vibration. Doolittle's grueling regimen of accelerated flying included loops at speeds up to 160 miles per hour (m.p.h.), yielding forces up to 6.1 times that of gravity (g). He flew single, double, triple, and quadruple barrel rolls, the latter at speeds as high as 160 m.p.h., producing 7.2 g. He maneuvered the PW-9 into power spirals at velocities up to 140 m.p.h., flown at full throttle banked up to 70 degrees, and resulting in 5.3 g. Doolittle tried a variety of miscellaneous maneuvers (rolls at the top of loops, Immelman turns, vertical banks, spins with engine throttled, and spins with full power), the most effective of which (vertical banks) produced 5.7 g. at an approach of 150 m.p.h.. He experienced flight in rough air, resulting in a maximum of only 2.2 g. Finally, he put the aircraft and himself through a series of hair-raising stunts in which he pulled suddenly out of dives, exerting gravitational forces ranging from 5 at 130 m.p.h. to 6.5 around 150 m.p.h. He achieved these results in 10 m.p.h. increments between 60 and 160 m.p.h. (Researchers also extrapolated the effects of pulling out of a 220 m.p.h. dive, a maneuver calculated to exert *14 g.* on the pilot). After the dive tests, engineers inspected the airplane's wings and found that "the veneer covering of the upper wing, on the under surface, had split from the trailing edge to the rear spar....In this particular construction there is no drag bracing between the spars; the veneer covering replaces it. The failure demonstrates clearly that the wing has deflected up and forward under the load." Doolittle concluded that pull-ups from dives posed the greatest danger (although barrel rolls at the same speed caused stresses only five or ten percent less). He pinpointed four elements affecting the extent of pressure loading on aircraft: the relationship between the diving velocity and the minimum speed, the degree of longitudinal stability, the damping due to pitch, and the time necessary to change the elevator's angle of attack from small to large. Doolittle warned that only in high performance military pursuit planes did high loads present extreme hazards; all other aircraft faced far less risk due to inherent limitations. Finally, on the PW-9 in question, he reported that although constructed to withstand dynamic loads well in excess of design specifications, in combat conditions where pilots routinely pulled out of dives at 185 m.p.h., the aircraft's wings would fail.[15]

The leaders of the Engineering Division required little time to analyze and reflect on the meaning of Doolittle's report. It revealed a significant but as yet unquantifiable peril to military pilots, suggesting the causes but offering no solutions. Until further research clarified the problem, the Air Service found itself unsure of its capacity to employ nimble, fast, dog-fighting aircraft in combat--or merely in mock combat for training purposes--with reasonable confidence of the safe return of pilots and their vehicles. Indeed, as the performance of military airplanes continued to improve rapidly, the danger became even more acute. Faced with potentially paralyzing uncertainties which lay beyond the competence of the McCook Field staff, the Chief of the Engineering Division enlisted George Lewis and the NACA to conduct the theoretical and experimental research necessary to understand the dynamics of pressure loads in flight. Hence, Lewis' and Victory's machinations during summer 1924 to obtain a special PW-9.

The official request for assistance arrived the third week of September in a memo from Major General Mason Patrick, Chief of the Air Service, addressed to NACA Chairman Charles Walcott. It envisioned a sweeping review of the pressure distribution problem, including an "extensive program of flight tests to obtain pressure distributions and accelerations for the purpose of determining the proper loading to be used in the design of airplanes." Further, because existing methods of computing stress loads had proven to be inaccurate, entirely new means of measurement needed to be developed and entirely new and comprehensive data needed to be gathered in order to arrive at design formulae applicable to the wide range of military aircraft, not just to specific ones. The universality of the results was underscored: "[t]he program should be sufficiently extensive to cover the determination of all flying loads likely to be critical, and should include accelerometer tests and pressure distribution tests on all surfaces."

---

[15]Anon., *Against the Wind: 90 Years of Flight Test in the Miami Valley* (Dayton, Ohio: Aeronautical Systems Center, 1994), 10, 11; James H. Doolittle, NACA TR 203, "Accelerations in Flight," (Washington, D.C.: NACA, 1925), 373-388 (quoted passage on 386-387).

The memo posed 18 difficult questions, which in themselves indicated the wide scope of the undertaking, the demand for fundamental inquiry, and the high importance the Air Service ascribed to the findings.

1. What are the maximum loads to which wings are subjected, and under what conditions of flight are they likely to occur?

2. How does the total load vary with respect to time and angle of attack while pulling out of a dive at high velocity?

3. What is the history of the change in pressure distribution on the wings while pulling out of a dive?

4. Will the maximum load be determined by the design of the airplane or by the physiological effects on the pilot?

5. What is the history of the acceleration and pressure distribution changes in other maneuvers that may be critical for some members? For instance, in the barrel roll and other maneuvers in which the loading is unsymmetrical?

6. What are the effects of changes in wing section, aspect ratio, [and] taper...upon the pressure distribution?

7. Do the variations above also depend on the angle of attack, or air speed?

8. What is the effect of the design of the tail surfaces on the maximum loads on the wings? For instance, does the use of balanced elevators permit greater loads to be obtained than with unbalanced elevators, and if so, how great is this effort?

9. Is the pressure distribution dependent upon any other variables, and if so, what is their effect?

10. Does the center of pressure vary along the span at any given moment, and, if so how?

11. Of what variables is the center of pressure a function, and what are the relationships?

12. To what loads are the ailerons subjected? When do they occur, and what is the effect of moving the ailerons upon the loads on the remainder of the wing?

13. When do the tail surfaces receive their greatest loads, and what are the relationships affecting their magnitude and distribution?

14. What loads are on the tail surfaces when the wings are subjected to their maximum load?

15. What are the tail loads when the rear spars are subjected to their maximum load?

16. What are the loads on the horizontal tail surfaces when the vertical ones are subjected     to their maximum load, and vice versa? What is the worst combination?

17. Are the air loads on the fuselage and chassis ever of importance and, if so, what is their magnitude under various conditions?

18. What are the accelerations of parts of the airplane other than the center of gravity in the various critical load conditions?[16]

When the request arrived in Washington, George Lewis showed it to Joseph Ames, now chairman of both the Executive and the Aerodynamics Committees. An experienced

---

[16]Mason Patrick to Charles Walcott, 18 September 1924, RA 138 File, LaRC Historical Reference Collection.

administrator, Ames realized the immense investment in time and resources represented by the Air Service inquiry and asked Lewis to canvas the Langley staff for suggestions and comments. An able young assistant aeronautical engineer named John W. "Gus" Crowley (M.I.T, mechanical engineering, 1920, soon-to-be chief of the Langley Flight Test Section and the future Associate Director for Research at NACA Headquarters) got the assignment to reply to Lewis and Ames and issued a short report after two months of investigation. Crowley realized the questions posed by the service could not be answered fully with flight research conducted on one aircraft type. But, upon meeting with the two Air Service representatives who drafted the memo for General Patrick, Crowley won their support for the expeditious approach: they agreed to confine the flight research to the PW-9 but to obtain "much information relative to most of the questions." After consultations with the NACA technical staff, Crowley proposed free flight tests to measure simultaneously the pressures on wings and tail surfaces during violent maneuvers and to record continuously the fluctuations in loads, airspeeds, accelerations, and so on. He predicted the need for wind tunnel investigations to supply answers to some of the questions but conceded these tests had not yet been formulated.

Meanwhile, the Langley staff negotiated appropriate modifications on their PW-9. Chief pilot Thomas Carroll met with George Tidmarsh, Boeing's Washington, D.C., representative, and asked for three essential design changes. To approximate the aircraft's original flying characteristics Carroll wanted to retain the aircraft's standard weight, achievable by removing the main gasoline tank and by replacing it with no more than 150 pounds of flight research instruments. His experience in the cockpit told him that the PW-9 really needed no more than a 15 to 20 gallon fuel capacity for the short flights common to flight research programs. Second, the two men agreed that the pressure orifices should be introduced on false ribs installed on the wing and tail surfaces, and Carroll suggested the manufacturer use aluminum rather than leak-prone rubber tubing in fabricating these new structures. Third, despite the punishing regimen to be imposed on this special aircraft, the NACA aviator declined to impose any numerical standard of structural reinforcement; rather, he asked the Boeing engineers to design such features as the tubular compression members, the wing box spars, and the flying wires "to a logical maximum." Indeed, Carroll felt that "the strengthening of the airplane is an engineering matter which the manufacturers...can best solve." The two men parted with the understanding that Tidmarsh would "hurry [the project] as much as possible." Boeing's Chief Engineer C.L. Egtvedt complied with Carroll's suggestions in a matter of days and returned to George Lewis a general plan for the modifications. Reducing the normal 75 gallons to 20, removing the main fuel tank, and taking off the plane's regular armament resulted in a total savings of 593 pounds. Adding 150 pounds of instruments still yielded a PW-9 443 pounds lighter than the production model. The reduction in weight alone--without any new strengthening or re-design--meant an aircraft with a 17 percent higher factor of safety, according to Egtvedt's calculations.

When Gus Crowley reviewed the Carroll-Egtvedt plan for modifying the NACA PW-9 he could not resist the supervisory engineer's temptation to qualify and to amplify the work of subordinates. He noted that with *all* of the additional weight--batteries as well as instruments, the tubing, and the false ribs--the total reduction amounted to only about 370 pounds. Crowley also advised Leigh Griffith and George Lewis not to be content with reducing stresses merely by reducing poundage. Whatever savings Boeing achieved "should be used to strengthen the whole airplane structure as much as possible as this airplane is to be used in particularly violent maneuvering." Thus, he recommended bracing the front truss (spars, wires, and fittings) and the rear fuselage. But in all other respects, Crowley accepted the outline agreed upon by Carroll and Tidmarsh. The NACA leadership also concurred and on December 2, 1924, Joseph Ames signed Research Authorization 138, "Investigation of Pressure Distribution and Accelerations on Pursuit Type Airplanes." The RA's short but encompassing statement of purpose underscored the project's fundamental importance to aeronautics: "To determine the proper loading to be used in the design of airplanes."[17]

---

[17]George Lewis to LMAL, 24 September 1924, RA 138 File, LaRC Historical Reference Collection; John W. Crowley, "Memorandum On the Letter From The Office of the Chief of Air Service on Pressure Distribution Tests," 24 November 1924, RA 138 File, LaRC Historical Reference Collection (first quoted passage); Thomas Carroll to George Lewis, 13 October 1924,

During 1925 the laboratory prepared itself for this auspicious undertaking, but not without some delays and frustration. First the principals needed to decide how to mesh the wind tunnel tests with the flight research. To answer Air Service questions six and seven (how did changes in wing section affect pressure distribution; and did the changes depend on airspeed or angle of attack), George Lewis corresponded with Dr. Max Michael Munk, Chief of the Langley Aerodynamics Division. A German emigré who took his degree with the legendary aerodynamicist Professor Ludwig Prandtl of Göttingen, Munk assumed the position of technical assistant to the NACA in 1920 at the age of 30. Brilliant, wildly prolific, yet shockingly arrogant, he added a theoretical depth to the laboratory during the six year interlude before he resigned from the laboratory. Munk suggested that the tunnel phase of pressure distribution research concentrate on the wing tips using thick and thin wing sections, tapered and non-tapered, cambered and non-cambered. He also suggested some new techniques for measuring pressure distribution and he proposed increasing the number of orifices on the aircraft surfaces. But where Munk was involved, dissent often raised its head. At the end of 1924 a young aeronautical engineer named Elliott Reid who ran the atmospheric wind tunnel locked horns with the German and declared himself "not in agreement with his [ideas about] the merits and importance of the...[Air Service] questions....Dr. Munk's suggestion of a new method of recording pressure distribution was thought decidedly impractical." After trying to compose his differences with Munk, Reid presented Leigh Griffith with an alternate proposal. Reid argued that the Air Service request had such wide compass that if undertaken fully the atmospheric wind tunnel would have to service just this one project. He therefore recommended the use of related data collected previously at Langley, at Göttingen, and at St. Cyr. He also rejected Munk's

---

RA 138 File, LaRC Historical Reference Collection (second to fourth quoted passages): C.L. Egvedt to the NACA, 27 October 1924, RA 138 File, LaRC Historical Reference Collection; John W. Crowley, "Memorandum On Proposed Pressure Distribution Tests On PW-9," 17 November 1924, RA 138 File, LaRC Historical Reference Collection (fifth quoted passage); RA 138: "Investigation of Pressure Distribution and Accelerations on Pursuit Type Aircraft," Joseph Ames, 2 December 1924, RA 138 File, LaRC Historical Reference Collection (sixth quoted passage).

proposal to study wing tip behavior because previous experiments revealed an absence of excessive loading there. Rather, the Langley tunnel could be employed to compare the disparities between pressure distribution on a model MB-3 airplane and a full-scale PW-9 at angles of attack encountered in flight. If the free flight and tunnel tests agreed, the model's proportions could be modified to complement and to augment the data collected in the flight research program. Once the direct comparisons ended, then the specific points raised in Air Service questions six and seven might be explored.

The actual wind tunnel program proved to be a victory for Reid. By February 1925 Lewis and Griffith had decided that first the direct comparison between full-sized and model tests would be run, following which the other questions would be conducted in accordance with the Air Service request. But in linking flight research and wind tunnel work, Leigh Griffith and George Lewis found themselves in a quandary: should a new Research Authorization be issued to accommodate the non-flying part of the project or should the two halves, which were inextricably bound together, be joined in one massive effort? Lewis, always the diplomat, arrived at the Solomonic answer. "Would it be well," asked the Research Director, "to have authorized by the Executive Committee an extension of or addition to Research Authorization No. 138, which we might designate as, say No.138A and 138B?" This system not only settled the problem of the pressure distribution project, but gave Lewis a bureaucratic method of expanding future Research Authorizations without watching them proliferate beyond his control. It also allowed for tidier record keeping. Griffith, to his credit, suggested making the wind tunnel work an *appendix* to the existing project to prevent new research from gobbling up the manpower and the resources of the primary project. These thoughts congealed on February 18, 1925, when the Executive Committee approved Research Authorization 138A and the precedent it represented. It bore the same title as Number 138 but added wind tunnel tests and described the whole range of ancillary assignments.[18]

---

[18]Once RA 138A got underway the Air Service announced that it preferred the model tests be run using a Douglas Observation (O-2) aircraft and after consultations with Gus Crowley and Elliott

The Lewis-Griffith formula for broadening existing Research Authorizations received its first test soon after its conception. In June 1925 George Lewis received a copy of a memo sent from the Office of the Chief of the Air Service (acting on behalf of the Engineering Division) to Charles Walcott and the Main Committee. It requested solutions to additional pieces of the pressure distribution puzzle. This time the service inquired about the loads placed simultaneously on the wing tips and the rear of the aircraft during accelerations in rolling maneuvers and pull-outs. In particular, military pilots wanted to learn more about stresses on the tail section while climbing out of dives and military engineers desired a "sounder and more rational" method for computing tail loads. This raised four new questions:

1. What inertia[l] forces on the rear portion of [the] fuselage should be considered as acting simultaneously with the maximum air load on the tail?
2. What air load on the tail should be assumed to act simultaneously with the maximum inertia[l] forces on the rear portion of the fuselage?
3. What is the relationship between the maximum inertia[l] forces at the center of gravity and at its tail?
4. What inertia[l] forces should be assumed to act on the rear portion of the fuselage in connection with the High and Low Incidence conditions on the wings?

Crowley and Carroll both felt these inquiries touched on important but poorly understood

---

Reid, George Lewis acceded to the request in December 1925. Elliott G. Reid to Leigh Griffith, 29 October 1925, RA 138A File, LaRC Historical Reference Collection; John Crowley to Leigh Griffith, 5 November 1925, RA 138A File, LARC; George Lewis to LMAL (Griffith), 24 December 1925, RA 138A File, LaRC Historical Reference Collection; Hansen, *Engineer in Charge*, 50.
     See Hansen, *Engineer in Charge*, 72-92 for a fascinating discussion of Max Munk s contributions to the NACA; George Lewis to LMAL (Griffith), 2 December 1924, RA 138 File, LaRC Historical Reference Collection; Max Munk to George Lewis, 29 November 1924, RA 138 File, LaRC Historical Reference Collection; Elliott G. Reid to Leigh Griffith, 8 December 1924, RA 138 File, LaRC Historical Reference Collection (first quoted passage); Elliott G. Reid to Leigh Griffith, 12 December 1924, RA 138 File, LaRC Historical Reference Collection; George Lewis to LMAL (Griffith), 15 December 1924, RA 138 File, LaRC Historical Reference Collection; Leigh Griffith to George Lewis 18 December 1924, RA 138 File, LaRC Historical Reference Collection; George Lewis to LMAL (Griffith), 4 February 1925, RA 138 File, LaRC Historical Reference Collection (second quoted passage); Leigh Griffith to George Lewis, 11 February 1925, RA 138 File, LaRC Historical Reference Collection; George Lewis to LMAL (Griffith), 26 February 1925, RA 138 File, LaRC Historical Reference Collection: RA 138A: "Investigation of Pressure Distribution and Accelerations on Pursuit Type Aircraft-Wind Tunnel Tests," Joseph Ames, 18 February 1925, RA 138A File, LaRC Historical Reference Collection.

phenomena and agreed the addition of two accelerometers to measure these forces posed no problems. Joseph Ames informed the Office of the Chief of the Air Service that the NACA intended to merge these inquiries with the on-going project and designate it Research Authorization 138B, "Investigation of Pressure Distribution and Accelerations on Pursuit Type Airplanes-Acceleration Readings on the PW-9."[19]

The project now lacked nothing but the necessary equipment. In some cases the undertaking's complexity spawned new types of instruments. For example, measuring angle of attack posed particular challenges. Gus Crowley proposed that the NACA begin a special Research Authorization to perfect a system in which a camera, mounted at a fixed angle, take photographs of the ground. Based on the real size of the objects pictured, researchers could deduce not only angle of attack but altitude, flight path approach, and so on. Crowley also concerned himself with the acquisition and installation of such standard test devices as manometers, airspeed meters, turn meters, accelerometers, and timers. Then there were the more essential items to procure. Because of delays by Boeing in completing the propellers for the PW-9, George Lewis consulted friends at the Bureau of Aeronautics who found a spare one for a PW-7, which the NACA accepted. Not only did they need a propeller. The aircraft required an engine and Joseph Ames, like Lewis, turned to the Bureau of Aeronautics for a loan. The Navy found a re-serviced D-12 and shipped it to Langley about a week after receiving the request. More troubling, the date of delivery of the airplane itself kept being postponed. Boeing's Tidmarsh first told Lewis to expect it first in mid June 1925, then on July 1; but it had not arrived by mid August. Then Lewis received word to expect it the third week in September, only to

[19]George Lewis to John Crowley and Thomas Carroll, 30 June 1925, RA 138 File, LaRC Historical Reference Collection; W.G. Kilmer to Charles Walcott, 24 June 1925, RA 138 File, LaRC Historical Reference Collection (first and second quoted passages); John Crowley to Henry J.E. Reid, 10 July 1925, RA 138 File, LaRC Historical Reference Collection; Henry J.E. Reid to George Lewis, 14 July 1925, RA 138 File, LaRC Historical Reference Collection; Joseph Ames to the Chief of the Army Air Service, 25 July 1925, RA 138B File, LaRC Historical Reference Collection; RA 138B: "Investigation of Pressure Distribution and Accelerations on Pursuit Type Airplanes-Acceleration Readings on the PW-9," Joseph Ames, 19 September 1925, RA 138B File, LaRC Historical Reference Collection.

learn in October that no exact date of shipment had yet been assigned. Finally, it left Seattle for Langley in mid-November and reached Hampton on January 7, 1926. Almost seven months elapsed waiting for the vehicle which would launch the NACA on its biggest flight research project to date.[20]

## A BREAKTHROUGH: THE PRESSURE DISTRIBUTION PROGRAM

To accompany an undertaking of such magnitude, the laboratory, quite accidentally, also received fresh leadership. Not long after their appointments in 1920 as Senior Staff Engineer and Chief Physicist, respectively, Leigh Griffith eclipsed Frederick Norton in rank. Fifteen years older than Norton and a trusted friend of George Lewis, Griffith became titular director of Langley with the title Engineer in Charge, bestowed by the NACA Executive Committee in 1923. Unfortunately, he then became embroiled with the imperious John Victory in a petty misunderstanding regarding correspondence policy. The war of words escalated during 1925. So bad did the situation become that Griffith, despite his closeness to Lewis and regardless of his able stewardship of the laboratory, left on an extended leave of absence, returned to California, and never returned to Langley again. Griffith's replacement proved to be a great success. The

---

[20]John Crowley to Leigh Griffith, 21 April 1925, RA 138 File, LaRC Historical Reference Collection; John Crowley to Henry J.E. Reid, 29 July 1925, RA 138 File, LaRC Historical Reference Collection; Thomas Carroll and John Crowley to Leigh Griffith, 13 May 1925, RA 138 File, LaRC Historical Reference Collection; Leigh Griffith to George Lewis, 14 May 1925, RA 138 File, LaRC Historical Reference Collection; Joseph Ames to Bureau of Aeronautics, 13 May 1925, RA 138 File, LaRC Historical Reference Collection; George Lewis to Thomas Carroll, 12 May 1925, RA 138 File, LaRC Historical Reference Collection; George Lewis to LMAL (Griffith), 11 June 1925, RA 138 File, LaRC Historical Reference Collection; Marsden Ware to the NACA (George Lewis), 31 July 1925, RA 138 File, LaRC Historical Reference Collection; George Lewis to Charles Monteith, 27 August 1925, RA 138 File, LaRC Historical Reference Collection; George Lewis to Thomas Carroll, 1 September 1925, RA 138 File, LaRC Historical Reference Collection; George Lewis to LMAL (Griffith), 30 October 1925, RA 138 File, LaRC Historical Reference Collection; George Tidmarsh to the NACA (Lewis), 31 October 1925, RA 138 Files, LaRC Historical Reference Collection.

new Engineer in Charge Henry J.E. Reid brought a quieting presence, a disciplined ego, and a firm commitment to solid research. An electrical engineer who graduated from the Worcester Polytechnic Institute, the 30 year old Reid had worked for five years in the Langley instrument section before his appointment, perfecting the velocity-gravity recorder for flight research. Because instrumentation was essential to so many of the laboratory's tasks, Reid had became known by many and respected by most. Indeed, despite his heavy administrative burdens he continued to contribute to his technical specialty, a commitment which counted for much among the NACA staff.

Henry Reid began his tenure with a project worthy of his skill and patience, one which held great promise for the NACA and for aeronautics as a whole. Because of its size, complexity, and potential impact, the flight loads research commanded his attention. The surprises were many. During the first inspection of the NACA PW-9, NACA test pilot Paul King discovered that Boeing delivered the plane without instruments. It lacked such essentials as a gas control valve, a throttle, a fuel gauge, and an oil pressure gauge. Once more, George Lewis importuned his friends at McCook Field and by the start of February 1926 most of the needed parts had arrived at Hampton. It still required several months for the PW-9 to be checked out fully and for all of its instrumentation to be designed, tested, and installed. By late September Lewis expressed concern about the delays. Reid responded on October 6, saying the flying tests were underway. They began with flights by Thomas Carroll and Robert Mixon not on the PW-9, but in a Curtiss JNS-1 airplane to determine whether the pressure measured at the orifices of various lengths of tubing differed significantly from the pressure readings taken at the manometer cells. The mechanics attached three sets of tubing (ones 5 feet and 15 feet; ones 5 feet and 25 feet; and ones 5 feet and 50 feet) to the wings of the Curtiss. Carroll and Mixon flew the aircraft with one set of tubes at a time. They made glides with power off up to 90 m.p.h. and pull-ups from the most gentle to the most severe. The results showed an experimentally insignificant difference between the longest tubes, the 5 foot tubes, and the manometer readings.[21]

Then began the preliminary flying tests on the PW-9. This machine had the look of
durability. Its short, thick fuselage, heavily braced pair of wings, big tires, and broad vertical tail
suggested a plane able to take hard use. The flight research program tested its ruggedness to the
limit. As it flew, all of the main surfaces--wings, horizontal stabilizer, elevator, vertical fin and
rudder--underwent moment to moment stress analysis at multiple points. During the many
flights the pilots subjected the aircraft to almost every conceivable condition of flight, including
the most violent maneuvering possible during dives, loops, barrel rolls, and pull-ups. The two
manometers allowed simultaneous, continuous recording at 120 locations on the airplane for four
minutes at a time--long enough to chart each maneuver completely. Accelerometers designed by
the NACA staff recorded the air speed at the wing tips, the tail, and the center of gravity during
pull-ups. The data showed that all three points experienced maximum speeds at the same time,
although the tail received the greatest stresses and the wing tips fluttered just after the point of
maximum acceleration. Good data from the research rolled in quickly. Gus Crowley observed
that "the tests are in progress and only partly worked up. To date no unusual developments in
the air pressures measured have occurred." He called the initial results of the accelerometer
readings "interesting developments." The cautious NACA Director, however, worried that early
success might lead to undue complacency or even to risk-taking; on viewing some of the flight
tests Lewis prohibited any pilot from performing maneuvers which achieved loads greater than 8
g, a force "unsafe for the pilot and the airplane."[22]

---

[21]Roland, *Model Research*, 1:35-87; Hansen, *Engineer in Charge*, 30-32, 421; Paul King to
Henry Reid, 14 January 1926, RA 138 File, LaRC Historical Reference Collection; George
Lewis to LMAL (Reid), 30 January 1926, RA 138 File, LaRC Historical Reference Collection;
W.G. Kilner to George Lewis, 2 February 1926, RA 138 File, LaRC Historical Reference
Collection; George Lewis to LMAL (Henry Reid), 25 September 1926 RA 138 File, LaRC
Historical Reference Collection; Reid to the NACA (Lewis), 6 October 1926, RA 138 File,
LaRC Historical Reference Collection; Thomas Carroll and Robert Mixon, NACA Technical
Note (hereafter NACA TN) 251, "The Effect of Tube Length Upon the Recorded Pressures From
A Pair of Static Orifices in a Wing Panel," (Washington, D.C.: NACA, 1926), 1-4.
[22]John Crowley, "Pressure Distribution and Acceleration Tests on a PW-9 Airplane," n.d., RA
138 File, LaRC Historical Reference Collection (first and second quoted passages); Henry Reid

Inevitably, positive results bred a desire to expand the research. Henry Reid proposed to the Main Committee in December 1926 that the examination of fuselage pressure deserved its own Research Authorization. Although earlier, but incomplete studies estimated the body might support 10 percent of the aircraft's total load, Reid felt designers needed to have accurate and full data about this phenomenon. Moreover, the PW-9's existing configuration for pressure distribution measurement equipped it uniquely for fuselage research once the completion of the wing and tail tests freed the two manometers for this purpose. The fuselage research also offered an opportunity to the wind tunnel section to study acceleration data available only in free flights and to derive from it more efficient fuselage shapes. Accordingly, the Executive Committee approved the extension of the existing work under the title of Research Authorization 138C: "Investigation of Pressure Distribution and Accelerations on Pursuit Type Airplanes-Fuselage Pressure Distribution." But this was not all. On the same day that George Lewis notified the lab about 138C, he gave the go-ahead to broaden 138A based on a request Henry Reid made in December 1926 to add three new tests to the wind tunnel program: positive and negative overhang, effect of tip shields on cross-span loading, and leading edge slots.[23]

Indeed, the wind tunnel work already began to pay dividends. Because the researchers were unable to perform some of the tests in the open laboratory of free flight, the tunnels answered many of the project's conundrums or offered valuable correlations to the flight test data. Paul Hemke, a Johns Hopkins Ph.D. in physics (who, with Elliott Reid resigned from the NACA in 1927 after repeated clashes with Max Munk) contributed research on the size of the

to John Crowley, 16 April 1927, RA 138 File, LaRC Historical Reference Collection (third quoted passage); Hansen, *Engineer in Charge*, 162.

[23]Henry Reid to the NACA (Lewis), 14 December 1926, RA 138C File, LaRC Historical Reference Collection; George Lewis to LMAL (Reid), 1 February 1927, RA 138 File, LaRC Historical Reference Collection; RA 138C: "Investigation of Pressure Distribution and Accelerations on Pursuit Airplane-Fuselage Pressure Distribution," Joseph Ames, 21 January 1927, RA 138C File, LaRC Historical Reference Collection; A.J. Fairbanks to Henry Reid, 9 December 1926, RA 138A File, LaRC Historical Reference Collection; Henry Reid to the NACA (Lewis), 9 December 1926, RA 138A File, LaRC Historical Reference Collection; George Lewis to LMAL (Reid), 1 February 1927, RA 138A File, LaRC Historical Reference Collection.

pressure orifices. Using the six inch NACA tunnel he experimented with the size and shape of

the openings to determine what effect these variations had on air pressure readings. He

discovered that the wider the hole, or the more rounded the edge, the greater the affect on air

pressure over the lifting surface. On a cylinder of one inch diameter an orifice of .06 inches

failed to affect substantially the pressure distribution over the cylinder. This discovery gave

flight research investigators their first clue about what constituted a good standard size for

pressure orifices. A second test report broadened the pressure distribution picture. Andrew J.

Fairbanks, a NACA engineer fresh out of Cornell University, issued a paper in April 1927 which

solved some of the mysteries of pressure induced specifically by the biplane configuration. He

mounted PW-9 wing models in the atmospheric wind tunnel and found that over the full range of

angle of attack (from -6 degrees to +24 degrees) the air pressure on the biplane--compared to the

monoplane--was "almost entirely restricted" to the areas over the lower wing and under the upper

one. Moreover, while it appeared that burbling, or the threshold of reduced aerodynamic

stability, occurred at the same angle of attack in an upper biplane wing as in a single wing, the

lower biplane wing burbled at a *higher* angle of attack than a single wing. Finally, the overhang

of the upper wing caused the lateral center of pressure in biplane wings to spread outward and (at

high angles of attack) forward compared to a wing tested as a monoplane.[24]

Elliott Reid's final research as a NACA employee appeared in print in September 1927.

In it, he pinpointed the importance of the whole pressure distribution problem for the NACA and

for the aeronautics community. "As...aerodynamics is, as yet, in a state of development, rather

than in one of refinement, it is natural that the steady motion of wings through the air has been

studied extensively while the essentials of the accelerated motions remain practically unknown.

The necessity of attacking the latter problem has been felt for some time; the necessity of

investigating the forces which act upon the wings and tail surfaces of modern airplanes during

[24]Hansen, *Engineer in Charge*, 93, 416, 417; Paul E. Hemke, NACA TN 250, "Influence of the Orifice of Measured Pressures," (Washington, D.C.: NACA, 1926), 1-10; A.J. Fairbanks, NACA TR 271, " Pressure Distribution Tests on PW-9 Models Showing Effects of Biplane Interference," (Washington, D.C.: NACA, 1927), 315-325 (quoted passage 325).

rapid maneuvers of which they are capable has focused attention upon this field of aerodynamics." Reid's research in this new avenue of inquiry involved the effects on airfoils at changing angles of attack. He devised an apparatus for the atmospheric wind tunnel in which an airfoil, set at a large angle of attack but able to rotate freely around an axis, measured oscillations on a recording cylinder. Reid thus discovered some of the characteristics of pitching airfoils in motion, versus those held in a steady position.[25]

These three important papers by aerodynamicists Hemke, Reid, and Fairbanks--the last of whom would also leave the NACA by the end of 1927--led George Lewis to wonder whether the wind tunnel research in support of the pressure distribution project should be terminated. Despite the loss of these three crucial employees, Henry Reid backed additional tunnel work under Research Authorization 138A. It materialized in a paper by Oscar E. Loeser, Jr., which broadened the research of Fairbanks by expanding the angle of attack envelope to a range of -18 degrees to + 90 degrees. Loeser's results provided fresh data to correlate with free flight tests in the pursuit of more durable aircraft designs. Again concentrating on biplane, he discovered that when angles of attack rose above the point of maximum lift, a reduction in upper wing pressure occurred due to the shielding action of the lower wing; he found a delay in burble on the lower wing due to the influence of the upper wing, as modified by angle of attack; he learned that the center of pressure on upper wings shifted outward and forward compared to distribution over monoplane wings and lower wings of biplanes; he ascertained that the overhanging portion of the upper airfoil had little impact on the lower one; and he reported a decrease in pressure on both wings (especially the lower) because of mutual interference in the region above zero and below maximum lift. Despite Loesser's contributions, however, with the publication of this report George Lewis stepped in firmly, determined to keep the project focused and on track. Against Henry Reid's wishes he sent a copy of Loeser's paper to the Air Service in partial fulfillment of the overall pressure distribution undertaking. Lewis then halted any more tunnel

_____

[25]Elliott Reid, NACA TN 266, "Airfoil Lift with Changing Angle of Attack," (Washington, D.C.: NACA, 1927), 1-17 (quoted passage, 1-2).

work on the PW-9 project by transferring all remaining tests to Research Authorization 249, devoted to biplane cellules.[26]

Still, Research Authorization 138 continued to be a mansion of many rooms, a project rich in ramifications. As one avenue of investigation disappeared, two others assumed roles of prominence. Late in February 1928. Lewis visited the laboratory and saw a Vought airplane being tested for cockpit pressure in the propeller research tunnel. As aircraft flew higher and faster during the 1920s not only the stresses on airframes, but the reduced accuracy of instruments in the cabin presented formidable dilemmas. This realization prompted Lewis to ask whether the recent cockpit experiments in the tunnel could be recreated in actual flight. Henry Reid told him "[i]t would not be difficult" and the Director's hunch about its importance was seconded by a Navy figure of much influence. Lieutenant Commander Walter S. Diehl, who headed the Bureau of Aeronautics liaison office with the NACA, actually shared office space with Lewis and his staff and became a familiar figure at Langley. Although a construction corps engineer, Diehl devoted his life to aviation. The two men developed a close partnership and Lewis relied on him not only as a conduit for Navy equipment, spare parts, borrowed aircraft, and funds for worthwhile research, but as a sounding board who possessed sharp technical and bureaucratic instincts. On the issue of cockpit pressures, Diehl exercised his influence. Lewis sent him the figures worked up by the propeller tunnel research and Diehl concluded that the loss of pressure caused "an appreciable effect. Why not," he asked, undertake "some more readings, perhaps including flight tests values, to be published as a note? I think the subject is worthy of some notice. There has always been a lot of conjecture but no testing to amount to anything on it." Attaching Diehl's comments, Lewis directed Reid to begin the research and to charge the

---

[26]George Lewis to LMAL (Reid), 30 August 1927, RA 138A File, LaRC Historical Reference Collection; Henry Reid to the NACA (Lewis), 1 September 1927, RA 138A File, LaRC Historical Reference Collection; Henry Reid to the NACA (Lewis), 13 March 1928, RA 138A File, LaRC Historical Reference Collection; George Lewis to Leslie MacDill, 14 April 1928, RA 138 File, LaRC Historical Reference Collection; Oscar E. Loeser, Jr., NACA TR 296, "Pressure Distribution Tests on PW-9 Wing Models From -18 Degrees Through -90 Degrees Angle of Attack," (Washington, D.C.: NACA, 1929), 335-353; Henry Reid to the NACA (Lewis), 7 August 1928, RA 138 File, LaRC Historical Reference Collection.

time and material required to Research Authorization 138. Reid launched the project a week later with preliminary tests flown in the recently overhauled Vought under conditions of high accelerations and high angles of attack. Manufacturers responded quickly and positively to these developments. Charles Colvin of Pioneer Instruments expressed a keen interest that the tests be conducted on closed-cabin aircraft with special attention to the instrument board. His company experienced serious difficulties trying to obtain accurate readings from their rate-of-climb indicators and their altimeters due to the ambient pressures in sealed cockpits. Efforts to reduce these ill-effects by adjustments in the instruments often failed due to the wild fluctuations in pressure encountered in flight, caused by the design of the aircraft themselves. By summer 1928, the importance of this branch of the pressure investigations became apparent. As the NACA PW-9 returned to the Langley hangars to retrofit its fuselage for the upcoming external pressure tests, plans for the following year included the acquisition of a commercial, closed cabin aircraft in order to pursue a full investigation of pressure inside the cockpit and the fuselage.[27]

The second possible line of new inquiry also involved interior pressures in aircraft flying high stress maneuvers, but this time the human machine became the center of focus. One of the pilots (disregarding Lewis' earlier order not to exceed 8 g in flying the PW-9, which appears to have been violated frequently) actually experienced *11 g* in acceleration maneuvers. Captain I. F. Peak of the Army Medical Corps examined him on landing and noticed blood pressure anomalies, leading the doctor to ask NACA officials whether they would allow him to study the physiological effects of high stress flight. Dr. Peak, also a pilot, proposed measuring the blood

---

[27]Henry Reid to the NACA (Lewis), 29 February 1928, RA 138 File, LaRC Historical Reference Collection (first quoted passage); Hansen, *Engineer in Charge*, 165; Walter S. Diehl to George Lewis, 5 March 1928, RA 138 File, LaRC Historical Reference Collection (second quoted passage); George Lewis to LMAL (Reid), 7 March 1928, RA 138 File, LaRC Historical Reference Collection; Henry Reid to the NACA (Lewis), 13 March 1929, RA 138 File, LaRC Historical Reference Collection; George Lewis to LMAL, 14 March 1928, RA 138 File, LaRC Historical Reference Collection; Charles Colvin to George Lewis, 24 May 1928, RA 138 File, LaRC Historical Reference Collection; George Lewis to Charles Colvin, 27 August 1928, RA 138 File, LaRC Historical Reference Collection; Henry Reid to the NACA (Lewis), 8 September 1928, RA 138 File, LaRC Historical Reference Collection.

pressure of one flier and also of himself to learn human the effects of short-term high acceleration; the consequences of medium to high accelerations over longer periods; and the biological outcomes of accelerations in inverted flight. No one involved in Research Authorization 138 had yet explored these factors at the frontier of human physiology, even though the fourth Air Service question posed to the NACA asked the essential question: would future design loads be governed by the limitations of the machine, or by the physical make-up of the man in the machine? Peak's offer created the opportunity to satisfy the NACA's military client and at the same time to delve into this unknown biomedical subject. George Lewis received a memorandum from the laboratory staff supporting the suggestion but he still approached it with caution. He discussed it with Colonel L. M. Hathaway, the Army's Chief Flight Surgeon, who "expressed great interest and indicated that any program proposed by Captain Peak will probably be approved" by the Medical Corps. Lewis also learned from their talk that no instruments yet existed to gather the data proposed by Peak. He therefore instructed Henry Reid to call a conference with Peak and the engineering staff to digest the technical questions and to issue a report describing the specific costs entailed, the instruments to be developed, and the research benefits to be derived. These hurdles appear to have dampered the initial eagerness to proceed. But more to the point, Lewis himself lacked enthusiasm for it, remaining as firm as ever in his conviction that the pressure distribution investigation should devote itself essentially to the formidable mechanical and aerodynamic problems confronting military aircraft. Still, raising questions about the human being in the machine represented an important first step toward recognizing the intimate relationship between high performance flight and the physical constraints of human endurance.[28]

As the year 1928 progressed Lewis' instinct to concentrate on the fundamental objectives of Research Authorization 138 appeared vindicated as the initial results started to unfold and the

---

[28]Elton Miller to Henry Reid, 23 June 1928, RA 138 File, LaRC Historical Reference Collection; Henry Reid to the NACA (Lewis), 25 June 1928, RA 138 File, LaRC Historical Reference Collection; George Lewis to LMAL (Reid), 2 July 1928, RA 138 File, LaRC Historical Reference Collection (quoted passage).

experimental lessons started to be revealed. Gradually, the project's engineers acquired a familiarity with the role of pressure distribution in at least some of the flight maneuvers. They also systematically compared the pressures measured in the NACA tests with the standard design guidelines. The wing spars represent a case in point. When loaded to 9 g, they revealed impressive agreement between the new data and the traditional criteria. Since the wing spars typically tolerated greater loads due to a standard thirty percent increase in bending moment, no design changes seemed appropriate. But the recent experiments demonstrated a far different situation for the leading edges of the wings. When the PW-9 pulled out of dives at 186 m.p.h. the pressures along the leading edge exceeded *400 pounds per square foot*. This load compared to a mere 200 pounds per square foot in the Thomas Morse aircraft pulling out at maximum dives of only 150 m.p.h. in NACA tests conducted but six years before. The result: the front edge of the wing needed to be strengthened and the forward wing spar positioned correctly to prevent failure in the generation of aircraft rolling off the assembly lines in the late 1920s. But some of the phenomena resulting from increases in speed had never been experienced before and consequently, no traditional design criteria could be applied. A pull-up from a dive at 186 m.p.h. occurred in 1.5 seconds, resulting in pressures which rose from 1 to 9 g. in about *one-half second*, almost constituting a shock load. The discovery of this virtually instantaneous spike in pressure posed problems unknown and unconsidered, ones whose solutions could only be surmised.

Indeed, the secrets of aircraft pressure distribution did not reveal themselves all at once or easily. In February 1928, a full day of inverted maneuvers by an Air Service lieutenant-- including treacherous upside-down barrel rolls from inverted to normal positions performed in a snap, as well as two upside down tail spins--resulted in experimental confusion when the pilot's notes failed to correspond with the altimeter, airspeed meter, and accelerometer readings. Then another puzzling phenomenon came to light. At the end of 1928 Henry Reid informed George Lewis of a discovery discerned wholly through flight research: the PW-9 aircraft consistently demonstrated that "normal force coefficients obtained in maneuvers, pull-ups for example, are

36

much larger than obtained from tests in steady flight or from wind tunnel tests." It suggested

flow conditions in accelerated flight at variance with those experienced in steady flying or in

tunnels, as well as a discrepancy between pressure distribution measured in flight versus the data

collected from model airfoils in the atmospheric tunnel. Reid recommended a comprehensive

review of these inconsistencies, arguing the prevailing interpretations of the entire flight research

program hung in the balance. He asked Lewis to either designate the new work a part of

Research Authorization 138, or because of its importance, to create a new RA expressly for this

project. The Aerodynamics Committee took Reid's proposal under advisement and decided to

allow the research to occur in the propeller tunnel under Authorization 138, provided it be

"carried on incidentally without interfering with the more important investigations on the

program." But one of the panel's most distinguished members, Edward Warner, did much to

deflate Reid's urgent appeal by suggesting the variation in data might be the consequence of high

angle of attack being arrived at more quickly in curvilinear flight. The answer, said Warner, lay

in varying the angle of attack at different rates. Whether right or wrong, his open skepticism

afforded Lewis the opportunity to once more rein in a protean project, keeping the attention

centered on the questions posed by the Air Service.[29]


## FLIGHT RESEARCH ACHIEVES FAME


While the pressure distribution experiments yielded undeniable research dividends for the

---

[29]"Pressure Distribution Tests on PW-9 Airplane," 19 April 1928, RA 138 File, LaRC Historical Reference Collection; Henry Reid to the NACA (Lewis), 13 March 1928, RA 138 File, LaRC Historical Reference Collection; Henry Reid to the NACA (Lewis), 6 December 1928, RA 138 File, LaRC Historical Reference Collection (first quoted passage); George Lewis to LMAL (Reid), 23 March 1929, RA 138 File, LaRC Historical Reference Collection (second quoted passage).

NACA, during the late 1920s the laboratory simultaneously pursued a wind tunnel project of even greater acclaim and magnitude. At the behest of the Bureau of Aeronautics and several aircraft industries, in 1926 Langley embarked on an investigation of radial engine cowlings to determine the degree to which covering the powerplant's mechanism might reduce drag without affecting engine cooling. Three years of empirical study, conducted mainly in the Propeller Research Tunnel under the direction of engineer Fred Weick, resulted in a revolutionary low-drag design. For this achievement, in 1929 the NACA won the first Robert J. Collier Trophy, awarded annually by the National Aeronautic Association for the most significant contribution to aviation research. (See chapter three for more on the NACA Cowling). Yet, in its own way, Research Authorization 138 also reaped large rewards for the NACA in 1929, a year in which Langley achieved national and international recognition for its structural loads investigations. After five years of perilous flying and sometimes uncooperative instrumentation, after painstaking and sometimes contradictory wind tunnel experiments, and after occasional bureaucratic battles, the work started to bear fruit. The NACA's chief clients and patrons--the military services--received first notice of the breakthroughs and benefited first. They learned about the discoveries from a beneficiary of the 1927 resignations of Hemke, Fairbanks, and Elliott Reid. Richard V. Rhode, a 25 year old University of Wisconsin graduate in mechanical engineering who arrived at the NACA in 1925, assumed the main responsibility for the PW-9 project under the supervision of Flight Test chief Gus Crowley. George Lewis began to disclose the project's findings in January 1929 when he transmitted to the Wright Field Experimental Engineering Section a preliminary paper by Rhode called "A Danger in Maneuvering Airplanes of Similar Type." The subsequent response by the Air Service to young Rhode's work, expressed in a condescending yet defensive tone, infuriated the NACA researcher and his colleagues. The reply rebuked him for writing his report for pilots, rather than for the nation's structural engineers. But Rhode refused to yield, telling his critics that all competent engineers would soon learn about these results, but the pilot needed to acquaint himself with the new data immediately because "he is the most interested person and should be educated in these matters to enable him to make more judicious use of the airplanes he must fly." The Air Service tried to dismiss as "entirely fanciful" Rhode's warnings about the inadequacies of existing aircraft design standards under extreme maneuvers. "If," replied Rhode, "the accelerations being induced on some of the present day fighters in their ground attack maneuvers are of the order of 7 g to 9 g...the danger would seem to be real." Finally, to the insulting remark that "it is questionable policy, and affects the morale of the pilot to call attention to dangers which may not exist except in fancy" Rhode reacted with scorn. "If a question may be permitted," he asked, "would it lower the morale of the motorist if he were told not to turn corners at 60 m.p.h.?" The Air Service soon recanted its position.[30]                    A more business-like atmosphere prevailed when George Lewis sent the Materiel Division a pre-publication copy of a second paper by Rhode. Entitled "Advance Data on the Tail Surface Loads and Pressures on the PW-9 Airplane," it received a warm welcome at Wright Field. The reply in late July confined itself to a few minor technical questions and ended with a respectful note of thanks: "Your further comments on the above points will assist us greatly in our attempt to set forth a rational system of design requirements for tail surfaces and will be very much appreciated." Rhode's answer betrayed none of the ill-feeling of a few months before. Indeed, the time had come for the parties to meet face-to-face. Plans were laid for an NACA, Army, Navy, and Department of Commerce conference to discuss the wing tip experiments being formulated at Langley during the Summer of 1929. Meantime, Major G. E. Brower, Acting Chief of the Wright Field Experimental Engineering Section invited Rhode to visit Dayton for two or three days and explain to the Air Service engineers the theoretical complexities and the appropriate tail loading requirements. The naval services also took notice of the data emerging from the Langley

[30]For a detailed and thought-provoking treatment of the NACA cowling see James R. Hansen, "Engineering Science and the Development of the NACA Low-Drag Engine Cowling," in *From Engineering Science to Big Science: The NACA and NASA Collier Trophy Research Project Winners*, ed. Pamela E. Mack (Washington, D.C.: NASA SP-4219, 1998), 1-27; Hansen, *Engineer in Charge*, 51-53, 123-128, 421; C.W. Howard to George Lewis, 14 February 1929, RA 138 File, LaRC Historical Reference Collection (second and fourth quoted passages); Richard V. Rhode to Henry Reid, 5 March 1929, RA 138 File, LaRC Historical Reference Collection (first, third, and fifth quoted passages).

Laboratory.  A U.S. Marine Corps officer wrote the NACA inquiring about pressure loads likely

to develop during maneuvers in a Curtiss Hawk aircraft carrying bombs on racks attached to its

lower wings.  Gus Crowley attempted an answer based on the years of acceleration research.

While he recommended a partial stress analysis to obtain a complete answer, he ventured that

"no dangerous loading condition should result from the use of bombs on the wings in high speed

dives."  During dives, the drag of the bomb and bomb rack flowed in the same direction as the

drag on the wings; during pull-outs the bombs acted in the opposite direction to the lift on the

wings but not so much as to present a danger.  William McAvoy, also asked to comment, called

attaching bombs to the underside of wings a "standard practice" although he admitted planes so

equipped never flew more than 120 m.p.h.  Personal experience impressed on him the

tremendous pressures at higher velocities and he warned his Navy counterparts about the dangers

of diving at 200 m.p.h. and finding it necessary in an emergency or in combat to pull-out with

the bombs still attached.[31]

The fame of the NACA soon spread beyond the military services.  Foreign experimenters

noticed the findings and sought out more.  Some British researchers who found discrepancies

with the NACA's results wrote to Joseph Ames requesting explanations of the acceleration tests

and inquiring about the American method of calculating load factors.  Ames promised to

authorize the Langley staff to analyze a British combat aircraft, fly it under similar

---

[31]George Lewis to the Air Service Material Division, 8 April 1929, RA 138 File, LaRC Historical Reference Collection; C.W. Howard to George Lewis, 8 May 1929, RA 138 File, LaRC Historical Reference Collection; C.W. Howard to George Lewis, 25 July 1929, RA 138 File, LaRC Historical Reference Collection (first quoted passage); Richard Rhode to Henry Reid, 7 August 1929, RA 138 File, LaRC Historical Reference Collection; Henry Reid to the NACA (Lewis), 12 August 1929, RA 138 File, LaRC Historical Reference Collection; G.E. Brower to George Lewis, 21 August 1929, RA 138 File, LaRC Historical Reference Collection; George Lewis to G.E. Brower, 11 September 1929, RA 138 File, LaRC Historical Reference Collection; George Lewis to LMAL (Reid), 16 October 1929, RA 138 File, LaRC Historical Reference Collection; C.W. Howard to George Lewis, 12 October 1929, RA 138 File, LaRC Historical Reference Collection; W.O. Brice to the NACA (Lewis), 21 November 1929, RA 138 File, LaRC Historical Reference Collection; John Crowley to Henry Reid, 16 December 1929, RA 138 File, LaRC Historical Reference Collection (second quoted passage); William McAvoy to Henry Reed, 16 December 1929, RA 138 File, LaRC Historical Reference Collection (third quoted passage).

circumstances, and compare it to similar American planes. Sometimes, apparent differences in findings--for example, when a Bristol aircraft experienced wing tip torsion but the NACA PW-9 did not--could be reconciled by checking the NACA data and correcting divergences in testing procedures. However, the investigators from abroad did not enjoy a monopoly on rising interest in the NACA. Journalists also wanted to learn about recent discoveries. Walter Raleigh of the Affiliated Press Service requested photographs illustrating "safety in aircraft construction." Lewis replied with a newsy letter about the aerodynamic loads project, describing in simple terms the data realized from flight research. "Information of this character," explained Lewis with evident pride, "makes it possible for the aircraft designer to design an aircraft structure with the confidence and surety of the designer of a bridge or any other structure where the loads to be imposed are accurately known."[32]

If the NACA research generated notice overseas, curiosity in the media, and respect among the military services, it caused a sensation in the aircraft industry. In significant part because of the pressure loads work, the Committee soon became a national clearinghouse for aeronautical knowledge. Jimmy Doolittle recognized this role when he apprised George Lewis in October 1929 of a recent brush with catastrophe while flying dive maneuvers in Cleveland, Ohio. At 4,000 feet he began his descent, eventually gaining speeds between 200 and 220 m.p.h. At an altitude of 2,000 feet, flying past vertical, Doolittle heard a loud snap and as he flew crossways to the wind, saw the wings disintegrate and a piece of fabric tear loose. The plane "slowed down appreciably" but he managed to land. After he climbed out he saw a piece of dural 8 to 10 feet long floating to the ground nearby and other remnants fluttering to earth All of the upper wing and one lower panel remained on the aircraft; the other panel had vanished. Doolittle attributed the failure to faulty materials, not to wing loading, because he had previously

---

[32]Joseph Ames to Richard Glazebrook, 26 February 1929, RA 138 File, LaRC Historical Reference Collection; Henry Reid to the NACA (Lewis), 8 May 1929, RA 138 File, LaRC Historical Reference Collection; John Crowley to Elton Miller, n.d., RA 138 File, LaRC Historical Reference Collection; George Lewis to Walter Raleigh, 17 July 1929, RA 138 File, LaRC Historical Reference Collection (quoted passage).

flown even more violent maneuvers without apparent strain and he learned from a mechanic that the wings and fittings were new. But when Lewis received a report by Doolittle on the incident, it raised alarms. On October 16, just a day before Doolittle's near disaster, Lewis had witnessed at Anacostia's Bolling Field the dive failures of not one, but two aircraft: a Martin single engine bomber and the Hall all-metal pursuit plane. The Martin had been instrumented by NACA engineers and flown by William McAvoy, Thomas Carroll's replacement as Chief Test Pilot. On its third dive "the front spar on the lower left wing failed" and tore away one-third of the fabric on the lower left wing. McAvoy landed safely. Then the Hall airplane, flying with a high load on the center of the upper wing experienced failure of the rear spar and a number of ribs. Doolittle's experience provided needed counterpoint to the NACA investigations of the Bolling Field events. Indeed, the Chief Engineer of the Glenn L. Martin Company specifically requested help from the NACA in restoring the ill-fated XT5M-1 bomber. The firm wished to rebuild the aircraft so it could withstand the extremely high pressures encountered in dives. Normally, Lewis declined to release specific research data to manufacturers before its formal publication; he regarded such information as untested until then. But because the Bureau of Aeronautics sponsored this aircraft, he agreed to let Richard Rhode prepare a brief memorandum relating some parallel flight test experiences of the NACA PW-9. Lewis also gave the Navy a copy of Rhode's paper, which included a table of maximum leading edge pressures and diagrams of the stress patterns on the PW-9. Rhode revealed in this November 1929 memo a choice piece of structural loads flight research: the leading edge pressures on the upper wing ranged from 150 to *500 pounds, "with the pressure increasing toward the tip."* [Author's italics] Experiments on torsional pressure on the ends of the wings predicted this phenomenon because "the outer sections [of the airfoil] operat[ed] at lower angles of attack than the inner...."[33]

[33]James Doolittle to George Lewis, 18 October 1929, RA 138 File, LaRC Historical Reference Collection (first quoted passage); Harold Brand to George Lewis, 17 October 1929, with attached statement by James Doolittle, RA 138 File, LaRC Historical Reference Collection; George Lewis to Edward P. Warner, 16 October 1929, RA 138 File, LaRC Historical Reference Collection (second quoted passage); George Lewis to James Doolittle, 25 October 1929, RA 138 File, LaRC Historical Reference Collection; George Lewis to LMAL (Reid), 25 October 1929,

Lewis made a similar exception for the Charles Ward Hall Aircraft Company, the development of whose XFH-1 pursuit plane also received Navy funding. Lewis instructed Rhode to compose a memo to the Hall Company using the PW-9 data to analyze the failure of the XFH-1. Perhaps with the confidence of youth, Rhode felt certain aspects of the PW-9 flight tests "easily explained" the Hall failure. He concluded after speaking with Charles Hall that "[t]he applied load distribution was different from the design load distribution in such respect that, while the applied load factor was within reasonable limits, certain portions of the spar were overstressed." Moreover, the NACA research demonstrated a tendency for "unusually high loads [to be] imposed on ribs in the area near the fuselage, which possibly would account for the failures of upper chord members aft of the rear spar." Still uneasy about distributing preliminary data, Lewis sent the memorandum to Mr. Hall, but on the condition it be returned to him in a week to ten days. Four days later Hall mailed the report back to Lewis with a cordial note confirming the NACA hypothesis. "I find myself...in agreement with Mr. Rhode," wrote Hall, "that the primary cause of failure was a local high peak of up pressure at very low angle of attack...." Finally, in response to Curtiss-Wright's Chief Engineer Dr. Theodore P. Wright, Lewis released another of Rhode's as yet unpublished papers related to the pressure distribution project (intimately related to Research Authorization 138 but placed under 209). Still cautious of disclosing preliminary results, Lewis *loaned* Wright a copy of Rhode's findings, which pertained to maximum loads attained on the horizontal tail surfaces of the F6C-4 aircraft during dives and pull-ups. He also made the Langley staff available for Wright's questions. Wright replied a with a flat statement: "elevator design loads have frequently been too small in the past and...we should immediately like to provide for adequate design standards...."[34]

---

RA 138 File, LaRC Historical Reference Collection; Lessiter Milburn to George Lewis, 23 October 1929, RA 138 File, LaRC Historical Reference Collection; George Lewis To Lloyd Harrison, 6 November 1929, RA 138 File, LaRC Historical Reference Collection; George Lewis to Lessiter Milburn, 6 November 1929, RA 138 File, LaRC Historical Reference Collection; Richard Rhode to Henry Reid, 31 October 1929, RA 138 File, LaRC Historical Reference Collection (third and fourth quoted passages).

[34]Richard Rhode to Henry Reid, 28 October 1929, RA 138 File, LaRC Historical Reference Collection (first and second quoted passage); George Lewis to Charles Ward Hall, 7 November

Universities, too, began to beat a path to the NACA as a result of its seminal flight research on loads. Professor Joseph Newell of M.I.T. asked Lewis to see the data on the PW-9 or F6C-4 aircraft relative to the loads encountered by wings and tails during terminal velocities. He also requested information demonstrating the ratios between terminal velocities in a dive versus high speed level flight. Perhaps unconsciously, Newell paid flight research a glowing compliment when he told Lewis that "a pilot's estimate as to how much faster an airplane could go in a dive than the maximum obtained in flight tests would be more satisfactory than any theoretical terminal velocities." Unfortunately, George Lewis could not return the favor. The NACA possessed no releasable data on the terminal speeds of airplanes, although the fastest dives to date (December 3, 1929) occurred at 280 m.p.h. Lewis did say the PW-9 research suggested terminal velocity approximated twice the top speed attainable in level flight (in the case of the NACA PW-9, a diving speed as high as 320 m.p.h.) However, Lewis invited Newell to review a pre-publication copy of an upcoming NACA report on the subject, to which Newell agreed eagerly, even asking for an approximate date when he might receive it.[35]

With the interest in this project mounting to a crescendo, at the end of 1929 Gus Crowley and Richard Rhode put a surprising request on their Christmas lists: just as the stock market began its fabled descent into catastrophe, the two engineers asked their bosses to hire several new employees. Rhode wanted assistance in order to devote his full energies to the pressure distribution data and to exhausting all of its utility. The results, said Reid in strong support of the request, "would be a credit to the Committee and of great value to aircraft designers." But the

---

1929, RA 138 File, LaRC Historical Reference Collection; Charles Ward Hall to George Lewis, 11 November 1929, RA 138 File, LaRC Historical Reference Collection (third quoted passage); George Lewis to T.P. Wright, 15 November 1929, RA 138 File, LaRC Historical Reference Collection (fourth quoted passage); T.P. Wright to George Lewis, 11 December 1929, RA 138 File, LaRC Historical Reference Collection (fifth quoted passage).
[35]Joseph Newell to George Lewis, 15 November 1929, RA 138 File, LaRC Historical Reference Collection (quoted passage); John Crowley to Henry Reid, 25 November 1929, RA 138 File, LaRC Historical Reference Collection; George Lewis to Joseph Newell, 3 December 1929, RA 138 File, LaRC Historical Reference Collection; Joseph Newell to George Lewis, 16 December 1929, RA 138 File, LaRC Historical Reference Collection.

Civil Service Register did not cooperate. Because the Great Depression had begun only two months earlier, the recent list of candidates did not yet reflect a tightening job market. "Experience is not so necessary," said Reid, "as a keen mind, good training in structures, and an interest in this kind of work." One new hire, Eugene Lundquist "proved to be the type of man desired." As Reid turned to George Lewis for support of the new positions and asked for advice about filling them, Edward H. Chamberlin, the NACA Chief Clerk, expressed little sympathy. Chamberlin claimed Crowley and Rhode had already surveyed the qualifications of 19 candidates, yet found none acceptable. But Lewis, although cautious, recognized important work and sided with Reid, Crowley, and Rhode. "I am in accord with...the desirability of expanding the work on pressure distribution and stress analysis," he decided on the last day of the year, "and approve the addition of one or two Junior Engineers to this section." The worsening economy soon made it all too easy to find highly qualified applicants to fill the vacancies.[36]

Almost six years after the NACA harnessed flight research to discover the limits of structural loading, published reports started to disseminate the findings worldwide. The first one appeared in February 1930 by Richard Rhode, entitled "The Pressure Distribution Over the Wings and Tail Surfaces of a PW-9 Pursuit Airplane in Flight." This report represented a new body of knowledge awaited anxiously by the aeronautics community. Rhode, with customary confidence, declared the existing rules of aircraft design "satisfactory...when applied to airplanes of conventional type and purpose." But in order to build airplanes strong enough to resist the increasing pressures of flight, yet light enough to be practical, "the engineer must have a thorough and accurate knowledge of the character of the loads that his structure must withstand." Rhode hastened to identify the greatest beneficiary of the NACA's labors; not the designers or manufacturers of airplanes, but the flying public which required the assurance of safe travel

---

[36]Henry Reid to the NACA (Lewis), 17 December 1929, RA 138 File, LaRC Historical Reference Collection (first, second, and third quoted passages); Edward Chamberlin to George Lewis, 21 December 1929, RA 138 File, LaRC Historical Reference Collection; George Lewis to LMAL (Reid), 31 December 1929, RA 138 File, LaRC Historical Reference Collection (fourth quoted passage).

before committing itself to commercial aviation.. He also suggested to the aircraft industry that air commerce had arrived at a crossroads in which further progress demanded from the aeronautical engineer a degree of professional competence almost unknown in other fields.

> It is perhaps needless to say that crashes resulting from structural failures in the air, even though relatively rare, have a particularly bad effect on the morale of flying personnel...and on the attitude of the public toward aviation, and must be eventually eliminated if confidence in the airplane is to become deep-rooted. It is manifest, therefore, that the structural design of airplanes must be put on an indisputably sound basis. This means that design rules must be based more on known phenomena, whether discovered analytically or experimentally, and less on conjecture.
> Thus, the present report attempts to portray the phenomena occurring on a pursuit-type airplane in maneuvers that it is called upon to perform, or...in special test maneuvers outlined to impose the same conditions of load that occur at the critical times in the more familiar maneuvers. To this end, pressure measurements were made on the right upper wing extended to include portions affected by slip stream, fuselage, and windshield, the left lower wing, and the tail surfaces of a Boeing PW-9 airplane, simultaneously, with accelerometer readings at the center of gravity, wing tip, and tail in the maneuvers above mentioned.[37]

The pilots who flew the PW-9 biplane subjected it to countless repetitions and variations of seven maneuvers, many of which Jimmy Doolittle pioneered in his pivotal flights which preceded Research Authorization 138: level flight, pull-ups, rolls, spins, inverted flight, dives, and pulling out of dives. The data gathered from the instruments and from the impressions of the aviators themselves suggested a whole new approach to aircraft design, one based on a reliable set of observations heretofore only surmised. Rhode cautioned that the load distribution of aircraft depended on the torsional rigidity of the airfoil structure. Fortunately, the effect of wing twist on the load distribution could be calculated satisfactorily. Perhaps surprisingly, the extreme forward position of the center of pressure on both the upper and the lower wings did not vary with acceleration and, at least on the upper wing, this center proved to be the same in full-scale and in the model wings. On the PW-9, and probably in other pursuit airplanes, the maximum force coefficient of the upper wing attained a significantly higher value in high angle

---

[37]Richard Rhode, NACA TR 364, "The Pressure Distribution Over the Wing and Tail Surfaces of A PW-9 Pursuit Airplane in Flight," (Washington, D.C.: NACA 1931), 687-688 (quoted passage, 687, block quote 687-688).

of attack than steady flight while the lower wing coefficients remained the same for pitching and for level flight. Furthermore, in powered high angle-of-attack maneuvers fuselage loads needed to be considered as a component of adequate engine support. The leading edges of the wings also experienced very heavy pressures--up to 450 pounds per square foot--during dives and pull-ups. Rhode also warned designers to be mindful of stresses on the rear spars which "may be greater than heretofore considered" and to increase tail-load specifications, especially at leading edges, to at least double the existing standards. The Materiel Division at Wright Field, which first raised the pressure distribution questions, received its copy of the Rhode report from George Lewis late in March 1930.[38]

But much more would follow. Only two weeks later the NACA published another influential piece of research by Richard Rhode, this time a Technical Note about "Pressure Distribution on the Tail Surfaces of a PW-9 Pursuit Airplane in Flight." Recent failures of tail sections in flight prompted him and the NACA to isolate this portion of the aircraft anatomy from the other pressure distribution tests and to publish the findings separately, as an interim measure before the completion of new design specifications . "It should serve as a guide," he wrote, "to those designers who wish to insure (sic) structural safety in their airplanes pending the formulation of more satisfactory design rules for tail surfaces." Rhode referred not only to recent air disasters involving tail plane failures as a reason to undertake the study, but to the underlying cause: the production of more aircraft, especially military models, capable of higher performance and greater capacity to maneuver violently. Once again, flight research offered the only reliable antidote to the problem. During the NACA flight tests, the PW-9 flew with 23 pressure stations on the right horizontal tail surface and 26 on the vertical, all of which took simultaneous stress measurements. Airspeed, acceleration, angular velocity, and control position were also recorded simultaneously. Before the NACA took to the skies, it was assumed that pressure loads of 45

---

[38]Rhode, NACA TR 364, "Pressure Distribution Over..A PW-9...Airplane," 697, 765, 771 (quoted passage, 771); George Lewis to C.W. Howard, 25 March 1930, RA 138 File, LaRC Historical Reference Collection.

pounds per square foot for the horizontal and 40 pounds for the vertical allowed safe margins of error. The data published by Rhode agreed with the traditional loading assumptions for vertical tail surfaces (although the flights at Langley showed the right barrel roll and pull-out from dives both could exceed safe values for vertical structures). But the horizontal surfaces presented a different situation. Especially in dives, but also in high speed barrel rolls, the standards proved inadequate and "the design requirements should be raised upwards." Even more significant, on pursuit planes the design factors for the leading edges of the horizontal stabilizer measured 135 pounds per square foot and 120 pounds per square foot for the fin leading edge. But these figures only represented *averages*, not taking into account the greater forces at specific points. At these spots in severe pull-ups the stabilizers "exceed the specified leading edge value by a very appreciable margin...." Even in less abrupt pull-ups the margin was too small for safe design. Thus, Rhode proposed doubling the specified leading edge load. Even at the fin, where the pressures in pull-ups were lower, the thinness of the PW-9's vertical stabilizer (compared to the thicker horizontal stabilizer) posed potential hazards. Eventually, Richard Rhode reduced the data on tail pressure to a simple design equation, one which became known in the Air Service Materiel Division and in the Aeronautics Branch of the Department of Commerce by the originator's name--the "Rhode Formula." Such shorthand references suggest the project's (and the NACA's) expanding influence.[39]

This influence spread by means other than formal publications. In a new role, the ubiquitous Edward Warner did much to bring the important stories of aeronautics to the general

---

[39]Successful formuli such as Rhode's commonly gained wide recognition for their originators and for the institutions with which they were associated. Theodore von Kármán's mathematical representation of the so-called Kármán Vortex Street won him fame and added to the reputation of his alma mater, Göttingen University.

Richard Rhode rehearsed his work on PW-9 tail loading two years earlier when he conducted similar research on a Navy F6C-4 Curtiss Hawk. His conclusions appeared as NACA TR 307, "The Pressure Distribution Over the Horizontal and Vertical Tail Surfaces of the F6C-4 Pursuit Airplane in Violent Maneuvers" (Washington, D.C.: NACA,1928); Richard Rhode, NACA TN 337, "Pressure Distribution of the Tail Surfaces of a PW-9 Pursuit Airplane in Flight," (Washington, D.C.: NACA, 1930), 1-7 (quoted passages, 1, 5, 6); George Lewis to Henry Reid, 23 March 1931, RA 138 File, LaRC Historical Reference Collection.

public. Warner joined the Main Committee of the NACA in 1929 and became editor of the popular magazine *Aviation* during the same year. In his NACA capacity he knew just about every project underway at Langley; in his role as purveyor of aeronautical information to the general public he needed articles. "It seems to me," he told George Lewis in mid-1930, "that enough pressure distribution work in maneuvers has been done at Langley Field so that we ought to begin to digest it for the benefit of the practical man who does not follow the laboratory reports in detail." He therefore asked Lewis to instruct one of the Langley engineers to submit a piece on "The Meaning of Pressure Distribution Tests to the Designer." However, neither Lewis nor Henry Reid favored exposing NACA research--particularly such hard won, costly, and valuable research--in the open literature. Both preferred to publish the NACA's most important findings through the medium of the Technical Reports. But Reid felt a brief paper or simple digest on pressure distribution research written for the general reader might be acceptable. Both men nominated Richard Rhode to be the author. But Warner had something else in mind; an essay "not in any sense simplified or popularized" but also not for professional researchers. Clearly, Rhode found himself caught between these senior figures who demanded conflicting articles. Rhode sent Warner an outline in August but admitted in a cover letter, "I am perhaps violating to some extent the policy of the committee to withhold publication of results until after the issuance of the report, but I see no way out of this if our article is to have any technical 'kick'...." George Lewis gave him a way out, instructing Rhode to delay the article until the NACA released all of its research findings. "The Place of Pressure Distribution Tests in Structural Design" eventually appeared in *Aviation* in February, 1931.[40]

[40]Roland, *Model Research*, 2: 430; Bilstein, "Edward Pearson Warner," 118; Edward Warner to George Lewis, 25 June 1930. RA 138 File, LaRC Historical Reference Collection (first quoted passage); George Lewis to Henry Reid, 28 June 1930, RA 138 File, LaRC Historical Reference Collection; Henry Reid to the NACA (Lewis), 2 July 1930, RA 138 File, LaRC Historical Reference Collection; George Lewis to Edward Warner, 8 July 1930, RA 138 File, LaRC Historical Reference Collection; Edward Warner to Richard Rhode, 10 July 1930, RA 138 File, LaRC Historical Reference Collection (second quoted passage); Richard Rhode to Edward Warner, 8 August 1930, with outline "Pressure Distribution Tests and Structural Design," RA 138 File, LaRC Historical Reference Collection (third quoted passage); George Lewis to

Extricated from this dilemma, Rhode pressed ahead with his heavy task of reducing the immense amount of flight research data to meaningful, compact analyses. He completed NACA Report Number 380, "Pressure Distribution Over the Fuselage of a PW-9 Pursuit Airplane in Flight," designed to "determine the contribution of the fuselage to total lift in conditions considered critical for the wing structure...." Conducted in spring 1929, the flight research again employed orifices attached to two manometers which provided continuous recordings of the maneuvers. Because of the complex shapes of the fuselage, some of the pressure points (for example, on the cockpit, nose, and cowling) needed to be altered in order to achieve reliable readings. In addition to the manometers, an airspeed meter, an accelerometer, a turnmeter, a control-position recorder, and a timer comprised the instrumentation. Because of the complex contours of the fuselage, Rhode had to be content with less precise data than in the wing or tail studies. Still, he gleaned enough to be able to eliminate the fuselage as a significant part of aircraft load bearing. In maneuvers consisting of steady flight, pull-ups, and rolls and spins, the PW-9's body accounted for a little less than three percent of total lift in low angles of attack and about four percent in high angles--an approximate compensation for the loss of wing surface represented by the width of the fuselage. Moreover, flight research showed that the fuselage not only bore little of the aerodynamic load, but contributed little structurally. Rhode suggested to his fellow aeronautical engineers that they simply ignore the lifting factor of the fuselage in their calculations in order to produce yet more conservative structural margins.[41]

Still not finished mining the riches of the loads research, during January 1931 Richard Rhode and Henry Pearson--a junior aeronautical engineer hired in 1930, probably one of the new positions approved by George Lewis for the project--published "A Method for Computing Leading Edge Loads." Although not mentioned in the report, the findings included tests on the Martin XT5M-1, as well as the PW-9. The Martin aircraft, designed under a Navy contract,

---

Richard Rhode, 11 August 1930, RA 138 File, LaRC Historical Reference Collection.
[41]Richard Rhode, NACA TR 380, "Pressure Distribution Over the Fuselage of a PW-9 Pursuit Airplane in Flight," (Washington, D.C.: NACA, 1931), 327-334 (quoted passage, 327).

experienced wing failure at Bolling Field in October 1929. At that time Rhode provided the Bureau of Aeronautics with preliminary data on leading edge failure; now the full story came to light. Originally, it had been designated a Technical Note but Henry Reid deemed it "of sufficient value" to launch it under the more polished and distinguished NACA Report series. Upon completion not only the Bureau of Aeronautics, but the Army and the Department of Commerce received early copies for comment and review. George Lewis heralded it to these recipients with the proud claim that a ' formula was developed which enables the quick determination of the proper design load for the portion of the wing forward of the front spar." Reid and others at the lab thought the results so important that they urged Lewis to call a conference of Army, Navy, and Department of Commerce representatives to discuss its implications for wing designs. But the NACA Director, "in hearty accord" with the idea in theory, overruled them because the Navy considered the results confidential. The article touched on high angle of attack and on nose-dives, the two areas of concern for wing pressure. For the high angles, the Army and Navy design rules appeared adequate; but the services' design factors for wing structures lacked sufficient strength to withstand nose-dives. Based on flight research and on variable density wind tunnel tests, Rhode presented a formula in which "theoretical rigor has been sacrificed for simplicity and ease of application." It provided a good degree of accuracy for monoplanes in nose-dives and could also be adapted for biplanes provided the requirements for the more heavily stressed lower wing exceeded those of the upper by about 30 percent.[42]

Finally, Rhode again encountered the same problem he experienced with the proposed article for *Aviation*. George Lewis directed him to present a paper on an aspect of his much-

[42]Hansen, *Engineer in Charge*, 420; Henry Reid to the NACA (Lewis), 22 January 1931, RA 138 File, LaRC Historical Reference Collection; Henry Reid to the NACA (Lewis), 9 March 1931, RA 138 File, LaRC Historical Reference Collection (first quoted passage); George Lewis to C.W. Howard, 18 March 1931, RA 138 File, LaRC Historical Reference Collection (second quoted passage); George Lewis to the LMAL (Reid), 21 March 1931, RA 138 File, LaRC Historical Reference Collection (third quoted passage); Richard Rhode and Henry Pearson, NACA TR 413, "A Method for Computing Leading Edge Loads," (Washington, D.C.: NACA, 1932), 249-257 (quoted passage, 249).

discussed research at the Society of Automotive Engineers in April 1931. But Gus Crowley reminded Lewis that with the meeting only two months away, Rhode would have to talk about work still in progress; in other words, Rhode would again find himself at odds with Lewis' prohibition against presenting any data not yet published in a NACA Report or Note. Lewis was unmoved. "It is desirable that Mr. Rhode present a paper at this meeting," because the able young engineer had important findings to discuss. At the same time, the Director warned it must "not contain any information which has not been released by the committee." Again enmeshed in the paradox of being ordered both to report, and not to report his conclusions, he submitted a paper on applied load factors to Lewis and the boss vetoed it. Ultimately, Rhode decided to present a condensed version of his latest Technical Note, published just as the conference met in April, and apparently freeing himself from Lewis' strictures. Yet, the Director still offered resistance. By "long-established precedent," Technical Notes were circulated only in the United States, but after conferring with Joseph Ames he conceded; the "Committee recognizes the particular interest in this Technical Note [and is] making this case an exception."

All of the fuss actually made some sense because Rhode's research really broke new ground. Co-authored with Eugene Lundquist and entitled "Preliminary Study of Applied Load Factors in Bumpy Air," it represented a tentative foray by the NACA into the open question of aircraft structures and weather. Perhaps more cautious with experience, Rhode declined to prescribe any design values based on these findings, realizing the pressure data and the weather factors rendered his advice inadequate for firm structural design decisions. But he and Lundquist did attempt approximate equations to account theoretically for the gusts causing unusual wing loading during bumpy or rough air. Meanwhile, in order to test loads and accelerations in flight and to build a base of empirical knowledge, NACA pilots flew the PW-9 and a Fairchild cabin monoplane outfitted with recording accelerometers and airspeed meters through turbulent meteorological conditions. All of the experiments occurred between September and December 1930 and between January and March 1931 over the Western United States. They flew routes from Salt Lake to Cheyenne, Oakland to Sacramento, Sacramento to Reno, and Seattle to

Portland. Most of the 94 flights experienced some turbulence, often quite violent. The pilot of a flight from Portland to Medford, Oregon reported on September 9, 1930 a bump at 11 a.m. so strong it "caused passengers to leave their seats." The authors also culled the existing literature of rough air flying. In the end, Rhode and Lundquist admitted the need for far more statistical flight research data about accelerations in these conditions. They also urged closer cooperation among aeronautical and weather agencies and sought instrumentation improvements (such as a combined airspeed meter/accelerometer capable of automatic operation to capture the relationship between velocity and acceleration). Finally, to further validate their rough air equations they recommended additional research on the impact of high velocity vertical currents on aircraft flying through gusts of air.[43]

## A FORETASTE OF THE HIGH SPEED CONUNDRUM

After the downpour of publications by Rhode in early 1931, Research Authorization 138 seemed destined for a dignified retirement. Indeed, after more than four years of the most intense flying the PW-9 itself seemed ready for withdrawal from service, requiring among other parts a new radiator which took more than a year to obtain. But surprisingly, flight research on pressure distribution actually staged a remarkable comeback. Its renewal began at the end of

---

[43]George Lewis to John Crowley, 19 February 1931, RA 138, LaRC Historical Reference Collection; John Crowley to George Lewis, 24 February 1931, RA 138 File, LaRC Historical Reference Collection; George Lewis to LMAL (Reid), 26 February 1931, RA 138 File, LaRC Historical Reference Collection (first quoted passage); Richard Rhode to Henry Reid, 27 March 1931, RA 138 File, LaRC Historical Reference Collection; H. R. Gilman to the Secretary, Society of Automotive Engineers, 27 May 1931, RA 138 File, LaRC Historical Reference Collection; H. R. Gilman to the NACA Secretary, 20 November 1931, RA 138 Files, LaRC Historical Reference Collection; George Lewis to John J. Ide, 1 December 1931, RA 138 file, LaRC Historical Reference Collection (second quoted passage); Henry Reid to the NACA (Lewis), 11 April 1931, RA 138 File, LaRC Historical Reference Collection; Richard Rhode and Eugene Lundquist, NACA TN 374, "Preliminary Study of Applied Load Factors in Bumpy Air,"(Washington, D.C.: NACA, 1931), 1-30 (third quoted passage, 22).

1930 when a loyal NACA supporter, Chief of the Bureau of Aeronautics William Moffett, asked

the Commander of the Anacostia Naval Air Station to stage a series of inverted flight

acceleration tests. The Admiral joined the NACA Main Committee in 1921 and remained on it

until his premature death aboard the airship *Akron* in 1933. While he brought no outstanding

technical capacities to the role, he graced the fledgling NACA with unerring political instincts,

powerful personal connections, and an unmatched zeal for the progress of naval aviation.

Moffett wanted to investigate, in particular, the effects on airplane structures of the inverted snap

roll (in which pilots pulled out of the inverted position during the second half of a loop, adding

one or two spins). He made it clear he wanted the tests to simulate normal flight, not abrupt or

violent maneuvers. The NACA agreed to collaborate on the Anacostia flight tests by installing

its instrumentation on the naval aircraft and by providing consultation on the results. In the

winter skies over Washington, D.C., a NACA recording accelerometer mounted on an F6C-4

aircraft measured performance during inverted pull-outs from dives (3 to 3.6 g), inverted snap

rolls (3.10 to 3.85 g), turns during inverted spins (2.12 g), and outside loops entered from the

inverted position (3.35 g). Henry Reid then asked George Lewis to use his good offices to

acquire the Navy's flight test results, "practically the only information known to exist on the

loads in inverted flight. We are anxious," said Reid, "to obtain these recently established data to

assist in establishing load factors for the inverted flight conditions." Gus Crowley had the

Bureau of Aeronautics report in hand by St. Patrick's Day, 1931.[44]

---

[44]William F. Trimble, *Admiral William A. Moffett: Architect of Naval Aviation* (Washington and London: Smithsonian Institution Press, 1994), 5-6, 255-267; Edward Sharp to the NACA (Lewis), 20 September 1931, RA 138 File, LaRC Historical Reference Collection; George Lewis to C. W. Howard, 11 November 1931, RA 138 File, LaRC Historical Reference Collection; Edward Chamberlin to C. W. Howard, 17 November 1931, RA 138 File, LaRC Historical Reference Collection; William Moffett to A. H. Douglas (Commanding Officer, Anacostia Naval Air Station), 6 December 1930, RA 138 File, LaRC Historical Reference Collection; A. H. Douglas to William Moffett, 27 January 1931, RA 138 File, LaRC Historical Reference Collection; A. H. Douglas to William Moffett, 5 February 1931, RA 138 File, LaRC Historical Reference Collection; Henry Reid to the NACA (Lewis), 27 February 1931, RA 138 File, LaRC Historical Reference Collection (quoted passage); George Lewis to LMAL (Reid), 13 March 1931, RA 138 File, LaRC Historical Reference Collection; George Lewis to LMAL (Reid), 16 March 1931, RA 138 File, LaRC Historical Reference Collection.

At the same time, the laboratory received data on a series of Bureau of Aeronautics dive and pull-out tests on F2B-1, F3B-1, and F4B-1 airplanes flown under service conditions by Navy pilots. There followed six months of review by Langley's Flight Test Section, a lag which caused increasing irritation in the Bureau of Aeronautics. "Since these service dive tests were made by Navy pilots flying Navy planes and the data is urgently needed...[for] decisions involving the structural integrity of Naval aircraft...a special effort should have been made to forward the flight path data requested." Richard Rhode finally delivered his report to the Bureau showing a series of curves plotting pressure loads from the moment of recovery during dives to the point of resumption of horizontal flight. Measuring forces as high as 14 g at 200 m.p.h., the data represented a better basis for structural design of service aircraft than the previous low speed pull-out tests by the NACA.

The delays in analyzing the Navy dive flights arose from commitments to the Army for related work. Because Gus Crowley felt the NACA got "a lot of extremely interesting [structural] data" from the Navy's dive tests, he proposed a parallel program using Army aviators to fly standard service dives and recoveries in order to learn about the structural loads encountered in such maneuvers. Crowley wanted the NACA Headquarters to alert the Chief of the Air Corps to the request "since there might not be the proper types of pilots on [Langley] Field and also because the work is dangerous and I feel should be done officially. " Unknowingly, Richard Rhode preempted his boss. While preparing one of his reports, Rhode spoke to an Air Corps Lieutenant about the differences between the NACA's measurements of accelerations in flight versus those of everyday military maneuvers. The young airman, assigned to the Air Corps Tactical School, suggested the NACA install equipment to record the practice patterns being flown during the training of combat pilots. Henry Reid seconded the plan, offering to provide the needed accelerometer and airspeed meter and to "arrang[e] the matter locally....". At first George Lewis agreed. But on the advice of a Langley major who recently lost a Boeing P12-C airplane in spin tests (and had been warned that all experimental work belonged at Wright Field), Lewis took the formal route suggested by Crowley. He requested from Air Corps Chief of Staff Major General James Fechet permission to undertake an "investigation [which] will not interfere in any way with the normal operation of the [Air Corps Tactical School] airplane." The NACA received approval in mid-April and initiated a flight research program.[45]

Because the flight tests occurred under Army auspices, Rhode and his associates coordinated the project closely with the military side of the air field. Accordingly, Captain Flickinger of the Tactical School supervised the NACA engineering staff as they installed an accelerometer and an air speed meter on a P12-C aircraft. By May 4 the modifications were finished and Flickinger approved the flying program. The series of tests recognized the extensive work already undertaken by the NACA on high-speed, violent maneuvers. It aimed instead for high angles-of-attack flown under regular service conditions; that is, sharp maneuvers at *moderate* speed designed to measure heavy load pressures. For instance, although the barrel roll stayed in the repertoire, the pilot performed it at only 90 or 100 m.p.h. in keeping with actual military tactics. The flights in late spring and early summer included moderate dives and pull-outs; short, steep dives and abrupt pull-outs (some simulating attacks on ground targets); climbing turns from dives; and push downs from shallow dives to imitate escape from pursuing enemies. Finally, Captain R. W. Clifton and the NACA's William McAvoy (flying the old PW-9) engaged in many staged combat exercises, trading offensive and defensive roles.. Each of the mock encounters were recorded for two minutes on NACA instrumentation. Richard Rhode

---

[45]George Lewis to Bureau of Aeronautics, 10 January 1930, RA 138 File, LaRC Historical Reference Collection; J. H. Towers to the NACA (Lewis), 12 January 1931, RA 138 File, LaRC Historical Reference Collection (quoted passage); Richard Rhode to Bureau of Aeronautics, 8 January 1931, RA 138 File, LaRC Historical Reference Collection; Henry Reid to the NACA (Lewis), 1 July 1930, RA 138 File, LaRC Historical Reference Collection; John Crowley to Elton Miller, n.d., RA 138 File, LaRC Historical Reference Collection (first quoted passage); Henry Reid to the NACA (Lewis), 28 February 1931, RA 138 File, LaRC Historical Reference Collection (second quoted passage); George Lewis to LMAL (Reid), 4 March 1931, RA 138 File, LaRC Historical Reference Collection; Henry Reid to the NACA (Lewis), 10 March 1931, RA 138 File, LaRC Historical Reference Collection; George Lewis to the NACA, 24 March 1931, RA 138 File, LaRC Historical Reference Collection (third quoted passage); W. G. Kilner to George Lewis, 16 April 1931, RA 138 File, LaRC Historical Reference Collection; George Lewis to LMAL (Reid), 18 April 1931, RA 138 File, LaRC Historical Reference Collection.

presented Henry Reid with the resulting data on August 11, in a report entitled "Acceleration Tests on an Army P12-C Airplane in Service Maneuvers," Illustrated ingeniously on a single chart, it plotted airspeed against acceleration on one axis and wing loading factors on the other. The simulated aerial duels showed the widest range of speed and wing loading by far, although the fabricated ground attacks also yielded a broad spread of loading factors. Pull-ups from attacks on enemy aircraft revealed a narrow but high band of pressure and pull-downs to escape pursuing enemies showed the smallest wing loads. Rhode pronounced the tests a solid success, with qualifications.

> These results constitute our most real information to date on load factors encountered in actual service conditions, with the exception of those encountered in the high-speed dives and recoveries...as executed in Naval maneuvers. While interesting and valuable, they can not be used alone to draw any final conclusions on design load factors, since they were obtained on one airplane only. However, it is exactly this type of data which will be obtained with the combined air-speed meter and accelerometer now being developed at the Laboratory, and from its records obtained over a period of time in a number of airplanes which are used in service maneuvers, it is confidently expected that the question of design of load factors for service airplanes can be definitely settled.[46]

The Langley Flight Test Section finally turned its focus on a phenomenon closely related to the NACA, Army, and Navy dive tests: the mystery of the terminal velocity of aircraft. Specifically, the NACA researchers wanted to ascertain the "structural margin of safety in the airplane in fast vertical dives." The question arose as early as July 1930 when Henry Reid queried Walter Diehl about the subject. When Diehl found himself unable to answer Reid satisfactorily, both men accepted the need for additional research. Thus, the 1931 flight test program at Langley Field featured the new Curtiss F6C-4 pursuit aircraft being flown, throttle wide-open, in dives up to 342 m.p.h. But related factors also received due attention. For instance, what role did engines play in the attainment of terminal velocity? What relationship existed between the airframe's structural safety and its powerplant's structural integrity in vertical dives? Henry Reid called George Lewis' attention to this line of inquiry by showing him

---

[46]Richard Rhode to Henry Reid, 11 August 1931, RA 138 File, LaRC Historical Reference Collection (block quote).

a letter drafted by Gus Crowley to several aeronautical engine manufacturers. An old engines man, Lewis assumed Crowley's role himself. He corresponded with three powerplant experts: George J. Mead, Vice President of Pratt and Whitney; Robert Insley, Vice President of Continental Aircraft; and Arthur Nutt, Vice President of Engineering at the Wright Aeronautical Corporation. The NACA Director asked them to predict the maximum safe rotations per minute (r.p.m.) of engines over the normal rated speed, to express any differences in safe speeds between radial and in-line engines, and to suggest the point at which inertial engine forces manifested themselves when operating in excess of approved velocities. Mead replied first, saying "very little work has been done along these lines" and although some Pratt and Whitney engines functioned at 50 percent over their proven r.p.m., under normal conditions their products were engineered to withstand no more than 20 percent over the maximum recommended r.p.m. He assured Lewis there would be no difference in safe velocities for in-line or radial engines but declined to comment on engine inertia since it "depends entirely upon the size of the engine and its normal operating speed." Robert Insley answered with much the same advice but Arthur Nutt claimed the problem could only be resolved by consulting the in-flight experiences of the military services.

George Lewis, meanwhile, decided to raise the question at the NACA Powerplants Subcommittee meeting on February 27, 1931, and in preparation asked the Flight Test Section to prepare Army and Navy dive data in tabular form. But Crowley and his staff did far more; Richard Rhode drafted for the meeting a paper suggesting the relationship between engine speed and terminal velocity for several aircraft. Rhode described previous tests of the PW-9 and the Navy dives in the F2B-1 and F4B-1 airplanes and declared a method had been achieved to calculate terminal velocities with a rate of error not higher than six percent. If this announcement did not raise eyebrows in the Powerplants committee, the rest of Rhode's remarks surely did. Apparently, engine speed in the dive affected greatly the speed of the aircraft's dive. In PW-9 flights, the engine at full throttle yielded a terminal speed of 280 m.p.h. compared to *326* m.p.h. with partial throttle. The NACA researchers discovered that a wide-open throttle in dive caused propellers to account for 30 percent of an aircraft's *total* drag. Yet, despite these revelations, Rhode admitted the nagging problem of engine speed during dives remained an open question. Even though the Langley staff could now design load factors in pursuit and dive-bomber airplanes to withstand terminal velocities, the engine speed still needed to be known. "[I]t would be as tragic," said Rhode, "for the engine to fly apart in a dive as it would be for the wings or tail to come off. Therefore, we have taken the position that the "terminal velocity" for which the airplane should be designed should be the velocity consistent with some engine speed which would not seriously stress the engine or propeller." Other than that piece of the puzzle, all of the other factors--controlling engine speed, calculating a terminal velocity for any particular engine speed, the time and altitude required to reach terminal velocity for any airplane--had been discovered.[47]

During 1931 the characteristics of terminal velocity and the implications for structural design became known and propagated. After the distribution of Rhode's paper at the Powerplants Committee meeting the Bureau of Aeronautics seized upon the promise of a simple formula to approximate terminal velocity and requested the NACA prepare and publish such findings. The Bureau made an unofficial inquiry in December 1931 when Lieutenant Commander R. D. MacCart told George Lewis "I am in great need of a standard method of calculating the terminal velocity of airplanes in that this is an important item in determining

---

[47]Henry Reid to the NACA (Lewis), 10 February 1931, RA 138 File, LaRC Historical Reference Collection; George Lewis to George Mead, 12 February 1931, RA 138 File, LaRC Historical Reference Collection (first quoted passage); Henry Reid to the NACA (Lewis), 16 July 1930. RA 138 File, LaRC Historical Reference Collection; George Lewis to LMAL (Reid), 22 July 1930, RA 138 File, LaRC Historical Reference Collection; George Lewis to Arthur Nutt, 9 April 1931, RA 138 File, LaRC Historical Reference Collection; George Mead to George Lewis, 17 February 1931, RA 138 File, LaRC Historical Reference Collection (second quoted passage). Robert Insley to George Lewis, 18 March 1931, RA 138 File, LaRC Historical Reference Collection; Arthur Nutt to George Lewis, 6 April 1931, RA 138 File, LaRC Historical Reference Collection; George Lewis to LMAL (Reid), 19 February 1931, RA 138 File, LaRC Historical Reference Collection; Henry Reid to the NACA (Lewis), 24 February 1931, RA 138 File, LaRC Historical Reference Collection; Richard Rhode to Henry Reid, 24 February 1931, RA 138 File, LaRC Historical Reference Collection (third quoted passage) and a second copy presented to the Powerplants Committee on 27 February 1927.

structural strength. [A]lthough I understand it will be published sometime, I would like to get a preliminary copy for immediate use. This information is desired for my own use in the Bureau." During the ten months between Rhode's announcement of his breakthrough and MacCart's letter, the Langley Flight Test staff added to their understanding of the problem, especially the operating conditions of the propeller. Lewis asked to see their present work, approved it, and relayed it to MacCart. "A Method For Calculating the Terminal Velocity of Airplanes" by Richard Rhode answered MacCart's plea. Rhode explained first the essential complications of approximating speeds at terminal velocity. First, as aircraft plunged toward the earth they encountered thicker atmosphere as they approached the ground, causing the vehicle to *decelerate* during the later stages of the fall. In addition, some aircraft lacked the capacity to climb to high enough altitudes to allow them to accelerate to terminal velocity. More important, an appreciation of terminal velocity required an understanding of the effects of the propeller because of its immense influence on aircraft drag. This factor, in turn, hinged on engine speed. Engines could be set to dead, idle, throttled, or wide open; the imponderable involved choosing the one which would elicit the least propeller drag. Rhode eliminated the dead setting because of its practical impossibility and the wide open because it might race the engine and result in its disintegration. He proposed the well-throttled position as the safest. Where did this leave the aeronautical engineer seated at his drafting table and mulling over the appropriate structural loads in dives? "[T]erminal velocity," suggested Rhodes, "should be the velocity which satisfies the requirements of consistent strength of airplane and power plant or which satisfies the drag equation when the airplane is offering its minimum drag and the propeller is offering a drag consistent with some safe [engine] r.p.m." At this point in the evolution of safe engine speeds, Army and Navy tests established ceilings of 2400 r.p.m. for small and 2,000 r.p.m. for large aircraft. Informed of the maximum engine velocities and the minimum drag coefficients of the airplane, engineers only needed to learn the degree of propeller drag to compute terminal velocity. Rhode arrived at a sample propeller figure by using as an example one nine feet long with a dynamic pitch-diameter ratio of 1.0 and a mean blade-width ratio of 0.1. Assuming the propeller's engine turned at the maximum 2,000 r.p.m. and the diving speed to be 240 m.p.h., the propeller accounted for 375 pounds of drag. Thus, armed with these facts--the propeller's drag, the maximum safe engine speed, the aircraft's minimum overall drag, and the velocity in dive-- manufacturers and the armed services could devise aircraft structures able to withstand the loads encountered in terminal velocity.

Of course, Rhode's conclusions remained to be ramified and tested under many conditions. In late February 1932, the Air Corps Materiel Division reported the results of flight tests which tried Rhodes' calculations on a P-12C airplane. Major C. W. Howard informed Lewis that the pilot flew the aircraft to 14,000 feet and dived to 4,000 feet before pulling out. At full power the engine turned 3,000 r.p m. and the plane reached 300 m.p.h.; with closed throttle it attained 260 m.p.h. at 2,600 r.p.m. Implicitly, the question arose why the Army could safely fly well above the 2,400 r.p.m. recommended in Rhodes' article. Rhode felt it necessary to respond and Lewis and Reid not only gave him the chance, but adopted word-for-word Rhodes' argument in Lewis' reply to Major Howard. In it, he revealed that more recent flights in the PW-9 showed substantial agreement with the values predicted in his essay on calculating terminal velocity. Indeed, these additional tests indicated engine speeds at full throttle "appreciably below' the figure of 2,400 r.p.m. referred to in the article. But Rhode conceded the Army's recent flights and other new evidence did suggest the maximum rate of permissible engine velocity exceeded 2,400 r.p.m., if only because the speed could not be curbed without propeller brakes. To address this matter and others, the NACA agreed to a Bureau of Aeronautics request for more dive tests using a Curtiss Hawk aircraft equipped with an air-cooled engine and a variable pitch propeller. These flights reassessed terminal velocities and engine speeds by employing a variety of propeller pitches during steep dives at full throttle. Meantime, the Navy adjusted its maximum engine r.p.m. up to 3,200 but urged its researchers to "[r]efer to [the Bureau of Aeronautics] regarding maximum permissible engine speed for your design."[48]

---

[48]George Lewis to C. W. Howard, 29 December 1931, RA 138 File, LaRC Historical Reference Collection; R. D. MacCart to George Lewis, 14 December 1931, RA 138 File, LaRC Historical

# A RECOGNIZED DISCIPLINE

But the additional studies attempting to augment Rhodes' equations with fuller experimental data did not occur under Research Authorization 138. After eight years Henry Reid, Gus Crowley, and Richard Rhode agreed mutually to close the pressure distribution experiments as a discrete project. The Authorization ended by order of the NACA Main Committee on April Fools' Day, 1932. But this long endeavor proved to be anything but foolish. First of all, its success did as much as any NACA activity to bring acclaim and reputation to this new institution. Henceforth, the military services, the universities, and the aircraft industries looked to the NACA for research leadership and innovation. Research Authorization 138 also left a distinct technical legacy. It not only clarified the mysteries of aircraft loading and underscored the structural limitations inherent in aircraft of higher and higher performance, but it presented aircraft designers a clear set of practical rules which resulted in flying machines capable of longer and safer service with far less likelihood of falling from the skies. No less important, this research won for the flight research practitioners a place beside the theorists and

---

Reference Collection (first quoted passage); George Lewis to LMAL (Reid), 17 December 1931, RA 138 File, LaRC Historical Reference Collection; Henry Reid to the NACA (Lewis), 23 December 1931, RA 138 File, LaRC Historical Reference Collection; George Lewis to R. D. MacCart, 28 December 1931, RA 138 File, LaRC Historical Reference Collection; Henry Reid to the NACA (Lewis), 23 December 1931, RA 138 File, LaRC Historical Reference Collection, with attachment: "A method for Calculating the Terminal Velocity of Airplanes," by Richard Rhode (second quoted passage); George Lewis to LMAL (Reid), 3 March 1932, RA 138 File, LaRC Historical Reference Collection; A. J. Lyon and C. W. Howard to George Lewis, 27 February 1932, RA 138 File, LaRC Historical Reference Collection; Henry Reid to the NACA (Lewis), 11 March 1932, RA 138 File, LaRC Historical Reference Collection; Richard Rhode to Henry Reid, 9 March 1932, RA 138 File, LaRC Historical Reference Collection; George Lewis to C. W. Howard, 18 March, 1932, RA 138 File, LaRC Historical Reference Collection (third quoted passage); George Lewis to LMAL (Reid), 29 March 1932, RA 138 File, LaRC Historical Reference Collection, with attachment: "Refer SR-55 Terminal Speed Calculations," (fourth quoted passage).

the wind tunnel experimentalists. Much, if not most of the insights gleaned by Rhode, McAvoy, and the other members of the Flight Test Section could not have been obtained by any means other than flying, often in perilous conditions. But the results did not flow merely from brave pilots; carefully designed experiments, ingenious instrumentation, and imaginative analysis proved the importance of aeronautics' open-air laboratory.[49] During the years following, flight research would show itself indispensable in designing pilot-friendly airplanes and in winning a World War.

---

[49]Henry Reid to the NACA (Lewis), 30 March 1932, RA 138 File, LaRC Historical Reference Collection; RA 138: "Pressure Distribution," 2 December 1924; George Lewis to LMAL (Reid), 18 April 1932, RA 138 File, LaRC Historical Reference Collection.

# CHAPTER 3

## Necessary Refinements:
## Flying Qualities Research

### A VARIED DISCIPLINE

Once flight research won its fame during the pressure distribution investigation, the NACA lost no time in applying its techniques to many different programs. Some of the new undertakings, like loads measurement itself, could be realized fully only by instrumenting and flying the aircraft themselves. Other projects, in contrast, involved multiple research approaches, flight testing being but one of several avenues. In part, the diversification of the NACA's techniques reflected a deepening experience with flight research. The sometimes perilous conditions under which pilots had collected pressure distribution and other data suggested the limitations of full-scale flying and implied the need for more sophisticated tools to conduct experimentation on the ground. A hiatus in large-scale construction occurred at Langley from 1921 (when Max Munk's Variable Density Tunnel received the go-ahead) until the authorization of the Propeller Research Tunnel in April 1925 (completed in 1927). But during the period 1925 to 1931 a virtual tidal wave of building resulted in no fewer than *five* new tunnels rising on the laboratory's broad expanses. Each of them compensated for deficiencies in flight research. The Propeller Research Tunnel was conceived to reduce the reliance on flight research for propeller data, a method which had proven to be time-consuming and costly. The Eleven Inch High Speed Tunnel (operational in 1928) allowed researchers to gauge aerodynamic effects at the approach of Mach 1 (the speed of sound), impossible to achieve with existing airplanes in free flight. The

Five-Foot Vertical Wind Tunnel (first in service in 1929) subjected models to simulated spins in order to analyze spin recovery without risk to pilots or to aircraft. The Seven by Ten Foot Atmospheric Wind Tunnel (opened in 1930) specialized in stability and control at the low speed range, often at velocities below those tolerated by full-scale machines. Finally, the mammoth Full Scale Tunnel (completed in 1931) allowed the next-best conditions to free flight by bringing the entire aircraft indoors and testing its characteristics under controlled conditions.[1]

Yet, even the projects which relied on these expensive new machines still required the services of flight testing. The famed NACA cowling investigation offers perhaps the foremost example of a multidisciplinary inquiry enriched by, but not dependent on flight research. (See chapter 2 for a cursory description of the NACA Cowling and the Collier Trophy). Cowling research originated with a request from the Bureau of Aeronautics in 1926 asking whether a covering over the front of radial engines might not reduce the degree of wind resistance encountered in flight. By this time, the Navy clearly favored the radial over the liquid-cooled engine. Lighter and leak-proof, the air-cooled radial also suffered two major (and related) disadvantages. It tended to overheat; and its large, round shape mounted just behind the propeller interrupted the airstream and increased drag. The problem, then, turned not just on designing a cowling to minimize turbulence around the engine; to be worthwhile it needed to channel the air to reduce the temperature of the powerplant. The initial responsibility for the undertaking fell to an able young engineer named Fred Weick. Selected personally by George Lewis just three years after he received a mechanical engineering degree from the University of Illinois, the 26 year old not only designed, but subsequently directed the Propeller Research Tunnel after its opening in July 1927. Lewis made an astute choice. Weick began the project by drafting a tentative research plan and, before bending metal, circulated it with due deference among industry leaders for advice and comments. Once the essentials had been agreed upon, he and his staff of engineers inaugurated the cowling investigation by positioning a J-5 Whirlwind

---

[1]Hansen, *Engineer in Charge*, 443-447.

engine in the tunnel and testing systematically the full range of cowling sizes, from those

shielding the entire engine to those offering little or no coverage. They arrived at the ten most

promising designs and assessed each for its capacity to cool the engine and to improve

aerodynamic efficiency. After much experimentation, cowling number 10 won the contest

Covering the entire front of the Whirlwind, it reduced temperatures by forcing air through a set

of slots and baffles onto the hottest parts of the engine. To everyone's astonishment, this model

also diminished drag by a factor approaching three. After November 1928, when the NACA

revealed these incredible findings to aircraft manufacturers, pilots Melvin Gough and William

McAvoy undertook the flight research phase of the project. A Curtiss Hawk AT-5A aircraft

borrowed from the Air Service and fitted with the number 10 over the same J-5 engine achieved

a top speed of 137 miles per hour compared to 118 miles per hour without the cowling, thus

yielding a 16 percent increase in velocity. But flight tests did not merely confirm the wind

tunnel data. They also showed that the size and shape of the opening which expelled the air at

the rear of the cowling assumed critical importance; the exit aperture needed to release the air at

a higher velocity and lower pressure than the air entering the cowling in order to allow the

maximum cooling effect. Finally, the test pilots gathered data comparing drag forces on a

conventional engine nacelle to the new NACA cowling. The results indicated a twofold increase

in efficiency with the improved design. Not surprisingly, in 1929 the NACA won its first Collier

Trophy on the strength of its cowling research. But much work remained to be done before the

program ended in 1936. Often assisted by flight research, it became increasingly dependent on

the theoretical labors of Langley's Physical Research Division, under the guidance of Max

Munk's successor, physicist Theodore Theodorsen.[2]

---

[2]For a full treatment of the evolution of the NACA cowling see James R. Hansen, "Engineering Science and the Development of the NACA Low-Drag Cowling," in *From Engineering Science to Big Science: The NACA and NASA Collier Trophy Research Project Winners*, ed. Pamela E. Mack (Washington, D.C.: NASA SP-4219, 1998), 1-27; Hansen, *Engineer in Charge*, 123-137, 424; John V. Becker, *The High-Speed Frontier: Case Histories of Four NACA Programs, 1920-1950* (Washington, D.C.: NASA SP-445, 1980), 139-140; William H. McAvoy, "Notes on the Design of the N.A.C.A. Cowling," *Aviation*, (September 1929): 636-638.

# FIRST INCARNATION: STABILITY AND CONTROL

The NACA cowling represented perhaps the most influential example of a project which enlisted flight research as one of a number of contributing disciplines. But during the mid-1930s Langley undertook a worthy successor to the pressure distribution work, one which employed flight research in a starring role in a program of fundamental importance. It involved the *flying qualities* of aircraft, defined by a leading researcher in the field as "the stability and control characteristics that have an important bearing on the safety of flight and on the pilots' impressions of the ease of flying an airplane in steady flight and in maneuvers." The first person to explain the underlying factors governing the stability and control of aircraft propagated his theories just after the Wrights flew over Kitty Hawk. But almost unbelievably, mathematician George Hartley Bryan arrived at his conclusions without knowing humans had flown; credible reports of the feat had not yet reached his native England. He initial foray into the subject occurred in 1903 when he read before the Royal Aeronautical Society a paper entitled "The Longitudinal Stability of Aeroplane gliders," a narrative based on his own experiments. It met with polite interest. The following year he revealed his solutions to the full problem of achieving dynamic control in aircraft. Bryan divided flying qualities into lateral and longitudinal groupings based upon degrees and types of oscillation produced by unstable motions. The complexity of his theory and the length of the accompanying computations prevented many aircraft designers from adapting his approach. Nonetheless, manufacturers in search of strong but light vehicles became intrigued with his ideas. He received due recognition after the publication of his volume entitled *Stability in Aviation* in 1911 and four years later won the Royal Aeronautical Society's Gold Medal. Engineering students still learn elementary stability theory essentially from Bryan's original formulation.

Nevertheless, during the 1910s and the1920s, stability remained uncharted territory to

4

practicing aircraft designers, one of the many imponderables of flight. Indeed, beginning with the original Wright Flyer, early aircraft lacked the property of inherent stability. Impressionistic "cut and try" methods enabled some manufacturers to arrive at satisfactory handling properties, although inferior flying qualities also caused many crashes. Indeed, no one knew what aircraft design factors yielded good flying qualities. The few conscious efforts to design stability often resulted in poor, or even dangerous flying qualities. Pilots and engineers soon appreciated the embedded dilemma: the better the stability, the less adequate the control. Only gradually did it become apparent that safe flight demanded the successful integration (and simultaneous collaboration) of these two essential ingredients.[3]

The NACA played an early and a central role in unraveling this conundrum. The Langley staff recognized from the beginning that for aeronautics to become a familiar part of American life, stability and control needed to be understood and mastered. Indeed, the second NACA Research Authorization, signed the day the laboratory opened in June 1920, launched a study on "Controllability Testing" led by Chief Physicist Frederick Norton (see chapter two) Joseph Ames and the Executive Committee nurtured high hopes for this project, expecting nothing less than "definite data" about controls leading to "definite quantitative standards for controllability." The Authorization instructed Norton to obtain "simultaneous records...of the acceleration, attitude, air-speed, and positions of and forces on all three controls...done in normal flight, in landings, and in stunting." Accordingly, Norton planned a series of free flights on the Curtiss JN4H and on the De Havilland DH4. The initial results, published in 1921 as NACA Report 120, "Practical Stability and Controllability of Airplanes," gave American aircraft designers their first systematic guidelines for producing airplanes with a satisfactory degree of stability and control. Still, in the context of the Committee's high expectations, Norton admitted

---

[3]W.Hewitt Phillips, "Flying Qualities From Early Airplanes to the Space Shuttle," *Journal of Guidance, Control, and Dynamics* 12 (July-August 1989): 449-450 (quoted passage 449); Longyard, *Who's Who*, "George Hartley Bryan"; W. Hewitt Phillips, *Journey in Aeronautical Research: A Career at NASA Langley Research Center*, Monographs in Aerospace History, Number 12 (Washington, D.C.: NASA Headquarters, 1998), 21-22.

frankly the limitations of his work:

> It should be realized...that the data on which these conclusions are based is rather meager and applies mainly to tractor airplanes with a single motor and that in some cases the results are obtained from one airplane, so that it can not be expected that this data will apply strictly to any airplane which is designed. Also, the conclusions will be modified as our information is increased. In fact, in the present state of the art it is quite impossible    to design at the first trial an airplane which is perfect in stability and control, but it should be possible, however, to design an airplane which is fairly satisfactory and from tests on this airplane to deduce what changes it is necessary to make in order to correct any given faults.[4]

Despite these qualifying remarks, Norton left no doubt about the direction in which further stability and control investigations ought to proceed. "Above everything else," he wrote, "the pilot and the designer should get together, as only in this way can a satisfactory airplane be evolved." Although preliminary in nature, Report Number 120 also provided vital data to aviators and engineers alike. Norton found, for example, that *longitudinal stability* improved when the area of the horizontal tail surface measured about 13 percent of that of the wing surface. It improved further with a flat bottomed tail section for low speed flight and with a tail section flat at the top for high speed flight. *Longitudinal control*, on the other hand, depended on such factors as designing a large elevator whose area accounted for as much as 45 percent of the total tail surface. This configuration yielded the greatest sense of controllability. To obtain the greatest feeling of "quickness and lightness" in the controls, Norton recommended small and lightweight elevators employing large gears between the stick and the elevator. For effective *lateral stability* he recommended a wing dihedral (that is, the upward or downward inclination of an airfoil, like the wing, from true horizontal) of three to six degrees; for *lateral control*, ailerons of between five and eleven percent of the area of the wing surface. *Directional Stability* for a fuselage of average length depended on having a tail fin (vertical stabilizer) whose area measured two percent of the aircraft's wing surface. *Directional control* for ordinary airplanes

---

[4]RA No. 2: "Controllability Testing," Joseph S. Ames, 11 and 28 June 1920, RA 2 File, LaRC Historical Reference Collection (first quoted passage); Frederick H. Norton, NACA TR 120, "Practical Stability and Controllability of Airplanes," (Washington, D.C.: NACA, 1921), 359-372 (block quote 371).

required a rudder about two percent of their wing area.[5]

Norton issued a second report the following year. Again employing the JN4H instrumented with an angular velocity recorder, a recording air speed meter, a control position recorder, and an accelerometer, he attempted to determine "what features of design lead to great maneuverability and controllability of the airplane." His flight test plan instructed the pilots to first fly steadily at a desired speed, then to activate all of the instruments by flipping a common switch. After doing so they moved each control to a definite angle as suddenly as possible and maintained position until the aircraft rotated through 90 degrees. They repeated this procedure at various speeds and with varied angles of control movement. Norton concluded from the tests that the maximum angular velocity and maximum angular acceleration were in proportion to the controls; that for any particular control movement both angular velocity and acceleration increased with airspeed, with the greatest rapidity just above the point of stalling; that "the time required to reach each maximum angular velocity is constant for all airspeeds and control displacements for a given airplane"; and that "a rough indication of general maneuverability" could be realized in the performance of a steeply banked turn in the minimum amount of time. Norton then presented simplified formulas for measuring the controllability and the maneuverability coefficients.[6]

Helpful as these early studies may have been, they only whetted the appetites of engineers, pilots, and aircraft manufacturers for practical, experimental, and theoretical knowledge of this paradoxical yet essential aspect of design. George Lewis took the lead to satisfy the demand. Based on his many Washington contacts, Lewis learned that the Post Office Air Mail Service would soon make inquiries about the optimal characteristics for a commercial aircraft. He informed Norton of this possibility in Spring 1922 and suggested research on

---

[5]Norton, NACA TR 120, "Practical Stability and Controllability of Airplanes," 371-372 (first quoted passage 359, second quoted passage 371).
[6]Frederick H. Norton and William G. Brown, NACA TR 153, "Controllability and Maneuverability of Airplanes," (Washington, D.C.: NACA, 1923), 537-552 (first quoted passage 538, second and third 537).

controls at very low flying speeds might be of value. Norton thought immediately of his own recent controllability tests and suggested to his boss a program which expanded his preliminary findings by repeating all of the flight tests on a Vought VE-7, De Havilland DeH-9, British Royal Aircraft Factory SE-5A, Fokker D-VIII, SPAD VII, and a Thomas-Morse MB-3. Lewis concurred and Joseph Ames, now both Executive Committee and Aerodynamics Committee chairman, won approval for Research Authorization 73: "The Comparative Stability, Controllability, and Maneuverability of Several Types of Airplanes." Realizing the 1922 NACA appropriation had been frozen at the 1921 ($200,000) level, Ames sought alternate sources of funding. With characteristic audacity he wrote to the Chief of the Bureau of Aeronautics three days after the Research Authorization opened, requesting $20,000 for this and two other projects which, he hoped, might interest the Bureau. Fortunately, the Navy did see their value and by June 1922 provided the necessary funding. The Langley shops, meanwhile, machined and assembled two single component turn meters as well as all the other parts required for the flight program except the gyroscope motors, purchased by the NACA for the purpose.[7]

Unfortunately, the project lost its most important ingredient soon after its start. Once again, Max Munk influenced the course of events. (See chapter 2 for more about Munk). During the design and construction of Munk's Variable Density Tunnel (VDT), the ill-tempered and opinionated physicist spent most of his time in Washington, D.C., advising from afar. But as Chief of the Aerodynamics Section, Norton found himself in a dilemma; responsible for the fabrication of this revolutionary piece of equipment, he still had to win approval from Munk for its design. Unfortunately, the German dismissed virtually everything Norton and his staff

---

[7]RA No. 73: "The Comparative Stability, Controllability, and Maneuverability of Several Types of Airplanes," Joseph Ames, 20 May 1922, RA 73 File, LaRC Historical Reference Collection; George Lewis to LMAL Chief Physicist (Frederick Norton), 4 May 1922, RA 73 File, LaRC Historical Reference Collection; Frederick Norton to George Lewis, 12 May 1922, RA 73 File, LaRC Historical Reference Collection; Joseph Ames to the Chief of the Bureau of Aeronautics, 23 May 1922, RA 73 File, LaRC Historical Reference Collection; George Lewis to Frederick Norton, 17 June 1922, RA 73 File, LaRC Historical Reference Collection; George Lewis to the Chief of the Bureau of Aeronautics, 10 January 1923, RA 73 File, LaRC Historical Reference Collection; Hansen, *Engineer in Charge*, 480-481.

suggested but offered few alternative ideas himself. Norton suffered through this frustrating process while the tunnel took shape. But he reached his limit after George Lewis, apparently unaware of the building hostility between the two men, sent Munk to Langley for extended periods to oversee the VDT's initial research program. Munk arrived in late 1922; Frederick Norton resigned from the laboratory in 1923 to work in industry and later in academia. He took with him all of the experience acquired during the initial stability and control work, as well as all of the general knowledge accumulated since he signed on as the laboratory's first employee in autumn 1918. Research Authorization 73 felt the results. Over the next four years much additional analysis of the problem occurred, but nothing again so systematic and coherent as Norton produced. Henry Reid sought to breathe life into stability research by soliciting from Navy pilots their impressions of the flying characteristics of various aircraft . The Bureau of Aeronautics complied by supplying raw data (handwritten pilot replies to a series of questions) from the initial flight trials of about 20 aircraft, subsequently reduced to standard forms by the Langley staff. The results left much to be desired. Most of the reports were sketchy and none reported any numerical information, only general comments about the "feel" of the controls. Aware of the resulting inadequacies and convinced of the NACA's declining interest in the project, the Bureau established its own performance test section late in 1926, although the Langley engineers and test pilots continued to offer advice about stability and control. Indeed, a December 1926 conference at the NACA's Washington offices attended by Navy representatives and by George Lewis, Jimmy Doolittle, Walter Diehl, and Thomas Carroll helped to define the continuing research problem, but resulted in no action; only a consensus that stability must be studied in tandem with control and that it remained a very stubborn but a very important research problem. With that, George Lewis canceled Research Authorization 73 in September 1927. Subsequent NACA reports suggested an ongoing interest in the subject, but no fully developed program emerged to rescue Frederick Norton's good beginning.[8]

---

[8]Hansen, *Engineer in Charge*, 30, 84-87, 481; David L. Bacon to Leigh Griffith, 26 October 1923, RA 73 Files, LaRC Historical Reference Collection; Leigh Griffith to NACA (Lewis), 28

A resurrection finally did occur, however, but almost ten years after the aborted conference in Washington. Once again, the protean Edward P. Warner emerged from a hectic career to influence the NACA. (For Warner's earlier influences, see chapters 1 and 2). Of course, he never wandered far, serving on the Main Committee all through the 1930s and as chairman of the Aerodynamics Committee from 1935 to 1941. In addition to his duties as editor of *Aviation* magazine, at President Roosevelt's request he joined the Federal Aviation Commission in 1935 and helped unscramble the air mail crisis resulting from the military's unsuccessful attempt in to provide airborne delivery to the nation. Then Warner returned to aeronautical engineering. United Airlines hired him as a consulting engineer and from late 1935 to1939 he drafted specifications and contract requirements for a daring new transport aircraft three times the size of the DC-3 and powered by four engines. After Douglas Aircraft won the project, Warner moved temporarily to Southern California where he and Chief Engineer Arthur Raymond designed of the DC-4E. Because of the unprecedented dimensions and the uncertain handling properties of this behemoth, Warner found himself reviewing a subject he first considered during his brief employment by the NACA, the same one later ramified by Frederick Norton: the vexing problem of stability and control. In this instance, the ingenious Warner

---

October 1923, RA 73 File, LaRC Historical Reference Collection; George Lewis to LMAL (Griffith), 31 October 1923, RA 73 File, LaRC Historical Reference Collection; Henry Reid to the NACA (Lewis), 20 November 1926, RA 73 File, LaRC Historical Reference Collection; George Lewis to LMAL (Reid), 14 December 1926, RA 73 File, LaRC Historical Reference Collection; Navy Aircraft Trial Reports, n.d., RA 73 File, LaRC Historical Reference Collection; Thomas Carroll to Henry Reid, 4 January 1927, RA 73 File, LaRC Historical Reference Collection; George Lewis to LMAL (Reid), 19 September 1927, RA 73 File, LaRC Historical Reference Collection.

Three NACA publications touching on the problem of stability after the cancellation of RA 73 include: Heinrich Hertel, NACA Technical Memorandum (hereafter NACA TM) 583, "Determination of the Maximum Control Forces and Attainable Quickness in the Operation of Airplane Controls," (Washington, D.C.: NACA, 1930), 1-26, a reprint of a German paper; Fred E. Weick, Hartley Soulé, and Melvin Gough, NACA TR 494, "A Flight Investigation of the Lateral Control Characteristics of Short Wide Ailerons and Various Spoilers With Different Amounts of Wing Dihedral," (Washington, D.C.: NACA, 1934), 381-394; and Hartley Soulé and William McAvoy, NACA TR 517, "Flight Investigation of Lateral Control Devices for Use with Full-Span Flaps," (Washington, D.C.: NACA, 1934), 209-219.

decided to pivot his investigation not on engineering data, but on the impressionistic but essential pilot descriptions of the flying qualities of a variety of transport airplanes. He sought the help not only of airline captains, but of the engineering staffs associated with manufacturers and operators, and of researchers employed by the NACA and other institutions. After surveying these sources, he transmuted the language of "feel" and movement into engineering terms which could be rendered, in turn, into design specifications. The preliminary results, transmitted to NACA officials in December 1935, represented the first attempt in America to define these critical design features.[9]

But Warner knew he had not solved the problem. As cargo, commercial, and bomber aircraft grew increasingly large and heavy during the 1920s and 1930, their controls became increasingly difficult to maneuver and often very slow to respond. Not only did it become *physically* exhausting for pilots to make these giants behave; nervous exhaustion began to grip the cockpit as aviators, in command of extravagantly expensive machines filled with more human beings than ever before, spent long flights fighting sluggish and unpredictable controls. Such conditions diminished pilot confidence and represented a threat to safety in emergencies requiring fast maneuver. In light of the complexity of the situation--one mitigated by the "feel" of the controls, not just their actual mechanical actions--Warner realized his first guidelines were imperfect at best. He pursued these unanswered questions with characteristic zeal. As the DC-4E underwent demonstration flights he continued to participate in the design process and occasionally joined the test pilots "for the purpose of observing stalling and other characteristics first hand." He also alerted the NACA's administrators and engineers to the purpose and the value of flying qualities research. Finally, in his new role as chairman of its Aerodynamics Committee, he attempted to involve Langley root and branch in his investigation, sending a request to Joseph Ames for Research Authorization 509, "Preliminary Study of Control

---

[9]Fred E. Weick to Edward P. Warner, 7 January 1936, RA 509 Files, LaRC Historical Reference Collection; Hartley Soulé, NACA TR 700, "Preliminary Investigation of the Flying Qualities of Airplanes," (Washington, D.C.: NACA, 1940), 449; Bilstein, "Edward Pearson Warner, ' 115, 119-120; Phillips, "Flying Qualities," 450-451; Phillips, *Journey in Aeronautical Research*, 22.

Requirements for Large Transport Airplanes." Warner asked the staff at Hampton to "obtain data for the determination of the requirements as to the flying qualities, particularly maneuverability and stability of transport airplanes and evolve a technique for making tests to determine these qualities." He envisioned a program which yielded dividends to aeronautical research in general and dividends to his own labors with Douglas in particular, one which started modestly and expanded over time. Warner's proposed Research Authorization instructed the laboratory to conduct "a simple and short series of flight tests...to determine the flying qualities in quantitative terms," along with the required instrumentation. Then he suggested flight trials of the resulting data, staged on one or more of the aircraft in the Langley inventory. Finally, Warner recommended a series of follow-on flights using borrowed transport aircraft to assess the problem in its entirety. With no apparent reservations, Joseph Ames approved the Research Authorization on January 14, 1936.[10]

## SECOND INCARNATION: FLYING QUALITIES

During the following six months the Flight Research Section geared up for the initial flight operations and arrived at some crucial assumptions. A brief Technical Note by test pilot Melvin Gough published during the same January confirmed Langley's renewed interest in the subject. A simulator constructed at the laboratory tested "the maximum forces a pilot can exert

---

[10]Edward Warner served as chair of the Aerodynamics Committee from 1935 to 1941. Joseph Ames assumed greater responsibilities, becoming chairman of the NACA Executive Committee in 1920 and remaining until 1937; and filling the top job of chairman of the NACA Main Committee between 1927 and 1939. Roland, *Model Research*, 2: 427, 439; RA No. 509: "Preliminary Study of Control Requirements for Large Transport Airplanes," Edward P. Warner, December 9, 1935, approved by Joseph S. Ames, January 14, 1936, RA 509 Files, LaRC Historical Reference Collection; Bilstein, "Edward Pearson Warner," 120 (quoted passage); Phillips, "Flying Qualities," 450; James R. Hansen, "Bigger: The Quest for Size," in *Milestones of Aviation: Smithsonian Institution National Air and Space Museum*, ed. John T. Greenwood (New York: Hugh Lauter Levin Associates, 1989), 168, 171, 172-173.

on the controls of an airplane...to obtain...systematic data upon which to base the location of controls within the cockpit and the design of the control surfaces." Gough concluded that pilots misjudged the all-important stick forces by as much as *50 percent*, guessing low for small forces and high for large ones, suggesting a yawning gap between the flier's expectations of handling and the actual effort necessary to produce a desired maneuver in flight. Meantime, Fred Weick, now a Senior Engineer and coming to the end of his service at Langley, confided to Warner that although his staff had spent much time reviewing his original (1935) design requirements, much remained to be done. Replying from Los Angeles, Warner urged Weick to press forward, but contented himself with the initial reports of Douglas test pilot Frank Collbohm. Flying a Lockheed Electra, Collbohm "found...that most of the requirements as we have set them up seem quite within the bounds of reason." Warner transmitted to the Flight Test Section improvements suggested by Collbohm's flight tests. He also sent the NACA researchers data collected during the first quarter of 1936 from a series of maneuvers conducted at Los Angeles Municipal Airport. A Douglas Sleeper Transport (actually, an enlarged DC-2) was put through its paces relative to take-off, maximum power, and single engine performance. Close scrutiny of these and other pieces of evidence and the outlay of "considerable time" by his staff led Henry Reid to decide in early May 1936 to allow the "active continuation" of the project, exactly on the three-tiered basis suggested by Warner.[11]

Reid received the fruit of the first phase of the Research Authorization two months later. Based on the experiences of the NACA pilots and engineers, Edward Warner's preliminary

---

[11]Melvin Gough, NACA Technical Note (hereafter NACA TN) 550, "Limitations of the Pilot in Applying Forces to Airplane Controls," (Washington, D.C.: NACA, 1936), 1 (first quoted passage), 11; Fred E. Weick to Edward P. Warner, 7 January 1936, RA 509 Files, LaRC Historical Reference Collection; Edward P.Warner to Fred Weick, 20 January 1936, RA 509 Files, LaRC Historical Reference Collection (second quoted passage); "Take-Off Performance," Douglas DST Aircraft, 7 January and 28 February 1936, RA 509 Files, LaRC Historical Reference Collection; "Air Speed Calibration/Maximum Power Tests," Douglas DST Aircraft, 28 February 1936, RA 509 Files, LaRC Historical Reference Collection; "Single Engine Tests," Douglas DST Aircraft, 25 March 1936, RA 509 Files, LaRC Historical Reference Collection; Henry Reid to the NACA (Lewis), 4 May 1936, RA 509 Files, LaRC Historical Reference Collection (third quoted passage).

suggestions, and new details supplied by Douglas Aircraft, the Langley experimenters unveiled a set of preliminary standards for handling characteristics which informed all of their subsequent research. Floyd L. Thompson--an Associate Aeronautical Engineer in the Flight Research Division and the future director of Langley--transmitted his colleagues' flying qualities requirements to the Engineer-in-Charge. Entitled "Suggested Requirements For Flying Qualities of Large Multi-Engine Airplanes," It described, in Thompson's words, an attempt to "crystallize ideas regarding what items are important and indicate wherein data are lacking concerning quantitative values." Those who actually conducted the research were more frank, calling its stated numerical limits "quantitatively unreliable, owing to the...lack of data concerning what constitutes satisfactory flying qualities." Nonetheless, as a first systematic attempt to provide pilots the handling properties required for predictable response and for safe flying, it exercised an enormous influence on future aircraft design and construction.

The researchers divided flying qualities into four categories: longitudinal stability, longitudinal control, lateral stability, and lateral control. An aircraft achieved *longitudinal stability* when, "with elevator free [it] shall be dynamically longitudinally stable throughout the speed range for all loading conditions." *Longitudinal control* occurred when aviators found it "possible to maintain steady flight [in pitch] at any speed from the...diving...to the minimum speed. This condition shall be met with any loading...and with any power condition...." Maintaining *lateral stability* required the same conditions as longitudinal stability; that is, with the elevator free, the aircraft needed to demonstrate lateral [side-to-side] stability throughout its range of speed and under all loading circumstances. *Lateral control* involved the complicated interplay between aileron and rudder forces. The Langley engineers decided that at 70 miles per hour with flaps down or 80 miles per hour with flaps up, the ailerons alone should be capable of banking the aircraft 15 degrees in 2.5 seconds; at 120 miles per hour or faster, the same maneuver should be accomplished in two seconds. Similarly, they determined that relying solely on the rudder during steady flight at 70 miles per hour with flaps down, at 80 miles per hour with flaps up, or at any speed above 80, it should be possible to affect a 15 degree change in heading

14

under the same time limits prescribed for ailerons. Finally, combining these two sources of lateral control, it seemed reasonable to expect the execution of a 45 degree banked turn in five seconds at 145 miles per hour with no more than 100 pounds exerted for rudder force and 75 pounds for either of the ailerons. The same force limitations applied in order to complete a 30 degree banked turn in four seconds at 200 miles per hour, and so on.[12]

Complementing these requirements, Thompson and his associates included for Reid a proposed flight program designed to verify their assumptions and to broaden the scope of inquiry. Called "General Program of Tests of Airplane Flying Qualities," it prescribed a series of pilot maneuvers keyed to the stated requirements. Thus, *longitudinal stability* would be investigated in two ways. With the elevator free, the aviator would trim the aircraft for a desired speed, push the stick forward to achieve a velocity five or ten miles per hour faster, then release the stick and record the oscillations as the machine returned to steady state at trim speed. With the elevator fixed, the pilot would return the stick manually to its original setting after experiencing the disturbance and hold it during the period of oscillations. *Longitudinal control* would be determined by free flight tests measuring the degree of force necessary to operate the elevator controls at different velocities, with varied tab settings, with power on, and with power off. *Lateral stability* measurements, on the other hand, required the research pilots to place the aircraft in trim at a desired speed; move the ailerons abruptly to obtain a 15 degree bank; let go of the controls; record the maximum angle of bank, maximum rate of roll, or maximum change in heading; and note the elapsed time between peaks of the resulting oscillations. Rudder-related disturbances in lateral stability would be determined by following the aileron procedures, except for a rudder kick designed to cause a change in heading of about 10 degrees. Finally, *lateral control* would be ascertained through several techniques. Pilots would be asked to fly in steady flight and at a variety of speeds and to apply abruptly the full aileron control, then to record the

---

[12]Hansen, *Engineer in Charge*, 46-47, 422; Floyd L. Thompson to Henry Reid, 14 July 1936, RA 509 Files, LaRC Historical Reference Collection (first quoted passage) with first attachment "Suggested Requirements For Flying Qualities of Large Multi-Engined Airplanes" (second, third, and fourth quoted passages).

maximum rate of roll or the time elapsed in attaining a specified angle of bank. Force exerted on the controls would also be obtained. To measure rudder control, the flier, again holding the aircraft steady at different speeds, would apply the full rudder suddenly and note the time needed to change heading 15 degrees, or note the rate of turn versus the passage of time. In order to learn the effectiveness of aileron combined with rudder, maneuvers would be undertaken to apply both at once and record the length of time necessary to achieve a bank of 45 degrees.[13]

Henry Reid recognized the seminal importance of these two memoranda and lost no time transmitting them to George Lewis and to the NACA's Aerodynamics Committee for approval. Meantime, during the same month, the NACA published a Technical Report even more important than the two papers just forward by Reid to Washington, D.C. Its author, engineer Hartley Soulé, assumed a leading role in the handling qualities project from its inception and possessed perhaps the best grasp of the subject of anyone at Langley. The thirty-one year old New York University graduate arrived in Hampton in 1927 and took his cue from his boss, Gus Crowley, Chief of the Flight Research Division. Crowley believed firmly in the primacy of free flight tests in evolving a set of practical standards for handling qualities. His unequivocal position, seconded by Soulé, was necessary in a laboratory where wind tunnels reigned supreme and their highly able practitioners (such as Fred Weick, John Stack, and Robert T. Jones) sought to employ them as the chief research tools in as many investigations as possible. Soulé's first report on the subject, entitled "Flight Measurements of the Dynamic Longitudinal Stability of Several Airplanes and A Correlation of the Measurements with Pilots' Observations of Handling Characteristics" re-opened the flying qualities program at Langley. Moreover, its techniques and results, although focused on smaller aircraft rather than transports, epitomized the formative period of Research Authorization 509.

In order to assess the degree of longitudinal stability expected in conventional airplanes, Soulé supervised tests on eight single-engine machines: the Fairchild 22, the Martin XBM-1 and

---

[13]Floyd L. Thompson to Henry Reid, 14 July 1936, RA 509 Files, LaRC Historical Reference Collection, with second attachment, "General Program of Tests of Airplane Flying Qualities."

the T4M-1, the Verville AT, the Fairchild FC2-W2, the Boeing F4B-2, the Consolidated NY-2, and the Douglas 0-2H. During the flight program, the pilots attained an altitude of 3,000 feet, obtained steady conditions at the desired speed, and achieved trim. To induce oscillations, they then lowered the aircraft's nose using the elevator and accelerated until reaching a speed of five miles per hour over the initial setting. Then the elevator was quickly returned to its original position for fixed runs and freed again for tests with no elevator control. Adjustable stops held the fixed elevator firm during oscillations. The results presented in Soulé's article suggested an undeniable relationship: the higher the speed the longer the period of oscillation. Indeed, at low speeds oscillations lasted for 11 to 23 seconds on the eight airplanes, at high speeds from 23 to 64 seconds. Perhaps most important for future work, an

> attempt was made to correlate the measured stability with pilots' opinions of the general handling characteristics of the airplane in order to obtain an indication of the most desirable degree of dynamic stability. The opinions of the two pilots concerning the handling characteristics of the airplanes apparently were not influenced by the stability characteristics as defined by the period and damping of the longitudinal oscillations.[14]

While Soulé's report attracted notice, the flying qualities project itself faced a period of quiescence while the lab's superiors in Washington attended to some bureaucratic considerations. Breaking with the tradition of disseminating research results to the outside world only after the NACA vetted them thoroughly and approved them for publication, George Lewis allowed regulators in the Bureau of Air Commerce in the Department of Commerce an opportunity to examine the two handling qualities memoranda *before* they passed the muster of the NACA Aerodynamics Committee. Lewis probably agreed because the Bureau, responsible for the nation's civil air regulations, needed to be aware of data with the potential to revolutionize the requirements for safe flight in large transports. However, the NACA Director did not change his spots entirely. L.V. Kerber, Chief of the Bureau's Manufacturing Inspection Service, asked if

---

[14]Hansen, *Engineer in Charge*, 90-91, 181-182, 422; George Lewis to Richard G. Gazley, 20 August 1936, RA 509 Files, LaRC Historical Reference Collection; Hartley Soulé, NACA TR 578, "Flight Measurements of the Dynamic Longitudinal Stability of Several Airplanes and a Correlation of the Measurements with Pilot's Observations of Handling Characteristics," (Washington, D.C.: NACA, 1936), 69, 70 (block quote, 69).

his office could retain these documents longer than the usual NACA ten day review period, impressing on Lewis their possible impact on the nation's air commerce, as well as on the existing aircraft strength requirements imposed by the Bureau of Commerce. Lewis not only refused, but insisted on their return without delay. These documents still needed to be circulated for comment to a number of the NACA stalwarts, including Dr. Albert Zahm, now with the Library of Congress; Dr. Lyman J. Briggs of the National Bureau of Standards; Lieutenant Colonel Oliver P. Echols of the Army Materiel Division at Wright Field; Walter Diehl of the Bureau of Aeronautics; and the most interested of all, Edward P. Warner.[15]

Warner found the two memoranda encouraging signs of the NACA's commitment to flying qualities research. But he saw an even clearer indication when George Lewis reported a meeting on Research Authorization 509 with W.C. Clayton, an aeronautical engineer in the Department of Commerce who coordinated the Bureau's design requirements with the commercial airlines and the aircraft manufacturers. They met in Washington in November 1936, after which Clayton traveled to Langley where he conferred with Henry Reid, Gus Crowley, Richard Rhode, and others. The visitor arrived at Langley's doorstep "to get a better understanding between the needs of the industry and the Committee's work in answering these needs." During the talks, Clayton offered to act as an intermediary between the NACA researchers and the industry during the handling qualities project. Reid and his lieutenants, hoping to obtain a Douglas DC-2 or a Boeing 247 for their experiments, accepted Clayton's role, especially after he promised to raise the issue of loaning commercial aircraft to Langley at an impending requirements conference with aircraft manufacturers and air carriers. Edward

---

[15]L.V. Kerber to George Lewis, 27 August 1936, RA 509 Files, LaRC Historical Reference Collection; George Lewis to L.V. Kerber, 28 August 1936, RA 509 Files, LaRC Historical Reference Collection; John F. Victory to L.V. Kerber, 7 October 1936, RA 509 Files, LaRC Historical Reference Collection; John F. Victory to Albert Zahm, 7 October 1936, RA 509 Files, LaRC Historical Reference Collection; John F. Victory to Lyman Briggs, 7 October 1936, RA 509 Files, LaRC Historical Reference Collection; John F. Victory to Oliver P. Echols, 7 October 1936, RA 509 Files, LaRC Historical Reference Collection; John F. Victory to Walter S. Diehl, 7 October 1936, RA 509 Files, LaRC Historical Reference Collection.

Warner, meanwhile, had already laid plans to visit New York City in early December and, when he got wind of the conference mentioned by Clayton, jumped at the chance to participate in it. But failing to win an invitation, he decided on December 1 to forego the pleasures of the Harvard Club and journey to Langley to meet the flying qualities investigators.[16]

Warner arrived the morning of December 3 at Henry Reid's office and found a number of the laboratory's leading lights awaiting him. Gus Crowley, Floyd Thompson, and Hartley Soulé from the Flight Research Division sat next to such wind tunnel representatives as the future West Coast laboratory director Smith DeFrance and the brilliant young aerodynamicist Eastman Jacobs. Among the figures present, Warner found perhaps the closest affinity to Jacobs, based on their shared technical interest. Almost immediately after graduation from Berkeley in 1924, Jacobs went to work for Langley and only months after his arrival developed an interest in high-speed aerodynamics. He found himself free to pursue this line of inquiry upon assuming the post of section head of the Variable Density Tunnel after Max Munk's celebrated and unlamented departure in 1926. On this day, however, Warner talked to Jacobs not about aerodynamics in general, but specifically about the need to press forward with the flying qualities research. Gus Crowley offered Warner some reassurance. He explained that a Stinson Reliant SR-8E cabin monoplane owned by the NACA would be ready the following week to begin flight tests to verify the methods of obtaining handling properties data. Once the trials finished on the Stinson, flight research on big transports would begin. But Hartley Soulé added a note of caution. "No flight routine had yet been settled for...tests [of the full-sized aircraft]. Such a routine, "he cautioned, "would depend upon test results obtained with the Stinson and further work, [the idea

---

[16]John F. Victory to Edward P. Warner, 7 October 1936, RA 509 Files, LaRC Historical Reference Collection; George Lewis to Edward P. Warner, 16 November 1936, RA 509 Files, LARC Files; William H. Herrnstein to Henry Reid, 13 November 1936, RA 509 Files, LaRC Historical Reference Collection (quoted passage); Edward P. Warner to George Lewis, 19 November 1936, RA 509 Files, LaRC Historical Reference Collection; Edward P. Warner to Leighton W. Rogers, 19 November 1936, RA 509 Files, LaRC Historical Reference Collection; Edward P. Warner to John F. Victory, 1 December 1936, RA 509 Files, LaRC Historical Reference Collection; George Lewis to Edward P. Warner, 1 December 1936, RA 509 Files, LaRC Historical Reference Collection.

being to] have a broad base at the start particularly." Warner also learned that the Stinson flight program would gather data using two types of instruments: those especially designed and installed by the Langley team, including control-position and control-force recorders, two turnmeters, an accelerometer, and an air speed recorder; and off-the-shelf motion picture cameras positioned to photograph the readings of the standard cockpit instruments. Warner expressed concern about the length of time required to install these instruments; Thompson estimated two days at most. Soulé raised a more fundamental question. Could the standard NACA control force device be adapted to the wheels and sticks of large cargo aircraft? Since no one yet knew which commercial or transport planes might be made available for the flights, it was decided to collect information on the control columns of all the likely candidates.

After the meeting ended, Eastman Jacobs conducted Warner to the flight section group where the participants, including test pilot Melvin Gough, discussed the visitor's observations about flying qualities. From an aerodynamisict's viewpoint, Jacobs found Warner's conclusions to be "essentially reasonable and definitely desireable." Warner made no secret to Jacobs and the others about his own objective for the NACA research: a quick, universal flight-check procedure by which the flying qualities of any type of commercial aircraft might be evaluated within a week. The NACA researchers liked this approach and recognized "its vital importance to the Laboratory, because a familiarity with new [aircraft] types will...get us out of the dark with regard to the practical effects of the application of new developments." Yet, adhering to the cautious NACA style, they urged Warner to await the preliminary tests on the Stinson and to use the resulting data to fashion his check-out procedures. Stimulated by this open discussion, Soulé followed it with a request to the Washington office for a finished, printed copy of Edward Warner's most recent specifications for four engine transports.[17]

---

[17]W.H. Herrnstein to Henry Reid, 3 December 1936, RA 509 Files, LaRC Historical Reference Collection (first quoted passage); Eastman Jacobs to Henry Reid, undated, RA 509 Files, LaRC Historical Reference Collection (second and third quoted passages); (Draft) Hartley Soulé to the NACA (Lewis?), n.d., and final version, Edward R. Sharp to the NACA (Lewis), 21 December 1936, RA 509 Files, LaRC Historical Reference Collection; Becker, *High-Speed Frontier*, 11; Hansen, *Engineer in Charge*, 418.

# TAKING FLIGHT

Once Soulé received and absorbed Warner's treatise--a document considerably more specific and more quantitatively exacting than his earlier attempts--he launched the Stinson flight research. The convening of the Aerodynamics Committee on January 19, 1937, afforded him the opportunity to inform his superiors of the progress of the flight tests and to raise some concerns. By this time the Stinson had been put through about half of its flying program, completing the longitudinal stability and control investigations in only five hours due to a limited number of power combinations. The lateral stability and control work required more time. The absence of trim tabs on the Stinson's rudder and aileron hindered the program's original intent of testing handling qualities in all three axes. Moreover, in order to mount the motion picture camera in the small cockpit the ground crew needed to rearrange the instrument panel before the flight maneuvers could begin. These preparations resulted in the successful filming of such standard instruments as the directional gyro, the artificial horizon, the turn-and-bank indicator, the air-speed meter, and the altimeter. However, the simplest instrument of all failed the technicians; a common stop watch affixed to the instrument panel could not be read by the camera because its second hand and gradations did not photograph well against its white dial.

Nonetheless, Soulé presented some impressive results at the end of this series of tests. The program lasted about seven weeks and ended on February 11, 1936, after 20 hours of flight time. The experiments, wrote Soulé on February 24, "were made for the purpose of determining the practicability of the flight program..., of developing the instrumentation essential to the flight tests proposed, and of making a start on the compilation of information on the flying qualities of existing airplanes." As the flight program progressed, only minor changes in the instrumentation suite proved necessary: a standard rudder force indicator was installed, along with a specially-

21

made device to record aileron and elevator forces from the Stinson's wheel. Until this time the laboratory could only gauge forces exerted on a stick. All of the NACA instruments operated according to expectation. The few deviations from the initial flight plan related to the aircraft's design limitations; it had just one propeller pitch and lacked aileron and elevator trim tabs. For the sake of simplicity, Soulé decided to limit the Stinson's performance to only one center of gravity. He also added some lateral stability maneuvers which checked recoveries from aileron and from rudder-induced disturbances.

For the most part, the Stinson flights seemed to substantiate the specifications for flying qualities proposed by Edward Warner. Pending additional review of the data, Soulé predicted some quantitative revisions to existing dogma. For instance, the assumption that longitudinal oscillation occurred for a minimum of 40 seconds was not borne out in the Stinson tests. Not surprisingly (in light of earlier findings), the period of oscillation rose and fell with the speed of the aircraft. But the Stinson investigation's most significant finding involved the future direction of NACA handling qualities research, concluding that the flight program demonstrated "the practicability of the specifications...[and that] the test program and instruments are sufficiently satisfactory to warrant...the continuation of the development work on a multi-engine airplane...." Soulé felt the next phase of the flying qualities program might begin in spring 1937, provided the NACA found an agency or a company willing to loan a large transport aircraft (or a cargo or bomber plane of comparable size and handling qualities) for the tests. Otherwise, all that remained were some minor adjustments in the instrumentation (converting the control-force recorder to an indicator) and some training for the NACA pilots on the big machines. Soulé predicted a 60 day flight program.[18]

---

[18]Soulé's final preliminary study on the Stinson tests appeared in September 1937 as a confidential memorandum report entitled "Measurement of the Flying Qualities of the Stinson Model SR-6E Airplane," Henry Reid to the NACA (Lewis), 11 September 1937, RA 509 Files, LARC Files; M.M. Muller to LMAL, 6 January 1937, RA 509 Files, LaRC Historical Reference Collection; "Specifications For Flying Qualities of Four-Engine Transports," n.d., RA 509 Files, LaRC Historical Reference Collection; George Lewis to LMAL (Reid), 6 January 1937, RA 509 Files, LaRC Historical Reference Collection; Henry Reid to the NACA (Lewis), 16 January

But where could the NACA turn for the needed testbed? Henry Reid had the short-term answer. He reminded George Lewis of some previous landing research conducted by the NACA for the Army Air Corps using a loaned Martin YB-12 bomber. The Air Corps had also expressed increasing interest in Langley's handling qualities research as larger and larger bombers and cargo planes began to enter the military inventory. The prospect of mutual benefit led Reid to suggest borrowing the Martin again for two months, beginning around April 1, 1937, to fulfill Soulé's flight schedule. Lewis proposed this solution to Lieutenant Colonel Oliver Echols, Chief of the Engineering Division at Wright Field. Never timid about asking for assistance, the NACA Director not only requested the YB-12 or a Martin B-10B bomber to conduct flight tests similar to those on the Stinson; because the Langley hangars were "already taxed to the limit," he also pressed Echols to house the aircraft "in one of the Air Corps hangars at Langley Field and [to service it] by Air Corps personnel....' The colonel agreed to provide shelter and maintenance for the B-10B aircraft, to make it available for the period indicated, and to include "any particularly desireable item" needed by the NACA researchers. Lewis, in turn, presented the Engineering Division with the general test plan and the suggested flying qualities requirements, both of which so impressed Echols that he asked the laboratory to treat as confidential all of the results of the Martin flight program. Finally, in compliance with Hartley Soulé's desire to measure not one center-of-gravity but several, Echols instructed his staff to forward both the specifications and diagrams of the B-10B, as well as two load schedules for the most forward and the most rearward center-of-gravity locations.[19]

---

1937, RA 509 Files, LaRC Historical Reference Collection; Hartley A. Soulé to Henry Reid, 3 May 1937, RA 509 Files, LaRC Historical Reference Collection; Hartley A. Soulé to Henry Reid, 24 February 1937, RA 509 Files, LaRC Historical Reference Collection (first and second quoted passages).

[19]Henry Reid to the NACA (Lewis), 26 February 1937, RA 509 Files, LaRC Historical Reference Collection (first quoted passage); George Lewis to Oliver Echols, 11 March 1937, RA 509 Files, LaRC Historical Reference Collection (second quoted passage); George Lewis to LMAL (Reid), 29 March 1937, RA 509 Files, LaRC Historical Reference Collection; Oliver Echols to George Lewis, 24 March 1937, RA 509 Files, LaRC Historical Reference Collection; George Lewis to Oliver Echols, 8 April 1937, RA 509 Files, LaRC Historical Reference Collection; Oliver Echols to George Lewis, 14 May 1937, RA 509 Files, LaRC Historical Reference Collection; Hartley

As the Langley technicians prepared the Martin bomber for its flights, some familiar visitors appeared at Hampton to make known their continued interest in the project. Still eager to shape events because of his commitments to Douglas, and also because of a paternal interest in the flying qualities, Edward Warner arrived at the laboratory in mid-June 1937. By then, the Stinson flights had received their last post-mortems and the Martin program had just begun. The discussions may not have been entirely welcome by the NACA engineers and pilots. Warner grilled Hartley Soulé about "the extent of the results, details of presentation, and the time required for the tests." He also questioned the precision of the recorded measurements and asked whether the same maneuvers flown under identical conditions really corresponded to one another. The intense Warner also offered some suggestions to the flight researchers. He advised them to fly pull-ups on the big transports with great care, duplicating exactly and consistently the normal flight paths of commercial airliners in order to pinpoint any delays between "the control movement and the upward motion of the airplane." Eleven days later W.C. Clayton, the Department of Commerce engineer who had come to Langley the previous year with the hope of disseminating the NACA's handling qualities research to the airlines and the aircraft manufacturers, returned to Hampton. Before launching a national tour of the industry, he wanted to find out how long the NACA required for each flying qualities investigation and which types of aircraft would be most beneficial for the NACA to borrow. Gus Crowley told him each series of experiments required about one month, or roughly 60 hours of actual flying time. In addition, the laboratory needed three weeks set-up time prior to delivery of any testbed and another week for the company pilots to familiarize the NACA's aviators with the idiosyncrasies of the planes on loan. The flight researchers told Clayton the most suitable candidates for their experiments included any of the large-size Lockheed machines, followed by the Boeing 247, followed by the Douglas DC-2.[20]

Soulé to Henry Reid, 3 May 1937, RA 509 Files, LaRC Historical Reference Collection; George Lewis to LMAL (Reid), 26 May 1937, RA 509 Files, LaRC Historical Reference Collection. [20]Hartley Soulé to Henry Reid, 15 June 1937, RA 509 Files, LaRC Historical Reference Collection (both quoted passages; John Crowley to Henry Reid, 26 June 1937, RA 509 Files,

Meanwhile, the Martin B-10B underwent its tests. The flight program occurred in early May and throughout June and required just 26 hours of flying time, half of what Crowley expected but closer to the quick assessment desired by Edward Warner. On the whole, the Martin experiments were "in essential agreement" with those on the Stinson. One technical fact complicated the investigation, however; because the cockpit lacked space for the observer to sit abreast of the pilot at the controls--and because the second seat had to communicate with the first by phone--the indicating instruments needed to be interpreted by the pilot during flight and relayed by voice to the observer. The awkwardness of the procedure led Floyd Thompson to conclude that "regardless of the system used for making measurements [flight research] is greatly handicapped when the observer does not have access to the pilot's cockpit." Nonetheless, all of the equipment worked satisfactorily but the control-force indicator which failed during sudden pushes or pulls. This instrument underwent modification so it could account for violent, as well as steady inputs.

Because of the Martin's greater range of flight settings than the Stinson--such as propeller pitch, throttle, flap position, and landing gear--the researchers limited the test plan to five regimes: high-speed, climbing, power-off, take-off, and landing. The most significant finding of the flight tests materialized during the longitudinal stability maneuvers. Flying with power on and weighted to achieve the rearmost center of gravity, the Martin demonstrated longitudinal instability. But stability returned with the power off and remained so during the forward center of gravity tests. The aircraft also exhibited poor dihedral stability, failing to level off quickly after lowering one wing. Moreover, when one engine was set at full power and the other at idle, the rudder tab failed to overcome the plane's change in heading due to asymmetric thrust. Nonetheless, Thompson realized that the importance of the Martin tests lay not in specific handling results but in the methods used to sample the aircraft's handling properties:

> [I]t is felt that the procedure has been fairly well perfected. Some further development of instruments and procedure will be required, but in general it is believed that from now on

the major point of interest will be the actual results obtained, rather than the perfection of procedure. [I]n machines wherein the observer has access to the pilot's cockpit, the complete program can be carried out in approximately one month. Some advance notice, however, is required to permit the preparation of instruments...[and] the control wheel installation should be made available at least two weeks in advance of the delivery of the airplane....[21]

The chance to weigh the evolving flying qualities requirements against still bigger and more complicated aircraft arose in an unexpected way and reflected the increasing impact of the NACA's flight research on the nation's air carriers. At a conference in Boston during summer 1937, George Lewis mentioned to United Airlines Superintendent of Engineering H.O. West a possible solution for stalling characteristics evident on their workhorse DC-3s. It involved a small instrument attached to the wing surface which informed the pilot of impending stalls. West followed up in August with a letter offering to make one such airplane available to Langley for tests, and attached a table showing DC-3 wing data. Lewis then wrote to R.D. Kelly, United's Supervisor of Research, asking the company to deliver the airliner to Langley sometime after the first week in September. A casual comment in Kelly's reply dismayed Lewis. The United executive mentioned, in passing, the establishment by his company of a Flight Research Group and asked whether Lewis knew of any experienced NACA engineers who might be interested in working for the giant air carrier. Kelly's question merely symbolized the growing reputation of the NACA in private industry, but Lewis, perhaps for good reason, did not take it benignly. Even though the NACA staff grew steadily during 1920s and the 1930s, George Lewis still found himself faced with a perpetual shortage of employees due to heavy turnover. Indeed, between 1919 and 1934 an average of 40 left each year; not a large number in itself, but *roughly one in seven NACA workers* in the year 1932. Moreover, Lewis continued to lead his institution with a strong personal imprint, more like a symphony conductor, as one historian points out, than a bureaucrat. These reasons explain why Lewis confessed himself "rather disturb[ed]" by Kelly's innocent inquiry and why he worried that when the United delegation arrived in Hampton to

---

[21]Floyd L. Thompson to Henry Reid, 14 July 1937, RA 509 Files, LaRC Historical Reference Collection.

deliver the DC-3 they might endeavor to lure one of Gus Crowley's men to Chicago. Lewis decided not to leave such job decisions to the locals and told Kelly with unconvincing naiveté, "I do not know at the present time of any man who is available and who has the qualifications you outlined...."[22]

This recruitment skirmish, which left the Langley ranks intact, did not endanger the DC-3 test program. Crowley, Soulé, Thompson, Jacobs, McAvoy, and Gough met at the beginning of September and decided to take advantage of a golden opportunity by folding the stall experiments into a broader program of low-speed flying qualities research. Since the aircraft would remain at Langley Field only five or six days, the team realized "qualitative observations" would need to supersede exact measurements in many instances. They agreed to first take the big machine on a preliminary flight in order to judge its overall handling characteristics as well as assess its performance at the minimum cruising speed. Afterwards, in conjunction with the United officials, they would agree on a flight plan and install only the essential instrumentation: a control-force indicator to discern elevator resistance; an air-speed indicator; a suspended air-speed head; and, to measure stall characteristics, one or more cameras to record the motions of thin black ribbons installed as tufts on one or both of the wings. Meanwhile, the staff requested a copy of the Bureau of Air Navigation's DC-3 Flight Report to become more familiar with the airplane's flying properties.[23]

---

[22]H.O. West to George Lewis, 19 August 1937, RA 509 Files, LaRC Historical Reference Collection; H.O. West to George Lewis, same date, RA 509 Files, LaRC Historical Reference Collection; George Lewis to R.D. Kelly, 20 August 1937, RA 509 Files, LaRC Historical Reference Collection; R.D. Kelly to George Lewis, 30 August 1937, RA 509 Files, LaRC Historical Reference Collection; R.D. Kelly to George Lewis, 27 August 1937, RA 509 Files, LaRC Historical Reference Collection (first quoted passage); George Lewis to Henry Reid (Personal), 31 August 1937, RA 509 Files, LaRC Historical Reference Collection (second quoted passage); George Lewis to R.D. Kelly, 31 August 1937, RA 509 Files, LaRC Historical Reference Collection; James R. Hansen, "George W. Lewis and the Management of Aeronautical Research," in *Aviation's Golden Age: Portraits from the 1920s and 1930s*, ed. William M. Leary (Iowa City: University of Iowa Press, 1989), 98, 101.
[23]George Lewis to R.D. Kelly, 20 August 1937, RA 509 Files, LaRC Historical Reference Collection; Hartley Soulé (?) to Henry Reid, 2 September 1937, RA 509 Files, LaRC Historical Reference Collection (quoted passage); Henry Reid to the NACA (Lewis), 1 September 1937,

During the final preparation for the tests, Lewis informed Kelly about the NACA's intention to install a leading edge spoiler on the DC-3 to combat its propensity for stalling. But he failed to mention the objective of also wringing some handling qualities data out of this investigation. Thus, when Kelly, two pilots, an engineer and a mechanic landed a DC-3 Mainliner at Langley Field on Sunday, September 26, for a week of tests, they only then learned about the covert flight program. The experiments lasted just six days, from September 27 to October 2. In essence, the United and the NACA traded favors. In exchange for the stall avoidance techniques, Langley learned through a combination of measurements in flight and discussions with the United pilots how such planes behaved and gained "a better appreciation of what the transport operators expect and are willing to accept " One indisputable and surprising fact emerged from the flying qualities tests, however brief; the DC-3's "longitudinal stability was poor." Still, Gus Crowley pronounced the experience of flying "this...latest and largest land transport machine now in use" a great success. Henry Reid felt the opportunity undeniably broadened and enriched the overall handling qualities inquiry.

> The Committee...benefited a great deal from these contacts, and particularly because of the fact that our pilots and engineers have been able to fly, handle, and observe some of the flying characteristics of this large airplane. This is the first time the personnel of the Laboratory staff has had such an opportunity and it is believed that it will be to the advantage of the Committee...if arrangements can be made to borrow such large airplanes as the need arises so that we may be kept in touch with current problems and may be in a position to so aid the industry.[24]

In exchange for this first taste of flight research on the new generation of multi-engined aircraft, the NACA provided United with invaluable short-term relief from the low speed stalls, violent events that happened with no pilot warning and that Soulé and the others found

RA 509 Files, LaRC Historical Reference Collection.
[24]George Lewis to R.D. Kelly, 3 September 1937, RA 509 Files, LaRC Historical Reference Collection; R.D. Kelly to John F. Victory, 22 September 1937, RA 509 Files, LaRC Historical Reference Collection; John Crowley to Henry Reid, 5 October 1937, RA 509 Files, LaRC Historical Reference Collection (first three quoted passages); Henry Reid to the NACA (Lewis), 6 October 1937, RA 509 Files, LaRC Historical Reference Collection (block quote); Hansen, *Engineer in Charge*, 112.

"definitely undesireable and likely to be dangerous." Because the stall warning indicator had not yet been perfected, the Langley engineers instead modified the DC-3, mounting sharp leading edges on the portions of the wings between the engines and the fuselage. A few miles an hour before the aircraft began to stall. these devices caused turbulent flow over the wings which in turn buffeted the tail section and resulted in a palpable sensation in the pilot's control column. Thus warned, airline captains could increase speed and avert disaster. R.D. Kelly felt United gained at least as much from its encounter at Hampton as Henry Reid did for the NACA.

> We have not completed our report of these tests as yet, but we plan to make immediate use of the information obtained by passing on some cf the highlights to our pilot personnel at once. We know this information will be very interesting to them and that it will give them a better knowledge of the characteristics of this airplane. Therefore, they will be able to take advantage of those characteristics which were found to be particularly good and to avoid those which were shown to be somewhat critical.[25]

Unfortunately, the warm feelings engendered by this collaboration proved to be short-lived. While George Lewis declared himself in sympathy with his staff's enthusiasm for continued partnership with United and even assured Kelly the NACA would be "more than pleased to conduct similar cooperative investigations in the future," in private he was nct so enthusiastic. Although Lewis, acting on Reid's prompting, did make an attempt to borrow another DC-3 (both from the Director of Air Commerce Fred D. Fagg, Jr., and from Arthur Raymond of Douglas Aircraft), nothing came of his efforts, telling in itself for a man who usually got what he wanted. His underlying assumptions about joint ventures--as well as about the sharing of equipment, of personnel, and of ideas-- became apparent when the indefatigable Edward Warner again raised the flag of the DC-4. During the last days of 1937, Warner informed Lewis of preparations to flight test Donald Douglas' great airliner. The size and cost of the Douglas behemoth assured extensive trials before any production decisions. Among these experiments, Warner wanted flying qualities to take precedence, regarding the DC-4 program as

---

[25]John Crowley to Henry Reid, 5 October 1937, RA 509 Files, LaRC Historical Reference Collection (short quoted passage); R.D. Kelly to George Lewis, 6 October 1937, RA 509 Files, LaRC Historical Reference Collection (block quote); Hansen, *Engineer in Charge*, 112.

an opportunity to finally elevate this critical part of aircraft worthiness from a speculative art to a quantitative science. But he stated conditions. Warner wanted to use the NACA's instruments and recording techniques; he wanted the tests to be conducted beside and above the Douglas factory; and he asked whether "one or two members of [Lewis'] staff, competent in the use of flight recording instruments [could] go to Santa Monica and remain there as part of the test crew for the duration of the test period?" Warner probably knew such requests might not meet with Lewis' instinctive agreement. He sweetened the proposition with an offer to compensate the NACA for its costs and he appealed to the Director's sense of past cooperation in the project. He also held out the likelihood that the NACA would be allowed to publish most or all of the results of the flight research program. But more important, Lewis "would be rendering service not merely to a single air line, but substantially to the entire air transport industry; since the companies involved in the DC-4 purchase...represent nearly 80% of the mileage flown and 90% of the [American] passenger traffic...."[26]

The Langley staff wanted to seize the opportunity. Gus Crowley felt the project would not only result in invaluable first hand knowledge about stability and control in large aircraft, but would gain the NACA much prestige. He urged appointing Langley's best minds and sending its best equipment to take full advantage of the situation, one which promised to be "different from any we have done or do on flying characteristics....[The] program that is set down will be modified from day to day in accordance with the findings on each flight. There will be frequent ...discussion of the results and their meaning. There will then be no opportunity, as is the usual case, to assemble all results and then analyze them as a whole." Crowley recommended Soulé

[26]George Lewis to LMAL (Reid), 11 October 1937, RA 509 Files, LaRC Historical Reference Collection; George Lewis to R.D. Kelly, 11 October 1937, RA 509 Files, LaRC Historical Reference Collection (first quoted passage); Henry Reid to the NACA (Lewis), 27 October 1937, RA 509 Files, LaRC Historical Reference Collection; George Lewis to Fred D. Fagg, Jr., 1 November 1937, RA 509 Files, LaRC Historical Reference Collection; George Lewis to Arthur Raymond (personal), 4 November 1937, RA 509 Files, LaRC Historical Reference Collection; Edward P. Warner to George Lewis, 28 December 1937, RA 509 Files, LaRC Historical Reference Collection (second quoted passage).

lead a team consisting of another engineer, as well as a technician from the instrument shop to calibrate the 15 assorted meters, gauges, recorders, and scopes required if the NACA participated. He did admit the Douglas project would impede the laboratory's own handling properties investigation; but no more, he thought, than if the work occurred at Langley itself. A few days later, Henry Reid put Crowley's name on the list of staff destined for California, knowing well his understanding of the NACA's procedures and his capacity to "forestall difficulties that might arise." In general, Reid endorsed Crowley's plan, calling it a way to "obtain a good deal of information about [the DC-4] and other work going on which would be of interest to the laboratory."

Lewis took a far dimmer view of Warner's suggestions. He assumed the Flight Research Division would be "crippled" by the loss of personnel for as long as a month. He knew the Committee could accept no compensation from Douglas for its labors; the funds would have to be paid directly to the Treasury. He guessed that "the information we do have and the instruments we have developed would probably become the common property and knowledge of those engaged in the tests, and we would lose much that we have gained in first studying this problem and developing a method and instrumental equipment for the study of the flight characteristics of airplanes." Finally, he wondered whether he could find among the staff such absolute loyalists that "the Committee's interest from every point of view would be their first thought."

Coincidentally, when Douglas Chief Engineer Arthur Raymond visited Langley on February 1, 1938, he too, expressed doubts about the NACA's collaboration in the DC-4 flights. He claimed the test vehicle lacked enough space to accommodate the Langley researchers along with the many others who wished to witness its initial flights from the cabin. Raymond's hosts did not find this reason persuasive and assured him if they were not on the flights the Committee would not participate in the tests. Raymond countered by saying the time for the NACA to conduct its research might be *after* the flight tests, when the airlines took possession of the DC-4s. The spirit of cooperation further diminished when Edward Warner sent a sharp and almost

31

condescending letter to Lewis reviewing Langley's informal report on its DC-3 flight program. He felt the wing stalling device installed by the NACA engineers failed to solve the big aircraft's underlying problem, that is, the fundamental aerodynamics of its airfoil. Warner apparently realized his words might seem impolitic to the NACA Director, but he made no apologies. "If [my statements seem] somewhat dogmatic in tone," he said, "that's to provoke an argument." In consultation with Soulé and Crowley, Floyd Thompson drafted a calm reply, but Warner again insisted on his main point. This bickering made George Lewis less likely than ever to agree to dispatch the NACA's men and equipment to the Douglas plant. Although Warner chaired the NACA's prestigious Aerodynamics Committee and was among the first to raise flying qualities as a subject of research, Lewis treated Research Authorization 509 like any other. He insisted on secrecy so long as the research continued; required his employees to adhere to a conservative, sequential process of experimentation; and allowed the release of the findings to the aeronautics community only through the NACA Reports, Memoranda, and Notes. At the cost of broader cooperation with private and public research entities, Lewis demanded the NACA always retain star billing for itself, a fact not lost on Arthur Raymond and the proud company he worked for.[27]

## A PAUSE TO REFLECT

[27]John Crowley to Henry Reid, 6 January 1938, RA 509 Files, LaRC Historical Reference Collection (first quoted passage); Henry Reid to George Lewis, 10 January 1938, RA 509 Files, LaRC Historical Reference Collection (second and third quoted passages); George Lewis to Henry Reid, 30 December 1937, RA 509 Files, LaRC Historical Reference Collection (fourth, fifth, and sixth quoted passages); Elton W. Miller to Henry Reid, 1 February 1938, RA 509 Files, LaRC Historical Reference Collection; Edward P. Warner to George Lewis, 6 April 1938, RA 509 Files, LaRC Historical Reference Collection (seventh quoted passage); Floyd Thompson to Henry Reid, 15 April 1938, RA 509 File, LaRC Historical Reference Collection; Henry Reid to the NACA (Lewis), 20 April 1938, RA 509 Files, LaRC Historical Reference Collection; Edward P. Warner to George Lewis, 26 April 1938, RA 509 Files, LaRC Historical Reference Collection.

During the fall of 1937 and winter 1938, those associated with Research Authorization 509 experienced a period of stock-taking. Langley flight research pilot Melvin Gough, one of the most able aviators in his field, made an important contribution during this introspective phase. Even though Gough earned a mechanical engineering degree from the Johns Hopkins University, after taking naval reserve training during the late 1920s he decided to trade his desk in the Propeller Research Tunnel Section for the flightline and the Flight Research Division. Gough flew many projects, including the recent handling qualities programs, and in October 1937 he delivered a lecture to the crew of the *U.S.S. Yorktown* about his recent experiences. He started out with a warning about the flying properties of low wing monoplanes. Because they possessed low drag and high wing loading, planes such as the DC-3 glided flat and landed fast, requiring flaps to raise the glide path and to induce lower landing speeds. The overall effect resulted in serious perils. A widened wing wake caused severe tail shaking and buffeting and reduced the power of the rudder. Longitudinal instability and decreased control effectiveness complicated safe flight. The pilot needed to be mindful that the plane's balance differed with the power off versus the power on. Moreover, while the wing and flap combination enjoyed high lifting capacity, the design also forced the wing tips to carry more load. This condition led to stalling at the wing tips and to a dangerous loss of lateral control. More important, Gough warned about "a general change in the "feel" of the airplane when the flaps are lowered." Because the old signs of impending stall no longer occurred. "the pilot finds it easier to stall unintentionally; so more and more he must resort to the mechanical interpretation of the air-speed meter rather than inherent feeling." For example, a steepening flight path in a glide at a steady altitude might result in a stall depending on the angle of attack. But pilots found angle of attack difficult to judge. Consequently, landings required constant vigilance, with pilots pushing the nose down farther than ever due to the steeper glide path and the danger of stall from an inadvertently high angle of attack. Gough left the clear impression with the Navy pilots that the new age of flying demanded a heretofore unknown acuteness of mind and body.

[O]ne should approach the modern airplane with the same enthusiasm and confidence as of old, but possibly with more caution, a more receptive mind, and greater expectancy. The airplane should be taken to altitude and its various stalling conditions observed and studied....Every shudder or shake or peculiarity should be carefully noted along with altitude and power changes. Every warning of an approaching stall should be so definitely fixed in mind that whenever again experienced a *lower angle of attack* will be automatically and instinctively sought. Possibly the greatest danger lies in steep slow glides, and turns. Avoid steeply banked turns at low speed particularly with flaps down. Once the danger zones are located, stay as far from them as possible ever after. Most of us probably heard the term "stall" first used in connection with an automobile...to note

the       cessation of activity. On the contrary, concerning an airplane, it signifies the beginning of rapidly occurring events leading to the end of all further activity.[28]

Another reason for reflection at this point in flying qualities research coincided with the hiring and transfer of Langley personnel. In 1937, a young man with great promise and a Masters degree in aeronautical engineering from the University of Minnesota arrived at the laboratory to work in flight research. Robert Gilruth first learned the NACA way under the tutelage of Hartley Soulé during the Martin B-10B project. But Gilruth's real mentor became Mel Gough who instilled an appreciation for the pilot's perspective and for the problems of engineering at the man/machine interface. Helped greatly by these two men, Gilruth got lucky the year after he started at Langley; Soulé decided to join the ranks of management, eventually becoming Chief of the Stability and Control Division. This left his young assistant Gilruth in charge of flying qualities research. Soulé retained a direct interest in the project and continued to make important contributions to it, but Robert Gilruth won the opportunity to carry the work to fruition. Meantime, just after the shift in roles, Henry Reid assembled Soulé, Floyd Thompson, Eastman Jacobs, and test pilot William McAvoy to discuss the direction of stability and control investigations at Langley. Gilruth, still a decidedly junior partner in the endeavor, did not attend. Soulé felt the project had reached the moment when greater emphasis should be placed on its theoretical groundings in order to achieve the ultimate objective of producing a set of specific

---

[28]Hansen, *Engineer in Charge*, 273, 417; Melvin N. Gough, "The Handling Characteristics of Modern Airplanes from the Pilot's Standpoint," talk given to U.S.S. Yorktown Squadrons VB-5 and VB-6, 7 October 1937, RA 509 Files, LARC Files (all quoted passages, including block quote).

design recommendations. Clearly, the commercial aircraft manufacturers awaited such practical guidance eagerly, but the importance of the research also manifested itself in military aircraft. Thus, a Boeing P-29A joined the ranks of aircraft undergoing handling qualities experiments at Langley. The P-29A had been added to investigate troubling stall phenomena associated with the newer, high performance aircraft and although researchers could not yet be sure whether wing design or longitudinal stability caused the problem, they succeeded in "greatly improv[ing]" the P-29A's handling qualities through a series of ad hoc modifications.[29]

The combination of Robert Gilruth's supervision and Mel Gough's piloting lead flying qualities research in new directions. The P-29A became only one of many aircraft added to the test docket. Having achieved an essential grasp of the flying qualities problem during the first two years of experiments, the Flight Research staff now broadened the horizons of the project and pursued the elusive goal of discovering quantifiable handling properties universally applicable to all aircraft. During the first half of 1938 the new team cleared the decks of past preoccupations and looked ahead to the new agenda. The widespread recognition of the importance of the subject left the laboratory struggling to keep abreast of questions from the military services, the air carriers, the manufacturers, and government regulators. Shortly after Melvin Gough delivered his lecture to the crew of the *U.S.S. Yorktown*, copies of it were requested by officials representing the Navy Bureau of Aeronautics and Carrier Division Two (of which the *Yorktown* was a part). This proved a challenge as Gough, who disliked writing, delivered the speech extemporaneously. Reproductions appeared only after he repeated his talk to a NACA technical audience with a stenographer present. Meanwhile, delays in the processing of motion picture film of the DC-3's flight tests at Langley caused genuine anxiety at United Air Lines. Sometime in January 1938, United's Research Supervisor R. D. Kelly saw George Lewis and asked for the film shot by the NACA almost four months before, during the week of flights

---

[29]Hansen, *Engineer in Charge*, 262, 265, 297, 416; John Crowley to Henry Reid, 6 January 1938, RA 509 Files, LaRC Historical Reference Collection; Robert T. Jones, "Report of Meeting on Stability and Control," 7 January 1938, RA 509 Files, LaRC Historical Reference Collection (quoted passage).

at Langley. The NACA Director became involved personally and prompted Henry Reid to expedite the editing in order to fulfill Kelly's request. But well into March Kelly wrote again, still anxious to receive the footage of the airliner in stalling condition. After George Lewis finally saw the contents of the reel he thought for a few days about how to honor the promise to share it with United, yet at the same time maintain control over its considerable technical value. He mailed a copy to Kelly with three restrictive provisos: it must be shown only to pilots and other United employees; Kelly must narrate the film personally since he participated in the events depicted; and he must return the reel to the NACA by April 10, only three days after its anticipated screening.[30]

In addition to negotiations over artifacts of past flying qualities research, disagreements emerged over some of the data itself. C.J. McCarthy of Chance Vought Aircraft wrote to Lewis about his company's program of longitudinal stability research. Based on Vought's own work, McCarthy expressed puzzlement about Hartley Soulé's observation that on the Stinson aircraft the shape of the elevator angle curves change during steady flight. Briefed by Gilruth, Lewis answered that Soulé meant to point out the difference between the elevator angle position measured from the stick (where the control system naturally experienced deflection under load) and the measurement at the elevator itself. In another example of manufacturer curiosity, an engineer at Stinson Aircraft called Hartley Soulé after failing to speak to George Lewis and asked to borrow the control force measuring wheel for Stinson's new model, or at least to obtain a loan of the drawing so the company could fabricate the instrument itself. More seriously, the

---

[30]George Lewis to LMAL (Reid), 19 January 1938, RA 509 Files, LaRC Historical Reference Collection; Henry Reid to the NACA (Lewis), 7 February 1938, RA 509 Files, LaRC Historical Reference Collection; B.B. Nichol to LMAL, 11 April 1938, RA 509 Files, LaRC Historical Reference Collection; Edward Sharp to B.B. Nichol, 19 April 1938, RA 509 Files, LaRC Historical Reference Collection; R.D. Kelly to George Lewis, 16 February, RA 509 Files, LaRC Historical Reference Collection; George Lewis to LMAL (Reid), 9 February 1938, RA 509 Files, LaRC Historical Reference Collection; Henry Reid to the NACA (Lewis), 14 February 1938, RA 509 Files, LaRC Historical Reference Collection; R.D. Kelly to George Lewis, 21 March 1938, RA 509 Files, LaRC Historical Reference Collection; Henry Reid to the NACA (Lewis), 23 March 1938, RA 509 Files, LaRC Historical Reference Collection; George Lewis to R.D. Kelly, 29 March 1938, RA 509 Files, LaRC Historical Reference Collection.

NACA found itself faced with a challenge to its flying qualities project by a fellow federal agency. The vice-presidents of United, American, and Transcontinental and Western Airlines composed a joint letter to the Bureau of Air Commerce's Chief of Aircraft Worthiness, L.V. Kerber, complaining that despite the NACA's research, "the flying characteristics of the DC-3 plane are entirely satisfactory for transport operation carrying passengers." In follow-up correspondence to Lewis, Kerber seemed to side with the air carriers. But the Langley response steadfastly supported its previous conclusions: at low speeds the aircraft risked a dangerous loss of lateral control, overcome only by use of the rudder and considerable pilot dexterity.[31]

Gilruth and Gough countered such criticism by scheduling more and more aircraft to undergo an increasingly rigorous and sophisticated schedule of flight tests. So much did they expand the repertoire of flying qualities research that George Lewis decided at the end of March 1938, to keep track of the deluge of work by issuing a completely new research authorization for every new aircraft added to the handling properties project by the Army or the Navy. Indeed, this bureaucratic adjustment suggested a real turning point in the program, one recognized by a contemporary in the Flight Research Division.

> It was realized that tests of a large variety of airplanes using improved instrumentation would be required to obtain more generally applicable flying qualities requirements....[W]ith Melvin R. Gough as the chief test pilot, [t]he technique for Gilruth's study of flying qualities was as follows. An airplane was fitted with recording instruments to record all relevant quantities such as control positions and forces, angular velocities, linear accelerations, airspeed, altitude, etc. Then a program of specified flight conditions and maneuvers was flown by skilled test pilots. After the flight, the data was transcribed from the flight records and plotted to show the relevant information, and the results were correlated with pilot opinion. The need to manually evaluate and plot each curve or data point helped to insure that unexpected results would not be overlooked.

[31]C.J. McCarthy to George Lewis, 24 February 1938, RA 509 Files, LaRC Historical Reference Collection; Robert Gilruth to Chief, Aerodynamics Division, 23 March 1938, RA 509 Files, LaRC Historical Reference Collection; George Lewis to C.J. McCarthy, 31 March 1938, RA 509 Files, LaRC Historical Reference Collection; Hartley Soulé to Chief, Aerodynamics Division, 30 June 1938, RA 509 Files, LaRC Historical Reference Collection; R.W. Schroeder, R.S. Damon, Paul E. Richter to L.V. Kerber, March 2, 1938, RA 509 Files, LaRC Historical Reference Collection (quoted passage); L.V. Kerber to George Lewis, 18 March 1938, RA 509 Files, LaRC Historical Reference Collection; Henry Reid to the NACA (Lewis), 26 March 1938, RA 509 Files; LaRC Historical Reference Collection.

Finally, reports were published on the individual studies.[32]


BEARING FRUIT


Among the first aircraft subjected to the new flight test regime of Gilruth and Gough, a

Boeing B-17 Flying Fortress arrived on loan from the Army Air Corps on July 5, 1938.  Because

of its size and power, because of its subsequent impact on civil and military design, it

represented an ideal successor to the highly influential DC-3 in the NACA's flying qualities

program.  The B-17 reflected Boeing's answer to an August 1934 Army Air Corps request for

proposals for a multi-engine bomber capable of transporting 2,000 pounds of ordnance over a

range of 2,200 miles at speeds up to 250 miles per hour.  Just one year later, a prototype B-17

emerged from its Seattle hangar, took flight, and required only nine hours to travel 2,000 miles

non-stop to Dayton, Ohio, averaging 233 miles per hour.  While the range and speed of the B-17

and the DC-3 differed only marginally, few contemporary machines matched the big bomber's

proportions.  Its 104 foot wingspan and length of 75 feet exceeded the Douglas plane by nine and

ten feet, respectively.  Its empty weight of nearly 34,000 pounds exceeded that of the DC-3 by a

factor of two.  Finally, although each of the B-17's engines developed the same 1,200

horsepower as the DC-3, the Boeing behemoth required four rather than two powerplants.

During its flight program, Gilruth and Gough prepared to test the B-17's handling

properties by installing the standard NACA instruments to collect simultaneous data for seven

separate factors of stability and control:


FACTORS                                                 INSTRUMENTS

---

[32]George Lewis to LMAL (Reid), 31 March 1938, RA 509 File, LaRC Historical Reference
Collection; William H. Phillips, "Flying Qualities from Early Airplanes to the Space Shuttle,"
*Journal of Guidance, Control, and Dynamics*, 12 (July-August 1989): 451 (block quote).

| | |
|---|---|
| 1. Air speed | Air speed recorder |
| 2. Time | Timer |
| 3. Force to operate three control surfaces | Control force indicator |
| 4. Position of three control surfaces | Control position recorders |
| 5. Position of elevator and rudder servo-control tabs | Control position recorders |
| 6. Angular motion about the three airplane axes | Angular velocity recorders |
| 7. Normal and longitudinal accelerations | 2-component accelerometer |

Unlike the hurried atmosphere prevailing during the week the NACA borrowed the DC-3 from United, this set of experiments occurred with comparative leisure. Over the course of sixteen days the NACA pilots flew ten flights and spent 20 hours in the air. On one of the early runs the B-17 flew to Wright Field and back in order to calibrate weight and center of gravity factors in the measurement of longitudinal stability. Thus prepared, the researchers loaded the aircraft with seven 300 pound bombs in order to vary the center of gravity. Including the seven person crew and full fuel tanks, the machine weighed 38,600 pounds. Gilruth chose two centers of gravity. One was positioned at about 27 percent mean aerodynamic chord of the wings (wheels up), reflecting a center of gravity far to the rear, aft even of the Army's permissible range. To compensate, researchers also conducted experiments with a more forward center of gravity. By hauling some of the bombs toward the cockpit during flight, they arrived at a mean aerodynamic chord of roughly 23.4 percent (wheels up), a middle center of gravity according to the Army specifications. The research pilots operated the aircraft in four conditions of flight: *cruising* (flaps up, landing gear up, engines set at 1,900 rotations per minute.); *gliding* (flaps up, landing gear up, engines throttled; *landing* (flaps down 58 degrees, landing gear down, engines throttled; and *take-off* (flaps up, landing gear up, engines set at 1,900 rotations per minute.).[33]

---

[33]Robert Gilruth and Melvin Gough, NACA Memorandum Report 36-150 for the Army Air Corps, "Measurements of the Flying Qualities of the Boeing B-17 Airplane," (Washington.

Once Gilruth and his assistants instrumented the B-17 and agreed upon its essential test program, Gough and the other pilots put the bomber through its paces. Their approach, routine for later generations of flight researchers, struck a leading contemporary engineer as "a notable original contribution by Gilruth." Relying on a growing library of past experiences, the team again simplified and verified the test procedures and, at the same time, assembled much new *quantifiable* data about desirable and undesirable flying qualities.

The engineers, pilots, and technicians associated with the project concentrated their efforts on acquiring information related to longitudinal and lateral stability and control, the classical foursome of flying qualities research. After several years of experience, the Flight Research Division defined an aircraft possessing longitudinal stability as "capable of flying by itself without deviating dangerously from a normal flight attitude or speed if...control is abandoned." To judge the B-17's handling in this respect, the pilot flew it in cruising condition at one, and then at the other center of gravity. After reaching a desired speed, he trimmed the airplane at all three axes. At this point he purposely disturbed the equilibrium by pushing the elevator control forward. Once the speed surpassed that set at trim by about ten miles per hour the aviator released the stick and recorded the resulting changes in airspeed and control position. This routine occurred again and again at speeds varying from 100 to 150 miles per hour. Gilruth explained the results: "The Boeing B-17 airplane was dynamically longitudinally stable under the above conditions, the motion being a damped oscillation; i.e., the airplane tended to return to steady flight." However, he also reported that the bomber demonstrated a predisposition toward spiraling. Minor adjustments of the rudder held this motion in check.

Nonetheless, this discovery led the investigators to shift their emphasis from longitudinal stability to the effects of elevator control. The Langley team sought data concerning the elevator angle and the control force required to achieve trim in steady flight at a variety of speeds. Flying the Fortress at the three trim tab settings under all four conditions of flight (cruising, gliding,

---

D.C.: NACA, 1939), 1-8; Sean Rossiter, *Legends of the Air: Aircraft, Pilots and Planemakers from the Museum of Flight* (Seattle, Washington: Sasquatch Books, 1990), 64-79.

landing, and take-off), the NACA research pilots conducted these maneuvers with the plane weighted for an aft center of gravity. Elevator influence over attitude proved to be "ample" in steady flight from the highest allowable speed to the stalling point. But not just in steady flight; the elevators also permitted three-point approaches and landings and held the B-17's attitude during take-offs. Moreover, the extent of elevator required to cause a stall in a glide (6.5 degrees) and in landing conditions (5.5 degrees) differed negligibly. But an undesirable handling quality emerged in tests recording elevator angles at different rates of speed. Apparently, shifting its position just one degree eventually slowed the huge machine from 170 to 115 miles per hour, suggesting that a light movement of the stick could result in wide swings in aircraft velocity. Here the broader problem of "feel" in the pilot's hands entered the calculations. Operating in trim at cruising speed, this particular airplane exhibited virtually *no* correlation between speed and control column force; thus, the pilot found himself--even in fair weather-- watching the airspeed gauge to attain constant flight conditions rather than flying by touch. Moreover, because only a slight movement of the control column could result in great changes in velocity, pilots attempting to fly at a constant speed often needed to make corrections not once, but several times in quick succession before achieving the desired elevator angle. Added to this burden, the stick frequently absorbed 50 pounds of force or more before the plane responded with any correction. Other aspects of longitudinal control proved more acceptable. In elevator-controlled pull-ups and push-downs, the B-17 reacted satisfactorily. During the high speeds demanded in pull-ups, very small changes in elevator angle caused significant reactions. Indeed, "the maximum allowable acceleration specified for the airplane could be obtained with an elevator movement of only approximately 7 degrees...[and] the evidence obtained indicates that the airplane can be maneuvered with the elevator to produce normal accelerations equal to those specified for the structure." During low-speed push-downs the "B-17 airplane was observed to be very responsive to the elevator....[T]he reaction to down elevator was immediate and powerful," essential for control during stalls and for holding attitude during landings and take-offs.[34]

The flight tests of lateral stability and control uncovered quite different results. The B-17 flew under the same four conditions of flight and used the same balance (27 percent mean aerodynamic chord) as during most of the longitudinal measurements. To produce a lateral disturbance, the pilots let go of the controls from a steady sideslip and found, to their surprise, that the ailerons failed to return to their prior positions. Thus, stability or instability became less the issue than ailerons not finding their trim setting on their own (although the oscillations caused by sideslip were damped quickly and effectively). Although the aviators did not think spiral stability crucial to the plane's flying qualities, the engineers felt an aircraft should return itself to normal flight attitude in the event of in-flight emergencies. Also, in rough air, lateral instability might result in consequential changes in course before the wings could be returned to the level position. Like spiraling tendencies, flight near the plane's stalling zone emerged as a cause of concern. As the aircraft approached the point of stalling while it decelerated, not until "sudden and violent" rolling instability occurred did the cockpit crew become aware of impending disaster. Clearly, the B-17 would be an excellent candidate for the NACA stall warning devices tested on the DC-3. On the other hand, the leveling qualities associated with the bomber's wing dihedral proved to be effective at all speeds, even counteracting sharp kicks at the rudder pedal. But aileron control turned out to be less satisfying. Pilots applied the ailerons variously and sharply at constant speeds to measure the effectiveness of rolling the airplane. They found a disturbing lack of feel caused by high control friction and by irreversible movement. Moreover, Gilruth and his colleagues discovered the ailerons to be relatively heavy and found their control cables tended to stretch, both of which reduced operating efficiency. Finally, the rudder control seemed less than adequate. After recording the effects of abrupt

---

[34]Phillips, *Journey in Aeronautical Research*, 22 (first quoted passage); Gilruth and Gough, NACA Memorandum Report 36-150 for the Army Air Corps, "Flying Qualities of the...B-17 Airplane," 9-22, 35 (all other quoted passages).

rudder displacements and the force required to maintain heading under asymmetrical power, the staff concluded the rudder, like the ailerons, weighed too much. Moreover, the pedal forces were too high for a good sense of touch and for adequate control.[35]

Despite the important data recorded during this flight program and the close working relationship between engineer Gilruth and pilot Gough, there developed some understandable differences between the disciplines they represented. Eventually, after still more experimentation and consultation, their viewpoints merged in a coherent set of handling requirements. Until then, not much unanimity existed among those who flew the airplanes and those who designed the flight research program. Indeed, pilot opinion about the flying qualities of particular airplanes diverged routinely from the recorded dynamic longitudinal and lateral motions. Just as the B-17 flight program ended, Melvin Gough informed the aeronautics community of the cockpit perspective on flying qualities. He pointed out that instability in itself did not necessarily mean an aircraft could not be operated successfully, provided the pilot had in hand controls sufficiently refined and delicate to compensate for the unstable tendencies. But, he admitted that "with control and stability both inadequate, the airplane is definitely dangerous." He felt the pivotal question really turned on *how much* stability. The more it prevailed, the lower the sensitivity in the controls, the rougher the flight, and the greater the pilot's burdens in mastering the aircraft. Gough admitted the need for more stability in existing aircraft design, estimating present models could safely possess twice the levels common in the late 1930s. Yet, he hastened to add some specific circumstances under which stability might and might not be welcome. In maneuvers requiring intense concentration for short periods such as during glides, landings, and take-offs, airline captains and their Air Corps and Navy brethren preferred light controls; on the other hand, cruising over long distances demanded good stability to relieve pilots of the exhausting task of constantly checking and adjusting attitude, heading, velocity, and level flight. But lateral stability remained an open question. While many felt spiral stability would be

---

[35]Gilruth and Gough, NACA Memorandum Report 36-150 for the Army Air Corps, "Flying Qualities of the...B-17 Airplane," 22-35.

43

"very desirable," no firm evidence existed to support it as a design objective. Indeed, said Gough, "[t]here is considerable difference of opinion as to the degree to which banking and turning should be automatically dependent upon each other, and to what extent their control should depend upon the pilot." Further complicating the objective of achieving universal flying qualities requirements, Gough reminded those quick to impose rigid standards that different types of aircraft required inherently different degrees of stability. "The important factors for the safe...airplane for the private owner," he wrote, "are entirely different in degree from those required for the airplane intended for the skilled military or combat pilot. Both requirements are at variance with the transport requirements, which consider the safety and comfort of the passengers under skilled guidance."[36]

Starting in 1940, Gilruth and the flying qualities team felt confident enough to begin to answer some of the contradictions expressed by Gough and to resolve the open technical questions with the publication of NACA Report Number 700, entitled "Preliminary Investigation of the Flying Qualities of Airplanes." Actually, its appearance in March represented a brief return to the field by the former flying qualities boss Hartley Soulé, who finally revealed the full details of his investigation of the Stinson aircraft. After more than four years observing the Stinson (and a dozen other vehicles) he presented--with the complete support of the cautious NACA leadership--a preliminary set of design requirements for the consideration of Boeing, Douglas, Lockheed, and all of the other manufacturers. Soulé presented this incarnation of handling properties research with a new degree of confidence and forthrightness, although he did admit candidly that the suggested numerical limits published in his report remained, as before, "quantitatively unreliable."[37]

For example, to attain satisfactory longitudinal control with the elevators, Soulé proposed

[36]Phillips, "Flying Qualities from the Airplane to the Space Shuttle," 451; Melvin N. Gough, "Notes on Stability from the Pilot's Standpoint," *Journal of the Aeronautical Sciences*, 6 (August 1939): 395, 398 (quoted passages).
[37]Hartley Soulé, NACA TR 700, "Preliminary Investigation of the Flying Qualities of Airplanes" (Washington, D.C.: NACA, 1940), 449-452 (quoted passage, 452).

five fundamental conditions, together constituting good flying qualities for this particular flight

regime. He followed these points with a specific set of procedures necessary for designers and

pilots to achieve the optimal relationships.

> Requirement.--The range of the elevator control shall be sufficient to meet the following conditions:
> a. With every setting of the trimming device, it shall be possible to maintain steady flight at any speed from the design probable diving speed to the minimum speed for any power condition, flap up.
> b. With every setting of the trimming device, it shall be possible to maintain steady flight at any speed from the placarded to the minimum, flap down.
> c. With the conventional type of landing gear, it shall be possible to make three-point landings and to hold the tail down while braking enough to give a     deceleration of 0.3g during the landing run down to a speed of 30 miles per hour.
> d. In the take-off run, it shall be possible to raise the tail off the ground by the time a speed of 30 miles per hour is attained.
> e. If a tricycle type of landing gear is used, it shall be possible to raise the nose wheel off the ground in a take-off run by the time a speed of 30 miles per hour is attained.

> Procedure for items a and b.--Measure the elevator angle at different speeds with different tab or stabilizer settings and different throttle positions.
> Procedure for item c.--Merely demonstrate the ability to make three-point landings. For the braking tests, run the airplane along the ground at a speed of approximately 50 miles per hour. Close the throttle and apply brakes to the maximum extent for which the pilot can maintain contact between the tail wheel and the ground. Record the air speed and the longitudinal acceleration as the airplane decelerates to less than 30 miles per hour.
> Procedure for item d.--Apply full throttle while holding the airplane with the brakes. Release brakes and attempt to raise the tail as soon as possible. Record speed at which the tail leaves the ground.[38]

Because Soulé's work represented the first attempt to prescribe definite requirements for

handling characteristics, it met with much praise, but also some criticism. Major H. Z. Bogert,

the Acting Chief of the Air Corps Experimental Engineering Section at Wright Field, thought the

report covered the subject in a "very thorough manner and [is] entirely satisfactory." Once the

NACA released more data covering a broader spectrum of aircraft, Bogert predicted

"specifications for flying qualities of future airplanes [will be as common] as specifications for

---

[38]Soulé, NACA TR 700, "Flying Qualities of Airplanes," 453.

structure and performance are...today."  On the other hand, John Easton, Chief of the Civil

Aeronautics Authority's Aircraft Section did raise objections, perhaps due to parochial concerns.

He felt take-offs, approaches, and ground handling failed to receive adequate coverage.  Soulé

defended his position convincingly, arguing "the ability to make three-point contact with zero

vertical velocity, to change in both directions at low speeds, to hold the tail while braking, and to

raise the tail for take-off ...add[ed] up to the take-off and landing qualities without the

requirement of a specific demonstration."  Moreover, the "variation of elevator force with throttle

setting and the ability to hold against a single engine on the ground at speeds above 50 miles per

hour also have direct bearing on the take-off and landing characteristics."[39]

Despite Hartley Soulé's essential contribution to the subject, flying qualities research still

awaited its signature expression.  It finally appeared in the form of NACA Report Number 755:

"Requirements for Satisfactory Flying Qualities of Airplanes" by Robert Gilruth.  Due to the

critical mass of data accumulated by Gilruth and Gough over the past few years, immense strides

were achieved in the short timespan between Soulé's pioneering report and the publication of

Technical Report 755 a year later.  Based on experimentation with 16 aircraft--most of which on

loan from the Army, but some borrowed from the aircraft industry and the airlines--Robert

Gilruth achieved the objective of a coherent, "easily measurable, yet fundamental" set of design

specifications first sought by Edward Warner more than five years earlier.  The publication of the

results could not have been better timed.  During summer of the previous year the government of

France capitulated to German attack and the Third Reich unleashed 1,000 aircraft and their

bombs on British targets from London to Scotland.  Six months after Gilruth's paper appeared

the Japanese joined the Italian and German governments in a tripartite pact; three months after

that the U.S. found itself at war.  Thus, the NACA's most definitive statement on flying qualities

---

[39]H.Z. Bogert to George Lewis, 3 June 1940, RA 509 Files, LaRC Historical Reference
Collection (first and second quoted passages); John Easton to George Lewis, 13 May 1940, RA
509 Files, LaRC Historical Reference Collection; George Lewis to John Easton, 12 June 1940,
RA 509 Files, LaRC Historical Reference Collection; W.H. Herrnstein to the NACA (Lewis), 21
May 1940, RA 509 File, LaRC Historical Reference Collection (quoted passages three and four).

received public dissemination early enough to have a decisive impact on wartime aircraft design. Indeed, it "formed the basis of subsequent military specifications for stability and control characteristics of airplanes." This time the author offered no apologies about the unreliability of quantitative data. This time, for instance, the NACA specified four separate, clearly defined categories of elevator control (steady flight, accelerated flight, take-offs, and landings). This time the requirements were sure, simple, and less time-consuming to verify.[40]

Indeed, the members of the Flight Research Division who worked with Gilruth probably surprised themselves with the gains realized in the flying qualities art between March 1940 and March 1941. The requirements for longitudinal control using elevators underwent revolutionary changes compared to those suggested by the Stinson tests. In *steady flight*, four simple precepts now prevailed:

1. Pilots were expected to be able to maintain minimum and maximum speeds.

2. Elevator control force needed to have the capacity in all settings to return the stick to trim.

3. Under the influence of different speeds, elevator control forces needed to be accompanied by push forces above the trim speed and pull forces below it.

4. Positive static longitudinal stability needed to be present during variations in elevator angle under the following conditions: with engines idling, flaps up or down, and speeds above the stall; with engines at power for level flight, flaps and landing gear down, and speeds above stall; with engines at full power, flaps up, at all speeds over 120 percent of the minimum velocity.

During *accelerated flight* Gilruth demanded from the elevator controls five essential characteristics.

1. To develop the maximum load factor or lift coefficient at any speed.

---

[40]Phillips, "Flying Qualities from Early Airplanes to the Space Shuttle," 451 (first quoted passage); Robert Gilruth, NACA TR 755, "Requirements for Satisfactory Flying Qualities of Airplanes," (Washington, D.C.: NACA, 1941), 49-57.

2.  To assume the various elevator angles during steady turning flight, reflected by a smooth curve at all speeds.

3.  To allow no fewer than four inches of rearward stick to alter the angle of attack in high maneuver airplanes.

4.  To permit normal acceleration proportional to the elevator control force during steady turning flight.

5.  To achieve a gradient in steady turning flight of 50 pounds per g in bombers and transports, less than six pounds per g in fighters, and for all aircraft a pull force of not less than 30 pounds to achieve the maximum load factor.

In *landings*, Gilruth defined good elevator control qualities as ones which sustained the aircraft off the ground prior to three-point landings, which restrained the machine from touching the ground until reaching its minimum speed, and which required no more than 50 pounds of force for wheel controls and 35 for stick-types to make landings.  Finally, during *take-offs*, Gilruth felt the elevators should be able to maintain the attitude of the plane from level to maximum lift after one-half of the necessary speed had been mustered.[41]

## FLYING QUALITIES GOES TO WAR

During the period between the publication of Soulé's and Gilruth's papers the U.S.

---

[41]Gilruth, NACA TR 755, "Satisfactory Flying Qualities of Airplanes," 50-52.

aircraft industry--already pressed by the demands of war production--showed a keen interest in applying the NACA's handling qualities research to the fighter, bomber, and cargo designs then under consideration. Among the many aircraft that benefited from this research, none attracted more attention than the P-51 Mustang. This aircraft originated with requirements established jointly by the British and French Air Ministries in the weeks before their respective countries faced the onslaught of the German forces. Just before the invasion of France and the Battle of Britain, in April 1940 a British Air Purchasing Commission arrived in the United States to procure an advanced aircraft to defend the skies over the United Kingdom. Because of the urgency of the situation, the commission first thought of existing warbirds such as the Bell P-39 and the Curtiss P-40. But North American Aviation of Los Angeles made a proposal which astounded the English visitors: the company committed itself to fabricating a prototype of an all-new aircraft, designed specifically to the French and British specifications, in only four months. North American won the go-ahead and the first flight of the XP-51 in October 1940 revealed an extraordinary machine, "an example of intelligent application of government research" which incorporated the latest NACA findings on laminar flow wings and on flying qualities. The wing project occurred under the auspices of Eastman Jacobs. After attending the Volta Conference on High-Speed Aerodynamics in 1935, the imaginative and daring Jacobs initiated studies on supersonic flow and activated design work for a nine-inch supersonic wind tunnel. He did so in the face of indifference, if not hostility, to supersonic research by NACA leaders. Nevertheless, Jacobs broadened these inquiries in 1937 when he and his wind tunnel associates opened an investigation on laminar flow over airfoils. The team scored a great success in 1938 when Jacobs' insight and persistence lead them to the conclusion that falling pressures could be achieved over most of a wing surface if they took the cross-section of an average airfoil and inverted its basic contours; that is, designed the nose to resemble the trailing edge, and the trailing edge to resemble the nose. Tests showed this method *halved* the drag over most of the wing surfaces. When North American's test pilots flew the XP-51 for the first time they were duly impressed by its speed in level flight (382 miles per hour), but they marveled at its

49

steadiness in even faster combat-related dive maneuvers. As most contemporary fighters approached Mach 0.7, the ill-effects of compressibility materialized; higher drag, loss of lift, the tendency for the nose to drop, and an increase in buffeting. But the XP-51's laminar flow airfoil minimized these perils, giving it great advantages over enemy aircraft in dogfights and in other wartime roles.

Yet, laminar flow and the capacity to achieve high speed with high stability did not constitute the NACA's only contribution to the North American designers. Flying qualities research continued unabated after the appearance of Gilruth's 1941 Technical Report, resulting ultimately in the flight testing of some 60 airplanes of all types. The accumulated knowledge proved to be of tremendous value to the XP-51's creators. A newly hired NACA pilot and aeronautical engineer named Jack Reeder remembered his initial impression of its handling qualities as "nearly ideal, particularly when compared with the other fighters of the period." Reeder flew the famous warbird many times afterwards and continued to be a great admirer of its flying qualities.

> I made some 43 high speed research flights in the XP-51 for various aerodynamic investigations. It was one of the most pleasant and exciting propeller-driven planes I have ever flown. It had nearly ideal handling qualities, and for the experienced pilot it had no vices. It had a desirable degree of static and dynamic stability about all axes, light but positive control forces, and it responded quickly and accurately to pilot control inputs.           Trim changes with power, flaps, and speed were small with low control changes. At           diving speeds, "compressibility" trim changes and buffeting were comparatively mild and recovery from high speed dives with longitudinal control alone was readily accomplished.[42]

Most of Reeder's flights occurred aboard an Army Air Forces (AAF) XP-51, testing its

---

[42]Maurice White, Herbert Hoover, and Howard Garris, NACA Memorandum Wartime Report 41-38 for the Army Air Forces, "Flying Qualities and Stalling Characteristics of North American XP-51." (Washington, D.C.: NACA, 1943), 38-41; John P. Reeder, "The Mustang Story: Recollections of the XP-51," *Sport Aviation Magazine* (September 1983; reprint, Langley Research Center, Hampton, Virginia, *NASA Facts*, April 1992), 1-3 (all quoted passages and block quote; page citations are to the reprint edition); Langley Research Center, Biographical Sketch of John P. Reeder, Headquarters NASA Historical Reference Collection, File Number 001774; Hansen, *Engineer in Charge*, 36-37, 111-118; Phillips, "Flying Qualities from Early Airplanes to the Space Shuttle," 451; Greenwood, *Milestones of Aviation*, 128.

flying qualities against the ever-evolving NACA standards. During this process the Langley researchers did uncover one flaw in an otherwise unblemished performance. The original requirements demanded an extraordinarily high roll rate, to be achieved at a speed of 400 miles per hour with the pilot exerting no more than 50 pounds of force on the stick. Reeder never attained more than 75 percent of the desired objective. Thus, the NACA initiated some modifications to improve this consequential aspect of combat flying. The flight researchers thickened and beveled the trailing edges of the ailerons in an effort to reduce the stick forces by causing "balancing pressure changes over the surfaces." Their solution worked. Not only did the XP-51 meet the British specifications, it now exhibited the highest roll rate of any front line fighter in the world--138 degrees per second compared to the FW-190's 119 and the Spitfire's 110. But this advantage was hard-bought. Reeder, Herbert Hoover (who joined the NACA in 1940), and the other test pilots underwent perilous flying conditions to prove the beveled ailerons, involving flight regimes at the edge of existing knowledge and experience. They jockeyed the elegant little fighter through one heart-stopping dive after another, attaining indicated airspeeds up to 492 miles per hour (520 miles per hour true airspeed). Yet, the experiments demonstrated more than the worthiness of beveled aileron trailing edges to improve roll rate. Coincident to these test flights, Reeder and Hoover reported a strange phenomenon. Robert Gilruth, now the chief of the Flight Research Division, learned from his pilots that in moments of favorable sunlight, as they pushed the Mustang downward into dives, the test pilots saw "the shadowy edges of shock waves cutting across the streamlines of their airplane's wings." Gilruth knew what this meant; a portion of the air flowing over the wings achieved velocities up to and even over the speed of sound. In this moment of realization, flying qualities intersected with laminar flow studies to produce a new avenue of flight research.[43]

---

[43]Reeder, "The Mustang Story: Recollections of the XP-51," 2-4 (first quoted passage); Herbert Hoover and Maurice White, NACA Memorandum Wartime Report 41-38 for the Army Air Forces, "Lateral-Control Characteristics of North American XP-51 Airplane...With Beveled Trailing-Edge Ailerons in High-Speed Flight, (Washington, D.C.: NACA, 1942), 1, 11-12; Phillips, *Flying Qualities from Early Airplanes to the Space Shuttle*, 451; Hansen, *Engineer in Charge*, 264 (second quoted passage).

## A NEW DIRECTION

Of course, the convergence of these two projects did not alone change the agenda of flight research. Factors both internal to the NACA and external to it brought about the reversal. Inside the institution, aerodynamicists believed high-speed flight to be much more than idle conversation over morning coffee in the Langley cafeteria. On the contrary, many of the lab's best theorists regarded it as a real eventuality. The national emergency embodied in the Second World War merely called forth the wherewithal to attack the problem frontally. At the same time, Big Power politics after the war legitimized the long-term cost and commitment required to sustain a program of this complexity. Thus, the existing state of scientific knowledge, the gains realized in wartime research, and the postwar anxiety about American defense all persuaded the NACA leadership to pursue a flight research program full of formidable engineering challenges, one which eventually attracted headlines because of its importance to national security.

# CHAPTER 4

## First Among Equals:
## Supersonic Flight

### DIVERSIFICATION

For the most part, George Lewis and his superiors on the NACA Main and Executive Committees concerned themselves with the technical advancement of aeronautics. But such experienced and worldly men as Chief of the Army Air Corps General Oscar Westover, Joseph Ames, Orville Wright, and Edward P. Warner also paid close attention to the international role of aviation and took due note of air power research conducted by other powers. During the mid-1930s, John Jay Ide, the NACA's intelligence officer in Paris, sent home urgent cables to the NACA leadership describing massive European building programs: a full-scale wind tunnel in Chalais-Meudon, an immense research complex in Guidonia, Italy, and a resurgence of aeronautical facilities all across Germany. Lewis apprised himself personally of the situation in 1936. During that summer he toured Germany and Russia to see their new installations and noted particularly the Deutsche Versuchsanstalt für Luftfahrt (DVL) near Berlin. In private moments back at his desk in Washington he still considered the Langley Laboratory to be second to none, a belief confirmed when the U.S. Senate passed a special appropriation of one million dollars for a new 20 by 25 foot propeller research tunnel, one that promised minimal scale effects. But motivated by his travels as well as a determination to *retain* the lead, Lewis canvassed Congress for funds to hire *500* new employees, in effect doubling the laboratory's complement. He also asked General Westover to chair a Special Committee on the Relation of the NACA to National Defense in Time of War. The subsequent report issued in 1938 recommended a second laboratory on the west coast or in the interior of the country, both to disperse the nation's aeronautical research establishment in the event of attack and to relieve the burdens of war work pouring into Langley. Some at the NACA also felt a second facility on the west coast would not merely serve the burgeoning aircraft industry in California and Washington, but would also act as a counterweight to the growing influence of the Guggenheim Aeronautical Laboratory at the California Institute of Technology (GALCIT), a dynamic research center directed by the brilliant and engaging physicist Theodore von Kármán. A second committee under the chairmanship of Admiral Arthur Cook, Chief of the Bureau of Aeronautics, proposed Moffett Field in Sunnyvale, California, a long-time naval airship station. After some opposition from the Virginia Congressional delegation, the House and Senate authorized construction in August 1939. In a show of gratitude for 20 years of stalwart service as chairman of the NACA Executive and Main Committees, the NACA named the new center for Joseph S. Ames, in failing health after suffering a paralyzing stroke in 1936.

Meantime, European engine advances also raised concerns in U.S. aviation circles. Right on the heels of the Moffett Field legislation none other than Charles Lindbergh took up the cudgels for a third NACA laboratory dedicated to propulsion. Somewhat diminished in stature because of his sympathy for Nazi Germany, Lindbergh nonetheless commanded respect on Capitol Hill. As chair of a Special Survey Committee on Aeronautical Research he warned that American engine technology risked being eclipsed by the advances of the Europeans. The high performance liquid-cooled powerplants designed for German, French, and British military aircraft threatened the sovereign status of the more efficient but less powerful air-cooled ones favored in America. Lindbergh wanted the NACA to reinvigorate its engine research--relegated by the Main Committee to the aircraft industry as early as 1916--by opening a facility dedicated solely to such investigations. The great aviator's name sounded bells in Congress and in June 1940, the month France capitulated to the German armies, monies were passed for the construction of the NACA Aircraft Engine Research Laboratory in Cleveland, Ohio.[1]

In their initial incarnations, both the Cleveland and the Northern California laboratories replicated the essential features of "mother Langley," and for good reason; the pioneers who first turned the keys in the new warehouses, hangars, and test facilities voluntarily transferred from Hampton to these distant outposts of the NACA. Half of the original 51 who opened Ames had arrived from Langley in 1940. Cleveland absorbed 150 Langley employees in 1941, including the entire Powerplants Division. With these individuals came the capacity to undertake flight research. Indeed, the earliest drawings of both facilities included the wherewithal to conduct full-scale flights. At Ames, the Flight Research Building--housing an immense eight acre hangar, a maintenance shop, and offices for engineers and pilots--opened in August 1940, only one year after Congress authorized the laboratory. Just as construction started, before a single aircraft taxied into the complex, the NACA headquarters issued its first flight test assignment to Ames: assume the de-icing work pursued at Langley since 1927. Project chief Lewis Rodert and his de-icing team (consisting of pilots William McAvoy and Lawrence Clousing) joined the initial cadre who journeyed West from Hampton, Virginia and they began their investigations immediately, a year before receiving a Research Authorization. The NACA ascribed such high importance to the icing hazards (which cost so many planes and crews during the war) that it approved the purchase of a twin-engine Lockheed 12 expressly for the purpose. It arrived at Ames in January 1941 after being outfitted at Langley with thermal heating elements embedded in its wings and tail. It embarked on its flight research program immediately. By 1943 an Army C-46 Commando underwent even more rigorous testing in the skies over California, equipped with the most advanced ice-protection system known and full instrumentation to record cloud

---

[1]Hansen. *Engineer in Charge*, 187-194; Virginia P. Dawson, *Engines and Innovation: Lewis Laboratory and American Propulsion Technology* (Washington, D.C.: NASA SP-4306, 1991), 1-15; Elizabeth A. Muenger, *Searching the Horizon: A History of Ames Research Center, 1940-1976* (Washington, D.C.: NASA SP-4304, 1985), 1-5; Edwin P. Hartman, *Adventures in Research: A History of Ames Research Center, 1940-1965* (Washington, D.C.: NASA SP-4302, 1970), 1-22; N. Ernest Dorsey, "Joseph Sweetman Ames: The Man," *American Journal of Physics*, 12, (June 1944): 148.

3

behavior.

In Cleveland, meanwhile, among the seven structures provided for in the original empowering legislation, a Flight Research Building (consisting of offices, a machine shop, and a hangar) opened as early as the end of 1941. It saw its first service, however, not as a shelter for aircraft, but as office space. When the laboratory's first director, Raymond Sharp, left Langley for Cleveland just after the attack on Pearl Harbor, he and his technical assistants had no administrative edifice to inhabit, so they established themselves in a local farm house and in the Flight Research complex. The hangar assumed its intended purpose when the Flight Research Division came into being in 1943. Strangely, for a time the engineers and pilots here found themselves engaged in much the same investigations as their colleagues at Ames. Under intense pressure to mitigate the losses attributed to icing on routes extending from the North Atlantic to Burma, the Army Air Forces (AAF) also enlisted the Engine Research Facility in the battle against the cold. The Clevelanders soon enjoyed an advantage over the Californians. An Icing Research Tunnel, constructed between 1942 to 1944 to take advantage of an immense refrigeration plant necessary for the new High Altitude Tunnel, offered a rare opportunity to study the effects of ice on aircraft in controlled conditions on the ground. Desirous of combining tunnel and flight testing, the AAF transferred a Lockheed P-38 Lightening to the Engine Laboratory to fly a program which evaluated the effect of turbosuperchargers on carburetor icing. By the end of the war, the Engine Facility employed a bigger icing staff than Ames. Just after the cessation of hostilities the Ohio investigators broadened their research with borrowed B-24 and B-25 bombers. These giants flew as far away as North Dakota and, respectively, conducted icing experiments on turbo-jet engines and on a variety of aircraft components. The logic of Cleveland's role in cold weather flying became inescapable; after Lewis Rodert moved to the Engine Facility in 1946 to become chief of flight research, all NACA icing research was consolidated under his leadership.[2]

---

[2]Muenger, *Searching the Horizon*, 9, 15-16, 19-20; Dawson, *Engines and Innovation*, 20, 25, 227-228, 241, 261; George W. Gray, *Frontiers of Flight: The Story of NACA Research* (New

## STIRRINGS AT HAMPTON

As the flight research staffs at Ames and at the Lewis Memorial Laboratory in Cleveland adapted quickly to the pressures of war work, the Langley flight research team found itself at a crossroads. The North American XP-51 dive tests (mentioned at the end of chapter 3) opened the possibility of a vast new aeronautical venture based on high speed flight. But the XP-51 did not only presage a quest for greater speed. The instrumentation packed aboard the little fighter "really wrapped everything together, tying [in] the ground facilities, wind tunnel and ground testing...into a focus point of a full scale airplane in which you could consider aerodynamic loads, stability control performance, everything...integrated into one complete [research] design." Indeed, one leading member of the team called it a "very complete flying wind tunnel." Actually, this description could not have been more complete or accurate. Precisely *because* transonic wind tunnel testing failed to yield the aerodynamic information necessary to design for high velocity aircraft, the participating NACA pilots found themselves flying extraordinarily dangerous missions in which all concerned held their breaths against the real eventuality of mid-air structural failure. No one knew the frustrations or the excitement attendant on supersonic research better than aerodynamicist John Stack, one of Langley's most celebrated figures. Arriving at the lab in 1928, by the early 1940s he became the NACA's leading exponent of high speed aerodynamics. While Stack possessed both the roguish charm and the hair-trigger temper sometimes associated with his parents' native Ireland, the MIT graduate also proved to be a highly able problem-solver, if not a theoretician. He apprenticed himself for a decade under the Variable Density Tunnels's section chief, Eastman Jacobs. To their mutual dismay, the two men

---

York: Knopf, 1948), 309-317, 325; Pearcy, *Flying the Frontiers*, 34-35.

discovered that at speeds approaching Mach 1 a "choking" of the airflow occurred in the throats of the laboratory's tunnels. Shock waves streamed from the models, careened into the tunnel walls, caromed back toward the rear parts of the models, and rendered hopeless all attempts at gauging the underlying aerodynamic phenomena. Unable to account for or to correct the problem, Stack admitted the hard facts: "[t]he laboratory approaches didn't look very promising. [S]o, where do we go? After some deliberation, free flight with men-instrumented airplane seemed the best and most direct way." Stack had first contemplated such a research airplane in 1933 and 1934 and even drew plans for it. But now, spurred by the intense interest in high performance engendered by the war and encouraged by the XP-51's crucial data, he again circulated the idea. Thus, flight research, a mature discipline with a generation of NACA practice to its credit, appeared to present the one hope of cracking the high-speed conundrum.[3]

An unusual confluence of institutions and personalities clustered around Stack's proposal. He presented his initiative to the NACA Headquarters in spring 1942. Stack needed George Lewis' approval and turned on the full force of his personality to get it. But even under the heat of Stack's high voltage campaign, Lewis only offered a tepid assent: Stack could begin work, but on a low-priority, back-channel basis. Stack understood the decision. The unobtrusive pursuit of high risk research had been a Lewis and a NACA hallmark for many years. Moreover, in the midst of the war the Director had few resources to spare. Stack accepted his support with gratitude and later gave Lewis high marks for being among the first and most noteworthy figures to back the project. But Lewis had distinguished company. As Stack assembled a small team of engineers to design a high-speed research airplane and as they actually drafted plans for a Mach

[3]De E. Beeler, interview by Richard P. Hallion, December 1976, 3, Hallion Papers, Dryden Flight Research Center (hereafter DFRC) Historical Reference Collection, (first quoted passage); De E. Beeler, interview by Richard P. Hallion, handwritten notes, 1 December 1974, Hallion Papers, DFRC Historical Reference Collection, (second quoted passage); Becker, *High-Speed Frontier*, 13-16; Hansen, *Engineer in Charge*, 256-258; John Stack, "History of the Rocket Research Airplanes," (statement presented at a meeting of the American Institute of Aeronautics and Astronautics, July 1965), 14-19, DFRC Library (third quoted passage, 19); Louis Rotundo, *Into the Unknown: The X-1 Story* (Washington and London: Smithsonian Institution Press, 1994), 8.

1 machine, the Army Air Forces started to take a keen interest. Intelligence from Europe suggested German scientists and engineers soon planned to unveil both rocket and turbojet propulsion for their combat aircraft. Breakthroughs such as these sounded familiar to General Frank Carroll, the Chief of the AAF Engineering Division at Wright Field. An extraordinarily able civilian engineer under his command had argued tirelessly for the feasibility of supersonic flight and urged the uniformed leadership to support high speed research. With the threatening developments in Europe, Ezra Kotcher finally won an audience. He convinced General Carroll to contact one of General Henry H. Arnold's most trusted personal advisors, Professor Theodore von Kármán of Caltech. The charismatic Hungarian, known as much as a *bon vivant* as an international authority on applied mechanics, retained personal ties to most of the consequential generals, admirals, scientists, and captains of industry involved in aeronautics. Kármán's opinion also carried great weight because he directed the only university-based rocketry program in the country, specializing both in sounding rockets and in small rocket canisters used to boost aircraft performance (called misleadingly Jet-Assisted Take-Off, or JATO). Kármán accepted Carroll's invitation and arrived at Wright Field on a Friday early in 1943. The general posed a simple question: could an aircraft be built to travel at 1,000 miles per hour?

> Here in this question was the culmination of all the theory and speculation on supersonic motion and flight in which I had been involved since almost the turn of the century. It was the first time that a practical question of this kind had been put to me. Had theory and technology arrived at the happy point where one could set a practical project into motion? Telling the General I would think about it, I returned to my hotel room in Dayton and arranged with Frank [Wattendorf, Kármán's friend, former student, and supervisor of construction of a ten-foot supersonic wind tunnel at Wright Field] to call in a few engineers from Wright Field. Spreading our papers on the floor, we worked all day Saturday and all day Sunday. On Monday I returned to Wright Field. In my valise was a preliminary design, with the main data on span, strength, and weight. I placed the figures before the General and his aides. Yes, I said, it is quite practical to build a plane that can fly at a thousand miles an hour.[4]

---

[4]Theodore von Kármán with Lee Edson, *The Wind and Beyond: Theodore von Kármán, Pioneer in Aviation and Pathfinder in Space* (Boston, Toronto: Little Brown, 1967), 233-234 (block quote); Stack, "History of the Rocket Research Airplanes," 19; Rotundo, *Into the Unknown*, 8-9; James O. Young, *Meeting the Challenge of Supersonic Flight* (Edwards, California: Air Force Flight Test Center History Office, 1997), 3-4; Hansen, *Engineer in Charge*, 259-260.

Kármán's favorable reply set in motion more eddies of activity. Early on, a distinction emerged between the NACA's research preferences and those of the Army Air Forces. John Stack wanted an advanced, highly instrumented turbojet aircraft capable of sustained flight in the transonic region in order to generate the maximum amount of data and thus break the code, as it were, of travel through this mysterious regime. Kotcher and Carroll sought a rocket plane capable of dashing through and well past the threshold between subsonic and supersonic speeds, of demonstrating the practicality of such flight, and of succeeding in its mission even if the vehicle needed to be launched not from the ground, but from a mother ship. Originating with these positions, events assumed a definite momentum. During July 1943 George Lewis recognized the scope and significance of the supersonic project and admitted the impossibility of Stack's conducting the work in typical "back of the envelope" NACA style. With a growing Army Air Forces commitment and the Navy showing signs of interest, Langley needed to acquit itself favorably and to take a leadership role. Consequently, the Committee directed the formation of a Compressibility Research Division at the laboratory and appointed John Stack to lead it. He and his engineers began by seeking more data. To augment the information already gathered during the XP-51 dives, Stack and his associates devised several ingenious techniques. One involved an Army B-29 Superfortress and missiles equipped with the Navy's most accurate radar tracking system. After being released from the bomb bay at 30,000 feet, the lead-packed missiles, implanted with specially designed NACA instruments, recorded the forces acting on the descending bodies as they achieved and exceeded the speed of sound. Not content with this data alone, John Stack sought the help of flight researcher Robert Gilruth. During the initial XP-51dive tests, Gilruth conceived of a simple way to circumvent the failure of the lab's wind tunnels at transonic speeds: merely mount a small airfoil vertically above the wing of a P-51D in the region of supersonic air flow, place miniature instruments in the fixture holding the airfoil to measure the direction and the forces at work, and ask the pilots to take the aircraft into steep dives. Gilruth counted on the well-known fact that while *airframes* experienced severe buffeting toward Mach 1, the air passing over the wings of high speed machines remained quite smooth.

8

Unfortunately, above the buffet boundary encountered around Mach 0.70, the scale model wing shook and the data proved of little use. In contrast, Langley engineer Henry Pearson conceived the idea of outfitting the P-51's standard airfoils with the most complete instrumentation used on any aircraft to date. His research succeeded in recording the transonic and supersonic air flows (up to Mach 1.4) required by Stack's Compressibility Research Division. Finally, a number of Langley technicians set up test stands on remote Wallops Island, Virginia, packed 40 pound rockets with the same instruments used in the drop-body tests, and launched them over the Atlantic Ocean to a height of 15,000 feet. These projectiles also reached Mach 1.4 as they streaked skywards, resulting in data on supersonic flight in the denser air found at lower altitudes.[5]

While the NACA undertook these research measures, the Bell Aircraft Company prepared, albeit unknowingly, to participate in the supersonic program. Under the direction of the exuberant Lawrence Bell, this company had just completed a grueling assignment for which it was personally selected by General Hap Arnold. The general liked Bell's enthusiasm and his firm's inventiveness. In just one year, Bell engineers closeted in their offices in Buffalo, New York, designed America's first jet-powered aircraft. Bell then fabricated the XP-59 Airacomet and chief test pilot Robert Stanley flew it for the first time over an isolated dry lake bed in the Southern California desert. While limitations in the power of its British-designed General Electric engines and unexpected aerodynamic shortfalls restricted flight to about 350 miles per hour, no one associated with the project--announced to the public in January 1944--doubted the capacity of Bell to produce exotic machines in short order. This deserved reputation and the timing of the XP-59 rollout left the New York manufacturer in an unmatched position to participate in an even more important investigation than the Airacomet. In mid-March 1944, the

---

[5]De Elroy Beeler, telephone interview with J.D. Hunley, 13 July 1999, DFRC Historical Reference Collection; Richard P. Hallion, "The Douglas D-558-1 Skystreak," 20 December 1971, unpublished paper, 1, Hallion Papers, DFRC Historical Reference Collection; Pearcy, *Flying the Frontiers*, 27; Young, *Meeting the Challenge of Supersonic Flight*, 4; Hansen, *Engineer in Charge*, 261-267; Stack, "History of the Rocket Research Airplanes," 21.

NACA called a meeting to discuss transonic flight with Army and Navy representatives. These sessions at Langley did not yield unanimity. Rather, the two different research tracks (traversing the sound barrier versus flying in the transonic region) emerged in open conflict. Stack attempted to win a unified, joint services approach to the problem based on his designs and on his conception of a long-endurance vehicle to gather data just above and just below Mach 1. But General Oliver Echols, by then the Army Air Forces Assistant Chief of Staff for Materiel, all but dismissed this approach, saying that during wartime the military services should not expend precious resources on non-military research planes. The Bureau of Aeronautics attendees, on the other hand, tended to side with Stack's objectives and the meeting ended without consensus.

Two months later the three parties convened again at Langley and achieved a compromise which papered over the disagreement but also pointed towards a solution. Known by this time as the Research Airplane Program Committee, its members agreed to launch the high speed investigations in two steps: "the first...us[ing] an airplane to obtain aerodynamic data to as high...flight speeds as could be obtained" (the Stack proposal); "the second...a high-speed flight research airplane...to reach the high[est] possible speeds and to have a [flight] duration on the order of from 10 to 15 minutes" (Kotcher's rocket plane). During the discussions, George Lewis realized this bifurcated approach meant the project would not be pursued in a fully unified framework and told the meeting that the NACA planned to release preliminary designs for a research airplane "to the Army *or the Navy* [author's italics]...as the NACA had no intentions of making a final design or constructing such an airplane." His assessment proved to be right. At conferences in July the Compressibility Research Division presented its turbojet design to the services, but the Army remained dissatisfied. Finally, in December all the parties assembled again in Hampton and Stack made a last plea for harmony, arguing his aircraft, unlike the rocket plane, offered direct military utility. The Army personnel left the meeting determined to have their way. By the end of the year Kotcher and his associates chose Bell Aircraft, fresh from the Airacomet development, to build a prototype called the XS-1 from specifications and engineering plans provided by the NACA. Apprised of the impending situation, Stack had already made overtures--supported by a trusted NACA ally, Walter Diehl--to the Bureau of

Aeronautics to sponsor the manufacture of the favored NACA design. The Bureau tentatively

selected Douglas Aircraft to design and fabricate the competing airplane early in 1945, pending

full approval in June of that year. Despite pursuing an active technical role in the development

of both aircraft, the NACA never wavered in its loyalties; Stack "displayed a strong preference

for the Navy airplane" and his staff extended themselves "in every way to assist in its

development." But Langley's compressibility chief did make one compromise; the Navy wanted

and won the point that the D(ouglas)-558 Skystreak would eventually evolve into a combat

aircraft, an outcome which ultimately proved to be chimerical.[6]

Fueled by a special Congressional appropriation, the rocket research aircraft program

took wing. The legislation designated the two services and the NACA as participating

organizations. Even though the Navy and Army paid for the projects, the Committee's federal

charter to supervise the science of flight won for it the preeminent role in drafting technical

specifications and in planning the flight test program. Yet, all of the parties agreed to the

sequence in which the high speed airplanes would be flight tested: first the Bell and Douglas

pilots would verify whether the performance satisfied contract specifications; then the military

aviators would press the machines to the limits of their flight envelopes; and finally, the NACA

cockpit crew would conduct the highly instrumented, incremental flight research for which the

NACA had become famous. While there continued to exist a sharp rivalry between the Army

and the Navy during the design and the development of their very different machines--a contest

---

[6]James O. Young, "Riding England's Coattails: The U.S. Army Air Forces and the Turbojet Revolution" in *Technology and the Air Force: A Retrospective Assessment*, ed. by Jacob Neufeld, George Watson, Jr., and David Chenoweth (Washington, D.C.: Air Force History and Museums Program, 1997), 3-39; Hansen, *Engineer in Charge*, 260-261; Meeting Minutes, NACA Committee on Aerodynamics, Washington, D.C., 24 May 1944, 8-9, Hallion Papers, DFRC Historical Reference Collection (first and second quoted passages); Young, *Meeting the Challenge of Supersonic Flight*, 5-10; Hallion, "D-558 Skystreak," 1; Becker, *High-Speed Frontier*, 91-92 (third quoted passage); Rotundo, *Into the Unknown*, 17-21; Carl F. Greene to the Director, Air Technical Service Command, 26 December 1944, AFFTC/HO Historical Reference Collection.

which even John Stack felt added vitality and momentum to the process--the letting of the contracts seemed to release a surprising degree of cooperation at the working level and even among the brass. Ideas circulated freely among the NACA, the Army, the Navy, and the contractor designers, technicians and pilots. So did the equipment. The Bureau of Aeronautics permitted Bell to employ in the XS-1 a Navy-sponsored rocket engine built by Reaction Motors; and the Army willingly revealed its air-launch techniques to Douglas and Navy engineers during planning for the advanced phases of the D-558. "And," said Stack, "they both turned...to a civilian agency to do the work...." Moreover, to be certain the two projects did not duplicate ends or means, the Bureau of Aeronautics retained close communications with Wright Field.[7]

During 1945 the NACA, the two airframe manufacturers, and the two services formulated their designs and put them to the test. Upon requests from either Bell or Douglas for advice or assistance, Stack instructed his team to respond quickly and thoroughly. Of the two aircraft, the D-558 advanced more slowly, in part due to the most recent findings on the comparative transonic qualities of swept wing versus straight wing aircraft. Based upon the research of Langley's brilliant, yet virtually self-taught aerodynamicist Robert T. Jones, not only did slender wings appear to be the most efficacious for high speed flight, but swept wings (discovered by Jones in 1945 independently of the German Adolph Busemann) appeared to reduce significantly the effects of compressibility. By the time the NACA and the Navy satisfied themselves about the value of swept wing--extensive wind tunnel experiments were conducted at the Caltech, the Southern California Cooperative, and the Langley eight-foot tunnels--they deemed it more practical to reserve the new configuration for the second phase of the D-558 project and use straight wings during the first stage. Douglas' Chief Engineer Edward Heinemann assumed primary responsibility for the Skystreak's design. Nevertheless, due to the unknown strength of forces in the Mach 1 range, Stack imposed on Heinemann, as well as on the Bell team, an

---

[7]Charles V. Eppley, *The Rocket Research Aircraft Program, 1946-1962* (Edwards California: Air Force Flight Test Center, 1963), 2; Hartley A. Soulé, "High-Speed Research Airplane Program," *Aero Digest* 63 (September 1951): 19-20; Stack, "History of the Rocket Research Airplanes," 25 (quoted passage); Hallion, "D-558 Skystreak," 4-5.

ultimate load of 18 g; that is, the aircraft required the capacity to withstand loads the equivalent of 18 times the force of gravity, a standard 50 percent higher than the capacity of existing fighters. The NACA-Navy team also expected the D-558 to fly as fast as Mach 0.89 while exhibiting satisfactory stability and control qualities at 10,000 and at 30,000 feet. Even though Stack favored the D-558, he still offered stiff criticism to Douglas at a mid-year design review, calling for more room for instrumentation, enlargement of the fuselage, and changes in the contour of the cockpit canopy.

As agreed upon by the NACA and by Ezra Kotcher, the XS-1 required rocket propulsion capable of sustaining powered flight for at least a two minute interval, during which time the machine would reach an altitude of 35,000 feet and develop speeds up to 800 miles per hour. John Stack expected the aircraft to accomplish these feats with 300 pounds of on-board instruments and 130 pounds of auxiliary equipment, all devised by the Langley engineers and all stuffed into every crevice of the little rocket plane. To avoid the complication of redesigning the straight, stubby wings of the XS-1, both Stack and Kotcher agreed not only to confine the swept-wing configuration to the D-558, but to further limit it to the model 2 aircraft. By the end of 1945 the Langley aerodynamicists had finished their wind tunnel work on the XS-1 and began to draw conclusions about its flight characteristics up to Mach 0.90. But if the aerodynamics started to come into focus, the rocket motors presented persistent problems. Stack blustered when Bell representatives threatened to resolve the difficulties by reducing the plane's period of maximum thrust by half and by lowering its flight ceiling. He reminded all involved why the NACA embarked on the program to begin with: unable to find transonic data in the wind tunnels, the Langley aerodynamicists looked to the world of full-scale flight. Despite such controversies, the XS-1 remained unencumbered by questions of basic re-design. Moreover, while the Skystreak needed to attain autonomous flight from its inception, the rocket plane faced a less daunting early program of air-launched, unpowered tests. Hence, the Bell team drove straight to an early finish, preparing the airframe (without rocket) for its first glide flight in January 1946. The D-558-1 took to the air almost 15 months later.

As the two teams readied their airplanes for flight research, Theodore von Kármán once again influenced supersonics. The cosmopolitan Hungarian and a hand-picked group of scientists journeyed to Europe under orders from Hap Arnold during summer 1945, in the final hours of the war. When they returned, Kármán wrote *Where We Stand* for the general. He broached the subject of transonic flight on the very first page and made it plain that the problem transcended the conflict about to be won. He envisioned a massive scientific undertaking involving "supersonic wind tunnels of large test sections...so that...a whole airplane...can be studied for optimum design." Kármán threw his extraordinary prestige behind a systematic and thorough investigation of "the very new horizon opened up by a velocity higher than sound [which] justifies the intensive research indicated. *We cannot hope to secure air superiority in any future conflict without entering the supersonic speed range.*" [author's italics].[8]

Emboldened by Kármán's prophecy, the XS-1 and the D-558 flight tests got underway. Bell's hot-headed and demanding Bob Stanley, no longer chief test pilot after being elevated to the position of chief engineer, sent his young and fearless replacement Jack Woolams on a pilgrimage to find a suitable flying site. Woolams thought first of the vastness of the Southern California dry lake bed where he had worked for eight months in the P-59 flight research program. But rainwater had accumulated on the high desert floor and the risk of intense downpours during January prompted Woolams to chose instead Pinecastle Field near Orlando, Florida. Temperate weather, a 10,000 foot runway, and adequate security won the approval of Stanley and his Wright Field sponsors. The Langley researchers made a game effort to conduct the glide flights over their home airstrip, but failed to persuade their partners. Still, the idea died

---

[8]Although the full-scale XS-1 never flew with-swept wings, Langley engineers tested this, as well as a forward-wing configuration in the Hampton wind tunnels. See the written comments on a draft of this chapter by DFRC aerodynamicist Ed Saltzman, DFRC Historical Reference Collection; Hallion, "D-558 Skystreak," 5-15; Young, *Meeting the Challenge*, 12-18, 20; Hansen, *Engineer in Charge*, 275-286, 288-294; Theodore von Kármán, *Where We Stand: First Report to General of the Army H.H. Arnold on Long Range Problems of the Air Forces with a Review of the German Plans and Developments*, (Washington, D.C.: U.S. Army Air Forces, 22 August 1945) 4-5, DFRC Historical Reference Collection (quoted passages).

hard in NACA circles; none other than Henry Reid regarded the Pinecastle flights merely as a prelude to future XS-1 tests over Langley. Still, the NACA gave the experiments unstinting support. Hartley Soulé, Mel Gough, and John Stack selected some of the lab's most able personnel to join the Florida contingent. To lead the group, Gough tapped Walter C. Williams, a young, tough-minded, and forceful aeronautical engineer from Louisiana who worked for him in the Flight Research Division. Williams had also collaborated with Soulé in stability and control and with John Stack on research airplane requirements. These experiences prepared him well for the critical challenges he and Gerald Truszynski (a radar specialist at Langley's Instrument Research Division) encountered during the XS-1 glide tests. Accompanied by three technicians and much telemetering gear and instruments, they journeyed south. The wisdom of Gough's choice--based mostly on the desire to select someone able to stand up to the autocratic Bob Stanley--proved itself almost from the moment Williams arrived in Orlando. The two men engaged in the first of many clashes of will, this one involving Stanley's demand to start the flights immediately versus Williams' insistence on the installation of the recording and radar equipment before plunging ahead.

Beginning on January 25, 1946, and during the following three months, Jack Woolams and the bullet-shaped plane dropped ten times from the belly of the B-29 mother ship. Even on its maiden flight he found it a delight to fly. It separated cleanly from the Superfortress, appeared aerodynamically sound, and at low speed (up to 275 miles per hour) handled beautifully in maneuver as well as in level conditions. He flew as fast as 400 miles per hour from the drop altitude of 25,000 feet. The only difficulties emerged on approach and landing. On the first flight, Woolams underestimated the steepness of his descent and landed 400 feet short of the runway. Another time the left landing gear retracted on impact, damaging the left wing. Clearly, these incidents taught that neither Pinecastle nor Langley were adequate for the more strenuous powered flights to come. Walt Williams noted the problems involved more than mere runway access or length (although these factors could not be underestimated). "One of the problems," said Williams, "was [Woolams] was launched above a scattered flight deck; a

15

scattered deck of clouds...maybe three-fourths, four-tenths cover. It was almost a standard condition at Langley." In addition, during the intense concentration of flying high performance aircraft, pilots might momentarily lose sight of the runway against the varied landscapes around both Pinecastle and Langley, a potentially fatal mistake at high speeds. Finally, the Pinecastle landing strip presented its own set of difficulties. On approach, aviators first saw a line of trees as the field came into view, with the consequence that even a fine test pilot like Woolams, flying at glide speeds, lost sight of the runway, failed to line up with it, and actually *crossed it* on his way to a hard, grass landing. After this experience, Woolams again recommended the Southern California desert. The reasons were compelling: a stable climate; greater isolation (for classified work and for avoiding populated areas); an almost endless expanse of dry lake for emergency landings; the confidence of having already flown the pathbreaking XP-59 flight test program there; the existing test base infrastructure (however makeshift) erected for the turbojet tests, including a flight line, equipment, and facilities; and a pool of military personnel for labor and for security. Regardless of residual hand wringing at Langley, the logic of the decision could not be denied. Bell representatives recognized the advantages, as did the Army Air Forces engineers, who persuaded the brass at Wright Field--probably General Frank Carroll himself--to launch the powered flights of the XS-1 under Western skies. Because "[it] was sort of a commitment that we were to work with the [Army Air Forces/U.S.] Air Force on X-1 from start to finish," Williams, his associates, and indeed the NACA itself followed the aircraft to its new destination.[9]

---

[9]Walter C. Williams, interview by Richard P. Hallion,13 June 1977, 9-15, 18, Hallion Papers, DFRC Historical Reference Collection (quoted passages); Rotundo, *Into the Unknown*, 49-89, 95; Richard P. Hallion, *On the Frontier: Flight Research at Dryden, 1946-1981* (Washington, D.C.: NASA SP-4303, 1984), 7-9; Hansen, *Engineer in Charge*, 296-297; Young, *Meeting the Challenge*, 18-22; J.D. Hunley to Michael Gorn (e-mail correspondence relating an interview between Hunley and De Elroy Beeler), 14 July 1999, DFRC Historical Reference Collection; De Elroy Beeler, interview with Michael Gorn, 23 April 1999, DFRC Historical Reference Collection.

## A DISTANT LAND

Between the end of the Pinecastle tests and the NACA's participation in the powered flights of the XS-1 six months later, the contractors found themselves pressed to meet their obligations. The least-tried link in the developmental chain--the XS-1's rocket motors--proved to be difficult as expected. Engine subcontractor Reaction Motors Incorporated passed acceptance tests on powerplants one and two, and during the summer of 1946 delivered them to Buffalo for test cell firings. Bell technicians encountered propellant valve failures in both, but project engineers were encouraged by the performance of the second one which proved to be remarkably durable on the shop floor, performing perfectly for an aggregate 1.5 hours over three weeks. The motors would continue to experience ups and downs during the interlude between Pinecastle and powered flight.

Meantime, the necessary parties began to assemble in the California desert. Jack Woolams journeyed West in March 1946 to prepare the ground for the Bell contingent coming soon afterward. He may have noticed some changes since his encounter with the Antelope Valley a few years before, but the fundamentals of the place remained unaltered. Between the 100 miles from mid-town Los Angeles to Palmdale lay the formidable San Gabriel Mountains, a barrier traversed over a two lane road which turned an otherwise straightforward drive into a four-hour ordeal. On first approach down the long descent to the floor of the Antelope Valley, the traveler discovered a barren panorama: a landscape flat, sparsely populated, and not just hot by day, but chilled at night. The terrain of the eastern Mojave Desert welcomed only the hardiest souls. The nineteenth century settlers who preceded the modern exodus consisted of miners who arrived at the time of the American Civil War. When they arrived at a rough crossroads called Mojave, they encountered nothing more than two buildings, both erected by Elias Dearborn in 1860: a stage coach station and a private home serving meals to those passing through. The miners surveyed and prospected and in 1873 found borax (sodium tetraborate, or boric acid and salt) to the northeast, in Death Valley. Uncommon until this discovery, the borax unearthed from the California desert became (and remains) the world's chief source of a mineral associated with washing powder and soap, pottery glazing, soldering, and mild antiseptics. A decade later a well-established but entirely makeshift route--made famous by the 20-mule teams which hauled the white powder on the first leg of its journey to markets across the globe--opened between the source in Death Valley and the town of Mojave in southeastern Kern County. The fortunes of Mojave improved further when W.W. Bowers discovered gold just south of the town in 1894. More good luck occurred with another gold strike, this time along the Borax Road at a mining camp called Johannesburg.. Starting in 1876, the Atcheson, Topeka, and Santa Fe, as well as the Southern Pacific railroads began to lay track in and around Mojave, establishing it as the railhead for the regional mines. The town solidified its position when untold quantities of borax began to issue from mines in Boron, a desert outpost east of Mojave on the Santa Fe line. Gradually, a few settlers began to join the itinerant miners. Effie, her husband Clifford, and his brother Ralph Corum bought 160 acres of land in 1910 on the western edge of Rodriguez Dry Lake, the biggest of the many dry lakes in the region and, indeed, the largest on earth. The Corum brothers built a home where the Santa Fe bisected the lakebed, opened a general store and a post office, drilled for water, and attracted other migrants to join them. They must have been persuasive men. Looking out over a shimmering and empty expanse measuring 12.5 miles by five miles at the longest and widest points, this tiny settlement of about 44 souls found itself perched on a hard sea of compacted silt commonly ranging in depth from 7.5 to 18 inches, but in some spots much deeper. The Corums wanted to use their own surname for the hamlet but when the post office protested that a California town called Coram already existed, they simply spelled the name in reverse and christened the settlement and, eventually (if temporarily), the dry lake as well. During the 1930s Muroc survived as a way station for the thousands of migrants from Oklahoma and Texas who streamed into California through Needles, trekked west as far as Mojave, and then branched south to Los Angeles or northwest to Bakersfield and beyond.

18

At the same time, the military value of the region became evident. The Army Air Corps, blocked by the Navy from using the Pacific Ocean as a bombardment range for its new generation of fighter and bomber aircraft, considered the Mojave Desert, located just over the San Bernardino Mountains from March Army Air Field. Disguised as Automobile Club representatives to avoid a cascade of land speculation, Hap Arnold, then the commander of March and two other officers journeyed to Muroc in 1933 to see the terrain for themselves. They returned to San Bernardino dazzled. Clearly, the isolated Muroc Dry Lake and its impervious surface promised the perfect field for aircraft operations, whether for bombing, for test flights, or for secret operations. Although legal title did not pass to the Air Corps until 1939, in September 1933 a detachment of March Field soldiers started laying out bombing and gunnery ranges on the eastern side of the great figure-eight shaped dry lake. The sound of repetitive gunfire and the occasional charge of explosives soon accompanied the appearance of aircraft from the other side of the San Bernardino Mountains as pilots tested their ordnance, their planes, and their firing skills. Lacking a mission other than target practice, the Muroc site remained under March Field jurisdiction for some time and the pilots and crew who flew the missions merely bivouacked beside their planes when they needed to stay overnight.[10]

The tempo accelerated during World War II. During summer 1941 Major George Holloman led 140 troops to the southwest quadrant of the lake and after erecting tents, undertook secret radio control tests of Douglas BT-2 trainers. After Japanese air forces devastated Pearl Harbor, American military planners realized the important security advantages of Muroc for the defense of the western U.S. Indeed, the 41st Bombardment Group's B-25s and the 6th

---

[10]See the written comments on a draft version of this chapter by Betty Love (a "computer" who later assumed a technical/engineering role) and also see the written comments on a draft version of this chapter by Ed Saltzman, both filed in the DFRC Historical Reference Collection; Rotundo, *Into the Unknown*, 96, 100-101; Young, *Meeting the Challenge*; Richard P. Hallion, "The Origins of Muroc AAFB," unpublished paper, 22 January 1972, 1-5, Hallion Papers, DFRC Historical Reference Collection; Hallion, *On the Frontier*, xiv-xv; Henry H. Arnold, *Global Mission* (New York: Harper, 1949), 136-137; Russ Leadabrand, *A Guidebook to the Mojave Desert of California, Including Death Valley, Joshua Tree National Monument, and the Antelope Valley* (Los Angeles, California: The Ward Ritchie Press, 1970), 27, 60-61.

Reconnaissance Squadron's aircraft arrived at Muroc on the afternoon of December 7, 1941. Two days later the 22nd Bombardment Group's B-26s and the 18th Reconnaissance Squadron's planes landed in anticipation of submarine patrol duty in the Pacific. Bombing practice continued as before, but now included a wooden facsimile of a Japanese heavy cruiser known jokingly as the *Muroc Maru*, constructed on the lake bed. The turning point for the region occurred when General Arnold, now Chief of the AAF, instructed his deputy, Colonel Benjamin Chidlaw to find a test site for the super-secret, jet-powered XP-59 aircraft. After a national search, Chidlaw selected Muroc. Consequently, Wright Field's Materiel Division established a flight test base on the northwest corner of the lake and dispatched Colonel R.P. Swofford to command. This high-profile project, pressed personally by General Arnold to close a menacing aeronautical lead opened by the Germans, caused drastic changes in the desert. Muroc ceased to be a satellite of March when the gunnery range became an autonomous Army post in July 1942.

The following month Bob Stanley of Bell arrived to fly the XP-59 but found just three structures standing against the vastness of Rogers (also known as Rodriguez and Muroc) Dry Lake: an unfinished portable hangar, a wooden military barrack, and a water tower. Freshly transplanted from Wright Field to command the same test site, Colonel Swofford took immediate action to accommodate Bell and the Wright Field personnel flooding in for the tests. By the end of 1942 he ordered on a high priority basis the construction of 20 by 48 foot hutments to house 100 men, a lavatory, an administrative building, a supply store, a recreation center, and a mess hall. Swofford also persuaded the Corps of Engineers to install a 10-mile-long, three-stranded barbed wire fence along the perimeter of the test base. Despite the recognized need for permanent quarters for the surge of incoming forces expected from Wright Field, those on the scene endured most of the winter of 1942-1943 with food supplied by Bell Aircraft and with shelter consisting of one Billeting Officers' Quarters barracks (with attached dining hall). These structures accommodated three officers, five enlisted men, and 40 Bell employees. Even during the following spring conditions improved only marginally when the hurriedly constructed hutments opened on the north base and the Wright Field technicians, mechanics, clerks, and carpenters streamed onto the compound. One mechanic described the prevailing situation.

> When I was stationed at Wright Field, I worked as a mechanic and one day I received orders to come to the Materiel Command Test Site at Muroc. They took four or five men from each of the hangars and sent them along, too. When I arrived at Muroc there were three hangars built, but only two were in use. Four or five P-59s were at the base undergoing tests and the base had actually been in operation a few months before I arrived [in September 1943]. I was in the second group of men to arrive at the field. The runway hadn't been built yet, there was no operations, no dispensary. When a man had to go on sick call a truck took him over to the [south] Air Base. There was a day room, but we had very little furniture and there wasn't much to do. The PX was only open for two or three hours a day and they sold only cokes and ice cream. I think there were only about 100 men after the first six months and retreat was the only formal activity held twice a week.

Because of a sense of shared adventure, morale proved to be quite good; but the psychological factor of isolation posed problems. To combat it, the commander authorized weekend leave for the soldiers in Los Angeles. A truck drove them into the city and at midnight picked them up at Hollywood and Vine for the long trip back, a cold journey in winter as the open vehicle negotiated the steep slopes of the San Gabriels.[11]

Despite its roughhewn qualities, Muroc improved somewhat under the pressure of war. It quickly gained a persona distinct from both March Field and from Wright Field. In November 1943 the bombing and gunnery range was designated the Muroc Army Air Field. The northwest corner of Rogers Dry Lake likewise assumed its own identity when it became known after

---

[11]Hallion, "Origins of Muroc AAFB," 5; Charles V. Eppley and N.A. Frank, "History of the Air Force Flight Test Center," 1 July to 31 December 1965, 6-7 (handwritten notes), Hallion Papers, DFRC Historical Reference Collection; Anon., "History [of the NACA Flight Research Center]," 1, n.d., in "Flight Research Background and History," DFRC Historical Reference Collection; Anon., *Ad Inexplorata: The Evolution of Flight Testing at Edwards Air Force Base* (Edwards, California: Air Force Flight Test Center History Office, 1996), 1; Young, "Riding England's Coattails," 21; Hallion, *On the Frontier*, xvii; R.P. Swofford to Benjamin Chidlaw, 9 December 1942, Air Force Flight Test Center History Office Historical Reference Collection (hereafter AFFTC/HO Historical Reference Collection); E.C. Itschner to South Pacific Division (Corps of Engineers), 4 July 1942, AFFTC/HO Historical Reference Collection; Benjamin Chidlaw to E.V. Schuyler, 12 October 1942, AFFTC/HO Historical Reference Collection; Malcolm Dodd to Ralph O. Brownfield, 29 December 1942, AFFTC/HO Historical Reference Collection; Personal Accounts of five soldiers who arrived from Wright Field to the Muroc Flight Test Base in 1943, AFFTC/HO Historical Reference Collection (block quote from Personal Account #1 by Sgt. John Novak).

August 1944 as the Muroc Flight Test Base. Accompanying these organizational developments, a new star in the flight research firmament launched its career at Muroc. During 1944, the XP-80 turbojet prototypes underwent intensive flight testing there. The product of the famed Lockheed "Skunk Works" of Burbank, California, the XP-80 advanced from concept to design to fabrication in a mere 143 days, a stunning feat even for the Skunk Works' extraordinary director Clarence (Kelly) Johnson. Its bigger version, the XP-80A achieved speeds of nearly 600 miles per hour in level flight and the production version P-80 Shooting Star rightfully claimed supremacy among the fighters of the world.

It also brightened the luster of Muroc as it gained laurels for itself. This renown manifested itself in accelerated base improvements. Early in 1944 five 20 foot by 96 foot prefabricated barracks and eight smaller 20 by 48 foot ones opened. So did a school house, a warehouse, and a fire station. Squadron administration buildings and others ranging from a dispensary to a latrine to a guard house soon followed. Yet, a number of the problems of everyday life persisted. The "hiring of civilian [support] personnel in this locality," wrote the commander, "is impossible," so services on base remained uneven. Moreover, the contract workers living at Muroc still experienced "undesirable conditions," as one inhabitant called them. But not everyone underwent the same discomforts. During the extraordinarily hot month of August 1945, salt tablets were distributed widely; but evaporative air coolers operated in only a few offices, offering the sole relief from excessively high temperatures. First Lieutenant Samuel Jacobs complained of the intense heat in crowded buildings but felt powerless to change the situation as the "endless red tape of procuring [the coolers] goes on while the men suffer...." Moreover, although no fewer than fourteen construction projects lay on the commander's desk, nearly all involved support of the mission, such as laying a runway and a taxiway, finishing two more hangars, and erecting a control tower. The living conditions ultimately raised questions about the future of Muroc. The base's reputation became known across the AAF, deterring some from serving there and resulting in short staffing. The situation emboldened the Test Base Director of Operations to admonish his superiors at Wright Field that the desert facility "now represents an investment of several million dollars. A return on this investment is expected by the Government. It can easily be realized by utilizing...the resources that are now available. This can be accomplished by the assignment of as many flight test employees" as the Air Technical Services Command could muster.[12]

The acute need for manpower manifested itself well before technicians uncrated the XS-1 at Muroc. By late 1945, only months after the end of World War II, the flight research program assumed a breadth no one could have imagined even as recently as the XP-59 experiments. A total of 31 projects awaited flight testing, including the P-80A, the XP-83, -84, and -86 fighters, and the XB-45 and -46 bombers. Researchers also wanted to measure the extent of noise in jet aircraft; to collect data on pressure distribution in such front-line aircraft as the C-47, the P-36, the P-51, and the P-80; to determine the maximum safe Mach numbers in dives of the latest fighters; and to measure helicopter vibration. Thus, the high speed research planes represented but two of many projects, although the national importance of the XS-1 and D-558 could not be denied. Acutely aware of the significance of the tests about to occur, the Langley contingent readied itself in spring and summer 1946 for the full program of experiments in the desert. But along with the technical preparations went a good deal of institutional adjustment. For an institution accustomed not only to being master of its own house but a jealous guardian of its own discoveries, the role of being only one partner in a large cooperative venture took some time to accept and to accommodate. For example, Army Air Forces press releases about the

---

[12]Anon., "History [of the NACA Flight Research Center]," 1, DFRC Historical Reference Collection; Eppley and Frank, "History of the Air Force Flight Test Center," 9, Hallion Papers, DFRC Historical Reference Collection; Anon., *Ad Inexplorata*, 3; S.J. Cook to Commander, Muroc Flight Test Base, 12 January 1944, AFFTC/HO Historical Reference Collection; John W. Morris, Jr. to Captain Chapman, 17 April 1944, AFFTC/HO Historical Reference Collection; Robert A. Kaiser to Laurence C. Craigie, 17 April 1945, AFFTC/HO Historical Reference Collection (first quoted passage); "List of Contracts on File At MFTB [Muroc Flight Test Base]," 15 January 1945, AFFTC/HO Historical Reference Collection; Samuel Jacobs, "Muroc Flight Test Base Unit History," 13 August 1945, AFFTC/HO Historical Reference Collection (second quoted passage); Muroc Test Flight Base Director of Operations to Director, Air Technical Service Command, 15 January 1945, AFFTC/HO Historical Reference Collection (third quoted passage).

Pinecastle flights trumpeted the achievements of Jack Woolams and Bell Aircraft but failed to even mention the NACA. As a result, sharp protests sailed from Langley to Dayton and the full role of Stack and his associates went into specially prepared War Department press kits. Despite such misunderstandings and bruised egos, the NACA did cooperate fully, even disclosing some of its research methods before publishing them. At the request of the AAF Engineering Division, the Langley staff agreed to participate in a Wright Field symposium in May 1946, timed just before the annual industry inspection of the Langley facilities. The program dealt with methods employed to collect both high speed and transonic data and featured Robert Jones and Robert Gilruth who talked, respectively, about the theoretical aspects of compressibility, stability and control at high speed; and the potential for rocket models to record high velocity information.

But the collaboration demanded by the high speed airplane program involved more than merely appearing at conferences. The NACA's designated research portion of the XS-1 flights required coordination not just with the Bell company, but with AAF representatives at Wright Field and at Muroc as well. The hard-edged Walt Williams and his staff pressed ahead, nonetheless, with his carefully organized instrumentation suite, one which his NACA antecedents would have recognized in an instant. He envisioned a program in two parts, designed to measure three factors: stability and control at high Mach numbers; aerodynamic loads on wings and tails through pressure distribution and strain-gauge techniques; and drag and performance data. The first phase would determine the operating boundaries of the aircraft and incrementally measure stability and control and aerodynamic loads up to the limiting conditions. The second would pursue more detailed renderings of loads using pressure distribution research. Finally, drag and performance would be recorded throughout the experiments. Williams decided to gather exhaustive sets of data for a series of designated speeds up to the margins of flight performance. Starting at Mach 0.83, then 0.86, then 0.89 he and his team would record complete stability and control and the associated loads "over each speed increment *before proceeding to the higher speed*." [Author's italics]. Pilots would be responsible for ten maneuvers at each

24

increment, including straight flight from launch to the realization of the desired speed point, steady turns at 1/2g increments up to the limit of buffeting or 5g, abrupt pull-ups to 8g, abrupt aileron rolls, abrupt deflection and hold of elevator controls, and abrupt deflection and release of rudder and aileron controls. Instrumentation consisted of the full NACA complement of devices to measure airspeed, altitude, acceleration, angle of attack, control forces, control positions, rolling velocity, sideslip angle, rocket chamber pressure, and strain at 12 points on the aircraft. Telemetering recorded airspeed, normal acceleration, and elevator and aileron position. Radar observed altitude and flight path. But, again, the NACA no longer worked solo; when the Langley team attempted to impose this full, complete, and rigid regime on the contractor flights as well as its own phase of research, Bell XS-1 Project Engineer Richard Frost resisted forcefully. "We do not foresee the need," he wrote to the leadership of the Air Materiel Command, "for delaying any flight tests, for instance, to permit detailed analysis of numerous data which the automatic instrumentation may have recorded the previous flight, nor delaying a flight because radar, or telemetering, or say, a multiple manometer were not functioning 100% since none of those items have any bearing on our contractual commitments."[13]

Wrangling over the respective roles of the contractor and the NACA persisted until and even after the Langley staff appeared at Muroc. In the meantime, final preparations went forward. Before his shocking death in an airplane accident on the eve of the Cleveland National Air Races, Jack Woolams laid the groundwork for the tests of the XS-1 at Muroc. He arranged for construction and delivery of two tanks: a large one to hold liquid oxygen, a smaller for liquid nitrogen. He oversaw the excavation of a loading pit for the rocket plane, a contrivance

[13]Frank N. Moyers, "Flight Research Program," 2 October 1945, AFFTC/HO Historical Reference Collection; Donald R. Eastman to Commanding General, Air Materiel Command, 27 June 1946, AFFTC/HO Historical Reference Collection; George E. Price to Jerome C. Hunsaker, 22 April 1946, AFFTC/HO Historical Reference Collection; George B. Patterson to Air Materiel Command Engineering Liaison Officer, 16 July 1946, AFFTC/HO Historical Reference Collection; Walter Williams to Chief of Research, 7 June 1946, AFFTC/HO Historical Reference Collection (first quoted passage); Richard Frost to Commanding General, Air Materiel Command, August 2, 1946, AFFTC/HO Historical Reference Collection (second quoted passage).

necessary so the B-29 mother ship could be wheeled over the Bell aircraft and the two could be attached at the bomber's belly. Woolams even succeeded in opening a rail spur by which cars carrying liquid oxygen could replenish the tank. Finally, despite the desperate lack of office space, he found what he could for the Bell workers and arranged to house some of them well off the base in Willow Springs, southwest of Mojave. Upon Woolam's passing, Bell replaced the irrepressible young aviator with another select flier, Charles "Slick" Goodlin, to undertake the acceptance tests of the XS-1. The Army, meanwhile, authorized NACA flight research pilots Mel Gough, Herbert Hoover, William Gray, Joel Baker, and Stefan Cavallo to fly (at the service's expense) several Army Air Forces cargo planes in support of the NACA mission.[14]

Despite steady progress toward launching the XS-1, the relations between the NACA and Bell continued to deteriorate as the date approached to ship the first prototype to Muroc. Walt Williams led a group of six to Buffalo on September 16 and 17, 1946, and at first all seemed cordial enough. Project Engineer Dick Frost permitted instrumentation specialists Paul Harper, Warren Walls, and Norman Hayes to begin stuffing the little fuselage with monitoring equipment. Meanwhile, Walt Williams, pilot Steve Cavallo, and engineer John Gardner followed Frost to the engine test stands where they saw an encouraging sight and heard encouraging news: the second engine already had been mounted on the XS-1 and the first one ran so impressively that Bell now pronounced itself "well pleased" with the powerplant. After leaving Gardner with a Reaction Motors representative, Williams and Cavallo sat down with Frost to review the test schedule. Frost opened with the assurance that "Bell's plans at present are all directed towards getting the XS-1 to Muroc as soon as possible," meaning shipped by 30 September. Then the discussion deteriorated. Just the week before, the Project Engineer refused to install rudder pedals conceived by the Langley engineers to measure the force applied by the pilot in maneuvers. This day in Buffalo he again rejected the instrument, saying Bell had never

---

[14]Rotundo, *Into the Unknown*, 96-97, 115; Donald R. Eastman to Memo for Files, 11 September 1946, AFFTC/HO Historical Reference Collection; Donald R. Eastman to Personnel with Operational Orders, 11 September 1946, AFFTC/HO Historical Reference Collection.

approved the modification and, "as far as he was concerned, the pedal-force recorders would not be installed...." Frost then asked what sort of data the NACA expected from the acceptance flights, to which Williams replied "complete stability and control data [and] the required aerodynamic load data...." Frost told Williams not to expect such an elaborate investigation, warning that up to Mach 0.80, no special flights would be undertaken. Bell Aircraft would concern itself only with Slick Goodlin's opinion about the aircraft's stability and control, as agreed upon by contract with the Army. Moreover, without so much as a courtesy copy to Langley, the prickly Bob Stanley had already sent the Materiel Division its acceptance flying plan, envisioning 16 to 18 flights after a series of unpowered glides with increasing increments of ballast. Although Stanley did overrule Frost about the pedal force instruments and agreed to put them on, he reinforced all else that had been said with even greater emphasis. He told Williams and Cavallo he had no more than 30 hours to perform all of his scheduled tests and "if the NACA requests for data could be worked into Bell's plans...some data would probably be obtained but *no interference would be allowed*" [author's italics]. Williams refused to be intimidated. He asked Stanley again and again for clarification about what Bell expected to achieve in its flight tests, but received no clear answer. The NACA representative stated his minimum demands: data on longitudinal stability and control in steady and accelerated flight, and on buffeting boundaries. He deemed these conditions "absolutely essential" for the NACA to continue its support of the XS-1 project. After his team completed rigging the XS-1 for the Muroc flights they returned home to Langley to prepare for the trip west.

Of all the assets Walt Williams assembled for this adventure, none assumed more importance than the confidence Henry Reid reposed in him. First, he equipped Williams with full control of the mission, informing officials at Muroc that as "the NACA representative in charge of the NACA personnel stationed at Muroc...Mr. Williams is authorized to make all necessary contacts and decisions for the NACA in connection with this project...." Reid also supported unequivocally the position Williams articulated at the turbulent meeting at Bell Aircraft: safety must take precedence over all other considerations, and work must be pursued in a thorough and orderly manner. "[B]efore asking anyone to proceed with the extremely hazardous flying a Mach number above 0.8," Reid observed, "everything [sh]ould be done to make certain that the airplane was satisfactory in all aspects in the speed range up to Mach 0.8. The test program was Langley's means of assuring itself of the airplane's satisfactory subcritical characteristics." The engines, completely new and untried in flight, required careful scrutiny for reliability. The degree of loading on the aircraft's surfaces needed to be understood. The landing gear failed twice in the Pinecastle tests, suggesting the need for further analysis. Reid felt the Bell criteria of Mach 0.80 and an 8g pull-out failed to lay the groundwork for safe flight at transonic speed. "Langley," he concluded, "does not want its pilots to undertake the research flying on the XS-1 following such limited acceptance tests as Bell proposes."[15]

Thus, instead of hopeful anticipation, a sense of wariness and anxiety pervaded the minds of Walt Williams and his team as they initiated the NACA's presence at Muroc Army Air Field. Williams and his associates knew that extreme circumstances might precipitate a complete withdrawal from the project. But even those who arrived with a positive outlook found their enthusiasm blunted by the conditions encountered at the end of the trip. The contrast was unsettling. While Tidewater Virginia could be notoriously hot and humid between June and August, those destined for Muroc left Hampton during early fall, the best season of the year. They left the changing colors of the thick stands of trees, the laboratory's solid brick structures, and broad sweeps of lawn more reminiscent of a college campus than a federal institution. They arrived at a place improved--but certainly not transformed--from the state Jack Woolams found it in 1943. Like the Bell technicians leaving Buffalo, many of the NACA people knew next to nothing about Muroc. Most of those who arrived by train approached not through the thriving

---

[15]Young, *Meeting the Challenge*, 25-28; Donald R. Eastman to Memo for Files, 11 September 1946, AFFTC/HO Historical Reference Collection; Walter Williams to Langley Chief of Research, 20 September 1946, AFFTC/HO Historical Reference Collection (first six quoted passages); Henry Reid to Commanding Officer, Muroc Army Air Base, 20 September 1946, AFFTC/HO Historical Reference Collection (seventh quoted passage); Henry Reid to the NACA, 26 September 1946, AFFTC/HO Historical Reference Collection (eighth and ninth quoted passages).

oasis of Los Angeles, but through the back gate; the blank, arid country of southeastern California. As one Bell employee remembered:

> We got off at Barstow. We transferred in Chicago to get on the Santa Fe that came through to Barstow....Barstow was just about like the end of the world when we got off there. We couldn't quite believe where we were. We thought maybe we were...going to be right near that city. They said, "No." They had a couple station wagons there that took us over [the present California] Highway 58 [then called U.S. Highway 446] down towards Muroc. It was getting worse all the time. Everybody said, "Where are you taking us?" There were no roads coming into the [northern edge of the] Base from Highway 58 at that time....There was a little dirt trail off of 58 that went across the sand dunes and down into the lakebed. When we got on the lakebed, the driver stopped there and we were all just kind of stunned by that huge expanse of dry lakebed, with all of its mirages and everything shimmering around. He said, "You see those two dark objects way out there in the distance a couple of miles? That's where you're going to live." I said, "That's where we are going to live, up here in this?" One guy said, "Would you mind turning this thing around and going back to Barstow so I can see if I can catch a train out of here?" I never anticipated living in a place like that. As we got closer to the Base, we could see the barracks and the hangar; we realized where we were going to be for the next year or so. It was very interesting, to say the very least.[16]

After this introduction, the catalog of discontent ranged from the trivial to the substantial. Some single employees arriving for the XS-1 experiments lived in the town of Muroc in a fire-prone Air Force housing area called "kerosene flats," named for the prevailing method of cooking and heating. On the rocket plane's proving grounds at the south base, other unmarried workers and engineers resided in hastily constructed barracks and found it necessary to install new windows in order to reduce the amount of sand blowing in by day and night. If they failed to make these modifications, they returned home from their shifts to find their beds so coated with wind-blown silt that all the bedding had to be stripped and shaken outside. In another effort to reduce the penetration of wind and sand into living quarters, local farmers hauled in bales of hay to wedge into the base of the buildings. They also interlaced the hay with thistle rope--thick cords spiked with stickers--to deter rattlesnakes from entering the living quarters. Indeed, more than one chef walking outside the base restaurant to dispose of garbage found himself face-to-face with coyotes or snakes. They either learned to handle a .22 rifle or they resigned the job.

---

[16]John W. "Jack" Russell, interview with unknown interviewer, Air Force Flight Test Center Oral History Series, April 1994, 5-6, DFRC Historical Reference Collection.

Hired in Los Angeles, these cooks rarely pleased their diners and usually quit without any prompting from the animal life. Armed security troops patrolling the perimeters posed yet another obstacle to a normal existence. The NACA contingent as well as the contractors worked all hours to prepare for the flight tests, treading back and forth on foot between the hangars and the barracks at no structured times. Turning a corner when they left the hangars--frequently around 2 a.m.--they occasionally encountered the chilling, metallic sound of a rifle mechanism being engaged and heard a disembodied voice telling them to freeze. Under such conditions, frayed nerves afflicted both the guards and of the workers and many feared accidental shootings.

To satisfy the demand for housing imposed as married Douglas Aircraft employees converged on Muroc to participate in a variety of flight research projects, the Bureau of Aeronautics tried the expedient of opening the Mojave Marine Base, abandoned and partially dismantled since the end of the war. Desperate to find temporary quarters for its staff too, the NACA asked Mel Gough to appeal to some of his Navy friends, who agreed to let the Langley workers lodge there temporarily. The resulting situation presented its own problems. One aircraft mechanic who lived there with his wife said when they first arrived, "the place was filthy." Appalled by the number of pests on the premises, he "went into Mojave and bought some stuff to kill the [them]. And I'm not kidding you--we swept them up in pans. We must have had a pound of them." Every morning, a mixed group of government and industry employees squeezed into a single station wagon and drove the 25 miles from Mojave to the flight test base. The wives of these men felt the difficulties more than their husbands. Before following their spouses west, most resided in cities and towns where the necessities of life lay close at hand. But if these women faced isolation in the eastern Mojave, many also forged close friendships with other families based on shared experiences. Their husbands, meanwhile, found both diversion and stimulation participating in the exciting projects to which they contributed.[17]

---

[17]Taped interviews with some of those involved in the early NACA flight research at Muroc suggest that for many, their lives centered not on housing or on leisure, but on their work. These engineers and technicians were young, excited by the wartime mission, absorbed in the complexities of their projects, and filled with the camaraderie and sense of shared objectives that

# THE SITUATION ON THE GROUND

Ignoring the inconveniences, Walt Williams and his Langley team concentrated on the

task before them and approached it in the traditional NACA way. Williams arrived at Muroc as

Engineer in Charge the morning of September 30, 1946, responsible to Flight Research Division

chief Melvin Gough, who managed the Muroc endeavor from his office in Langley. Late that

evening two more of the Langley contingent appeared in the persons of engineers William S.

Aiken and Cloyce Matheny. Williams, Aiken, and Matheny were met by Instrument Engineer

George P. Minalga and Telemetering Engineer Harold B. Youngblood who had already reported

for duty. This initial cadre of five became identified as the NACA Muroc Flight Test Unit. All

of them, like everyone assigned to this military camp, faced the same fundamental obstacle: an

acute shortage of housing. Willow Springs had been overrun by renters from Bell. The

---

often develop among people functioning in comparative isolation. They cared little about where
or how they lived and what they did or did not eat. If they needed an adequate meal, a relaxing
swim, or wanted to see a movie, the NACA workforce found a welcome at the base Officers'
and Non-Commissioned Officers Clubs. But for them, whatever hardships--and equally,
whatever recreations--existed in the early days of Muroc essentially paled in comparison to the
thrill of life on the job. See Don Thompson interview with Michael Gorn (by telephone), 11
March 1999, DFRC Historical Reference Collection; Clyde Bailey, Richard Cox, Don Borchers,
and Ralph Sparks interview with Michael Gorn, Palmdale, California, 30 March 1999, DFRC
Historical Reference Collection; De Elroy Beeler interview with Michael Gorn, 23 April 1999,
Santa Barbara, California, DFRC Historical Reference Collection.
Hallion, *On the Frontier*, 11; Russell, interview, April 1994, 7-8, 16-17; Betty Love interview
with Michael H. Gorn, Palmdale, California, 10 April 1997, 3, DFRC Historical Reference
Collection; Williams, interview, 13 June 1977, 25-27; Donald Borchers, interviewed by Curt
Ascher, Lancaster, California, 16 December 1997, 15-16 (quoted passage), DFRC Historical
Reference Collection; De E. Beeler, interview, December 1976, 20; Hubert Drake and Gerald
Truszynski, interviewed by Dr. J. Dill Hunley, Edwards, California, 15 November 1996, 7-8,
DFRC Historical Reference Collection.

31

unoccupied Navy (i.e., Marine) quarters in Mojave seemed destined for Army Air Forces personnel. Walt Williams drove the entire territory from Tehachapi, an hour northwest of the base, to Lancaster but found nothing available. The NACA's top official on Muroc spent his first nights sleeping in marginal conditions. "I am in a shack," the excitable Louisianan told Mel Gough, "with three Northrop mechanics." He had no office and no phone. To make matters worse, the shipment of the XS-1 had been postponed due to delays in refurbishing its B-29 mother ship; the five Langley men might all have stayed in Langley at least another week. Williams chafed and grumbled at this turn of events, as he did at all postponements. Indeed, he personified impatience. One close friend described Williams' habit when he came to visit with his family. "When he'd come driving up to the house he'd hop out. And he would leave his wife in the car. And she'd have to open the door and grab the baby and come in afterwards." He found work for Aiken and Matheny calibrating the strain gauges and instructed Minalga to set up his instruments. But if the wait lasted longer than two weeks, he fretted about finding enough work to put everyone's time to good use.[18]

The succeeding days brought both encouragement and annoyances. On October 2 he reported to Gough that he now occupied a single room in one of the dormitories and had found a good ranch house in Palmdale divided into apartments, one of which he was promised upon first vacancy. Williams deemed the rent high but calculated his Langley per diem would cover not only this expense, but gasoline to cover the daily 80 mile round trip. He also made progress on infrastructure needs, obtaining a NACA post office box in Muroc and completing paper work for office furniture. Having done all he could for the moment, he and his comrades assisted Minalga in his preparations. But Williams still fussed and complained. He called the cool, cloudy, and windy weather "nothing to brag about" and thinking again about the rocket plane said, "[o]nly one day so far this week would have been suitable for an XS-1 flight...." He visited the

---

[18]Unpublished DFRC Chronology, n.d., DFRC Historical Reference Collection; Hansen, *Engineer in Charge*, 297-298; Walter Williams to Melvin Gough, 1 October 1946, AFFTC/HO Historical Reference Collection (first quoted passage); Borchers, interview, 41 (second quoted passage).

establishment of the legendary Florence (Pancho) Barnes, former aviatrix and stunt pilot and now the proprietress of a large parcel of land on which she operated several businesses: a restaurant and bar that attracted many stationed at Muroc; a motel of some 12 units with a swimming pool; a ranch on which Barnes raised pigs and other livestock; a farm where she grew alfalfa; and an airstrip with a small hangar. In rare instance of understatement, Williams described this independent and flamboyant woman as "quite a character." He was not so demure about a viewpoint often heard on the air field; that the NACA team deserved no more consideration than contractors, even though they represented an independent government agency. Williams and some of his subordinates objected strongly to this mistaken impression and lost no time dispelling it. He also worried about personal details: how would his per diem be paid? By the NACA or the Army? Had the local Citizen's Bank received his paycheck, as he instructed the Langley payroll office before leaving Hampton?[19]

During the second week in Muroc Williams, in concert with his staff, began to solve some of his important problems, allowing him to forget about the trivial ones. When Republic Aircraft withdrew from Muroc after completing a major project, Williams persuaded the base housing officer to reserve the contractor's barracks for the NACA arrivals, a significant victory since the Republic accommodations bore the dubious distinction of being the best on the base. He solidified his own housing situation by taking the expensive but "very nice" Palmdale apartment for $28.50 a week, telling Mel Gough, "Well, I didn't come out here to make money." The NACA staff also moved into its own office, equipped with telephone, on October 8. These events occurred just in time. Williams expected Langley pilots Joel Baker and Jack Reeder to arrive that very day, and greeted engineers Charles Forsyth, Beverly Brown, and John Gardner; Instrument Technician Warren Walls; and Crew Chief Howard Hinman on the ninth. Another

---

[19]Walter Williams to Melvin Gough, 2 October 1946, AFFTC/HO Historical Reference Collection; Walter Williams to Melvin Gough, 3 October 1946, AFFTC/HO Historical Reference Collection (both quoted passages); Williams, interview, 28; Walter Williams interview with John Terreo, Computer Sciences Corporation, Edwards, California, April 1994, 11, DFRC Historical Reference Collection (in a series entitled *The Legacy of Pancho Barnes*, published by the Air Force Flight Test Center).

reason why Williams dwelled less on inconsequential matters was the appearance on the evening of October 7 of his nemesis Bob Stanley. Not only did Bell's chief representative arrive; so did the XS-1 and the B-29. After their first encounter at the Bell factory, Williams and Stanley braced themselves for further confrontation now that the essential ingredients of the flight program were on the ground at Muroc. One observer of Williams called him, "a hell of a [smart] guy, [but] he's a bull in a China closet...." One of Stanley's admirers described him as a man of supreme self-confidence, "a whiz at everything...." But Stanley also thought nothing of humiliating his subordinates; if he deemed a mechanic incompetent, he might tear the tools from his hands and, in the presence of others, finish the job himself. "You didn't tell Bob Stanley anything," his friend recalled. Thus, Williams and Stanley represented the perfect rivals.

They met the next morning, at which time Stanley announced his intention to launch the XS-1 early on October 9. Williams returned fire, saying "we [are] not ready and could not possibly be ready by tomorrow" since instrumentation specialist Walls had not yet arrived on base. Stanley argued that delaying the initial Muroc flight set a bad precedent for the entire program, but Williams demanded the program begin only when the NACA and Bell *both* felt satisfied with the preparations. Stanley then upped the ante, charging the debate really turned on "who (NACA or Bell) would dictate the program during the contractual flight tests." To settle the conflict, Stanley and Williams sequestered an Air Materiel Command representative as a witness and telephoned Dayton to determine the AAF's wishes: did Wright Field want instrumentation on all test flights or not? The reply gave the NACA a clear sense of its importance in the XS-1 program. The voice on the line saw nothing to prevent Bell conducting the initial flight on the timetable planned by Stanley. But, "if something did happen to the XS-1 without the telemetering installed it would be very embarrassing for the [Air Materiel Command] as well as the Bell company." All but admitting his bosses' inflexibility, Dick Frost told Williams privately that once Stanley returned to Buffalo he would "see that things worked out

34

better for [the NACA]."[20]

Tensions remained between the two combatants but the immediate cause of the controversy resolved itself. Stanley ordered the B-29 and XS-1 into the skies on the 9th of October for a glide test, even though some NACA instruments awaited installation and others required check outs. Everything seemed fine as Slick Goodlin waited for the bomber to achieve a safe altitude before lowering himself by ladder into the tiny rocket plane. But a malfunction in the B-29 cabin pressure regulator resulted in a dangerous buildup of exhaust from the XS-1's nitrogen-driven attitude gyro. Emergency releases failed to work so the B-29's cabin door had to be jettisoned and although it was secured by a lanyard, damage resulted to the door itself, to the door frame of the XS-1, and to the egress ladder. The big aircraft landed with Goodlin trapped in the research aircraft. Williams took full advantage of the subsequent delays necessitated by repairs. He rallied his forces to make the most of the opportunity, putting them on overtime and night work so that when the Bell technicians ended their daily assignments to fix the recent damage, the NACA team followed close behind to add the last instruments, complete the check outs, and make the calibrations. Before the second attempted research flight of the XS-1, the Muroc Flight Test Unit staff had set up all of its essential instrumentation. In part, they finished the job quickly because of the aircraft's compact size and uncomplicated interior design. Just 30 feet 11 inches long (less the nose boom) with a 28 foot wing span, the Bell machine weighed only about 7,000 empty and its "testing tools...were very, very simple...." Two big tanks, which held oxygen and alcohol/water, and eight nitrogen spheres took up most on the interior space. Only the rocket engine offered real difficulties. Time ran out, however, before Williams' crew could wire the telemetering system, designed to transmit a few key flight factors in case the

[20]Walter Williams to Melvin Gough, 4 October 1946, AFFTC/HO Historical Reference Collection; Walter Williams to Melvin Gough, 8 October 1946, AFFTC/HO Historical Reference Collection (first, second, sixth, and ninth quoted passages); Unpublished DFRC Chronology, n.d., DFRC Historical Reference Collection; Young, *Meeting the Challenge*, 27; Beeler, interview, 14 (third quoted passage); John W. Russell, interview, 7-8 (fourth and fifth quoted passages); Walter Williams to the Langley Chief of Research, 10 October 1946, AFFTC/HO Historical Reference Collection (seventh and eighth quoted passages).

aircraft failed to land safely.

Delighted to capitalize on Stanley's impatience, Walt Williams barely suppressed his pleasure when he recounted how "the Bell Company tried to make a flight today [October 9], but ran into a little trouble. I think it was the usual case of going off half cocked." Nevertheless, still bent on fulfilling the contractual obligations "with the NACA getting as little [data] as possible," Stanley drove his technicians to get the B-29 back in the air, using a sledge hammer himself to fix the bomber door. The Bell mechanics mended the broken ladder but, under Stanley's impossible timetable, were unable to solve either the pressure regulator problem or the broken manual release. One of the crew showed a grim sense of humor when he handed Slick Goodlin a screwdriver in case escape from the XS-1 became necessary. Unlike the late Jack Woolams, who showed zeal for the supersonic project, Goodlin reportedly expressed only tepid interest in the flight program, no great love for the aircraft, and little enthusiasm for the objective of reaching Mach 0.80, other than to do so without delay. The pilot took the first step toward that goal in a glide flight which began around 3 p.m. on October 11, 1946. He and the XS-1 dropped uneventfully from the B-29 and accomplished some stalls at 130 miles per hour with flaps and gear down. Approaching at 180 but touching down at 140 miles per hour, he rolled at least 10,000 feet before slowing to a halt. This first successful flight not only instilled a sense of confidence in the XS-1 operation, but made the discomforts of Muroc a little less aggravating and proved the wisdom of the site selection. Goodlin had both praise and complaints for the XS-1. He liked the overall handling qualities of it but felt the lightness of the controls caused him to overcompensate and suggested engineering some additional friction in the system. Also, the brakes failed to operate properly, hence the long ride on the ground. Worse than that, Williams and his engineers ended up with almost no data from the flight. Someone turned on the NACA instruments far too early, a full eight minutes before the drop, leaving only the first 30 seconds recorded on film. Because the telemetering equipment still awaited installation, Stanley's rush to get the XS-1 into the skies resulted in a useless flight from the NACA viewpoint.[21]

After weathering Stanley's impetuous behavior for a month, Williams got the first signs of relief. In mid October Hartley Soulé met with Air Materiel Command officials Colonels R.S. Gorman and George Smith, and Mr. J.H. Voyles in Dayton to clarify the NACA's role in the XS-1 project. Soulé left with all Williams could have hoped for. The Army agreed to hold Bell to that part of its contract which stipulated that satisfactory flying characteristics as high as Mach 0.80, thus allowing the service to require a longer and more complete contractor program which satisfied the NACA's demands. Also, since the Materiel Command intended to transfer the aircraft to the NACA, the Army representatives agreed that the NACA contingent must be satisfied with the plane's performance before acceptance. Soulé, in turn, promised that Williams and his cohorts would decide whether to approve the machine in fewer than 20 powered flights. A few days later, Gorman and Voyles arrived at Muroc and offered further reassurance to Williams by saying the AAF would not accept the XS-1 until the NACA concurred. Bob Stanley, Dick Frost and Slick Goodlin then received instructions from Gorman and Voyles to permit the NACA to collect "as much data as possible...during these tests....Bell should make it possible for the NACA to have their instrumentation ready for every flight." Stanley refused to cave in, but did admit he had been pressing hard because of contractual obligations and predicted (with surprising self-awareness) that when he returned to Buffalo, Williams and the NACA group would have sufficient time to accomplish his mission. Indeed, a sign of greater

---

[21]U.S. Air Force, "Air Force Supersonic Research Airplane XS-1," Report Number 1, 9 January 1948, 2-3, DFRC Library; Walter Williams, "History of the Rocket Research Airplanes," (paper presented at the American Institute of Aeronautics and Astronautics meeting, 28 July 1965), 3, Hallion Papers, DFRC Historical Reference Collection (first quoted passage); Drake and Truszynski, interview, 6-7; Joel Baker to Herbert Hoover, 9 October 1946, AFFTC/HO Historical Reference Collection; Walter Williams to Melvin Gough, 9 October 1946, AFFTC/HO Historical Reference Collection (second and third quoted passages); Walter Williams to Melvin Gough, 10 October 1946, AFFTC/HO Historical Reference Collection; Joel Baker to Herbert Hoover, 10 October 1946, AFFTC/HO Historical Reference Collection; Walter Williams to Langley Chief of Research, 11 October 1946, AFFTC/HO Historical Reference Collection; Walter Williams to Melvin Gough, 11 October 1946, AFFTC/HO Historical Reference Collection; Walter Williams to Melvin Gough, 14 October 1946, AFFTC/HO Historical Reference Collection.

cooperation manifested itself during this discussion. Slick Goodlin extended an olive branch by offering to confer with the NACA group before the flights and review the data with them afterwards. Williams knew Goodlin meant what he said; despite his ambivalence about the XS-1, the Bell pilot unexpectedly visited the NACA office a few days earlier and offered to fly the maneuvers desired by Williams and his staff.[22]

As it turned out, events overtook the negotiations with Bell. For all of Stanley's incredible will to complete the acceptance tests swiftly, his company found itself compelled to shut down flight operations not because of tardy behavior by the NACA, but for technical reasons. He and two-thirds of the contractor staff returned to New York during the third week in October to await completion of modifications on the XS-1. First, the fuel tanks required flushing out. Partially filled with water for ballast during the initial glide flights, they had been contaminated by dirt which threatened to clog the entire system. Second, the controls for the dome pressure regulators in the B-29 needed to be transferred to the rocket plane itself so the pilot could load and unload the nitrogen domes himself, rather than relying on the existing, cumbersome system in which two men fueled the plane from the bomb bay. Meantime, the ground testing of the Reaction Motors rockets ceased when 10,000 gallons of the wrong type of alcohol arrived at the Muroc loading docks. The break in action allowed the B-29 to be flown to Oklahoma City for routine maintenance inspection. But once there, it waited in a hangar for parts until mid-November. Everything took longer than expected; an expected hiatus of two or three weeks more than doubled in length. Bob Stanley did not return to Muroc until November 27.[23]

---

[22]Hartley Soulé to Langley Chief of Research, 15 October 1946, AFFTC/HO Historical Reference Collection; Walter Williams to Langley Chief of Research, 18 October 1946, AFFTC/HO Historical Reference Collection (quoted passage); Walter Williams to Langley Chief of Research, 21 October 1946, AFFTC/HO Historical Reference Collection; Walter Williams to Melvin Gough, 15 October 1946, AFFTC/HO Historical Reference Collection.
[23]Walter Williams to Melvin Gough, 17 October 1946, AFFTC/HO Historical Reference Collection; Joel Baker to Langley Chief of Research, 25 October 1946, AFFTC/HO Historical Reference Collection; Walter Williams to Melvin Gough, 9 November 1946, AFFTC/HO Historical Reference Collection; Walter Williams to Hartley Soulé, 27 November 1946,

During the interregnum, Walt Williams struggled to find patience and to maintain the momentum in the NACA hangars . He, Warren Walls, William Aiken took the opportunity to investigate the D-558 program in Douglas Aircraft's El Segundo, California, plant. They drove to the Santa Monica office of the NACA's Western Coordinator, Edwin Hartman, who escorted the party to the nearby factory. Upon inspection of the mock-ups, Williams saw that the Douglas engineers had left adequate space for instrumentation and for telemetering equipment in the Phase I design, but a good deal more room in the Phase II compartments. The NACA visitors were pleased to see most of the standard NACA recording instruments being installed in a configuration similar to the XS-1, with one suite containing a twelve-channel oscillograph for strain-gauge recordings and the other package consisting of two 60-cell manometers. Two differences with the XS-1 also came to light: the Douglas planes would take measurements directly from the control system rather than the pilot's controls and would automatically record all data on a specially made 30-channel Miller Oscillograph. If anything, Walt Williams thought the general instrument management more flexible than in the XS-1. On the other hand, the assembly of the first D-558 had not progressed as far as he expected; its fuselage still lay in three separate pieces. The number two aircraft trailed the other slightly on the production line. Douglas officials predicted mid-December for completion of the original test model, mid-January for shipment to Muroc, and first flight about one month later. Williams assured his hosts the NACA would "undoubtedly still be at Muroc when they came out and would be interested in following the Douglas tests," to which the Douglas representatives expressed an eagerness to join forces. Motivated perhaps by John Stack's original vision of the transonic program, as well as by the recent struggles with Stanley and impatience with delays, Williams seemed gratified by the spirit of cooperation in El Segundo and "left [Douglas] with the impression that D-558 was based on more sound engineering than XS-1. The whole thing seems to be on a more business like basis. We are getting a better research vehicle there even though it does not have the speed

---

potential."[24]

The good feeling vanished soon after he returned to Muroc and experienced increasing frustrations. Williams felt stymied during the break in the XS-1 project and at the same time found himself with time on his hands. As a consequence, both old and new administrative problems, while real and pressing, received more time, attention, and emotional involvement than they might have otherwise. Starting in mid October he asked Langley time and again to send two women to collate the data soon to be recorded from the XS-1 instrumentation. Known as computers, the women who dominated this highly specialized profession possessed great patience and significant mathematical skill. They extracted engineering data from traces recorded on rolls of film; plotted calibration curves; and calculated Mach number, altitude, the control derivatives, loads, and other parameters of the test aircraft. Williams ran afoul of his superiors when he insisted one of the women also perform his clerical duties. He apparently withdrew this demand because in December 1946 Roxanah Yancey and Isabell Martin left Hampton to join the Muroc team as computers.

Other problems proved less simple to solve. Perhaps in an attempt to monitor more closely the work of its distant operating unit, Langley directed the Ames Laboratory to designate someone to act as a liaison between Williams' team and Ames. Consequently, on a Friday in October Louis H. Smaus of the Ames Instrument Development Section drove the few hours from Northern California and appeared unannounced at Muroc. Williams gave him a cold welcome. "I don't see what purpose he can serve. We have a telephone and an airplane." Smaus returned to Sunnyvale almost immediately, but not before Williams and his staff relieved him of the government station wagon in which he arrived. But this did not end the attempt at fraternal West Coast relations. Acting on instructions issued jointly from Langley and from NACA Headquarters, another Ames official offered to help Williams. He wanted to send one

[24]Walter Williams to Melvin Gough, 25 October 1946, AFFTC/HO Historical Reference Collection (both quoted passages); Walter Williams to Melvin Gough, 28 October 1946, AFFTC/HO Historical Reference Collection; Walter Williams to Langley Chief of Research, 29 October 1946, AFFTC/HO Historical Reference Collection.

aeronautical engineer immediately and volunteered the Ames personnel pool for any vacancies Muroc needed to fill in the future. Once again, Williams rebuffed the overture, saying his group had been "set up as a self-sufficient unit to handle the XS-1 project and, at present, there was no need for additional personnel. [I]t was decided that no personnel from Ames would be sent to Muroc on the XS-1 project for the present, and Ames participation in the program will probably...consist of occasional visits to Muroc." But such independence may have had a price. Floyd Thompson, Langley's Assistant Chief of Research, dispatched an able young engineer named De Elroy Beeler to Muroc to manage the XS-1 flight loads program. Beeler arrived at Muroc in January 1947, soon became Williams' chief assistant, and within a year assumed the role of Head of Engineering. He and Beeler each managed their own staffs and reported separately to Hampton. While Williams remained in charge, he no longer ran a "one man show," accountable only to his own inclinations.[25]

On the other hand, Walt Williams did bear the consequences arising from the shortcomings of Muroc housing, one of the most serious challenges to morale experienced by his staff. He struggled with all his power to ameliorate the situation, suspecting that it did not represent a passing hardship; the tests scheduled for the XS-1. the D-558, and other research aircraft implied a long-term NACA commitment to Muroc. He faced problems on several fronts.

---

[25]For a discussion of the female computers and the art of collecting instrumentation data before the age of electronic computing, see Sheryll Goecke Powers, *Women in Flight Research at NASA Dryden Research Center from 1946 to 1995*, Monographs in Aerospace History Number 6 (Washington, D.C.: National Aeronautics and Space Administration, 1997), 3, 12-13, Appendix B: Data Reduction and Instrumentation Before Digital Computers, 45-64; Walter Williams to Melvin Gough, 14 October 1946, AFFTC/HO Historical Reference Collection; Walter Williams to Melvin Gough, 15 October 1946, AFFTC/HO Historical Reference Collection (first quoted passage); Walter Williams to Melvin Gough, 25 October 1946, AFFTC/HO Historical Reference Collection; Walter Williams to Hartley Soulé, 27 November 1946, AFFTC/HO Historical Reference Collection; Walter Williams to Melvin Gough, 23 October 1946, AFFTC/HO Historical Reference Collection; Walter Williams to Langley Chief of Research, 29 November 1946, AFFTC/HO Historical Reference Collection (second quoted passage); De Beeler, interview, 14-16; Joe Weil, interview by Richard P. Hallion, NASA Historical Interview, 18 August 1977, DFRC Historical Reference Collection (third quoted passage); Hallion, *On the Frontier*, 13, 15, 262.

Kern County authorities threatened to close the abandoned Marine Base, raising anxiety among the NACA couples living there. "The apartments at Muroc Homes," wrote Williams with customary candor, "are dumps. I am going broke at Palmdale. Other fellows are not feeling too good about being away from their families but don't feel they can put up with housing conditions here." Williams wanted experienced employees, but these individuals tended to be married men who would neither tolerate long absences from their wives and children nor subject their loved ones to unfriendly conditions, such as an outbreak of food poisoning which swept through the NACA ranks during this period. If anything, the situation worsened toward the end of 1946. The Base Housing office stopped accepting applications from NACA employees, even for the apartments equipped with kerosene cookstoves and heating. Williams finally advised his friend Mel Gough that in light of Langley's apparent decision to maintain "a large group out here for a very long time, judging from the airplane[s] they are getting involved in," the NACA Committee and headquarters should express their displeasure to the AAF brass about the existing state of affairs. "I hate to keep harping on the housing situation," a frustrated Williams told Gough, "but it is the one thing that keeps the people from being happy out here." One bright spot emerged when the base announced authorization to construct 100 unfurnished housing units suitable for married couples. But because the land lay just outside the base property line, it would have to be purchased, a fact which some locals discovered and which triggered land speculation. Moreover, Williams knew by now not to believe the optimistic housing projections of the Army. Still, he allowed himself to be hopeful when the Air Materiel Command ranked Muroc at the top of its construction priorities and requested Williams' estimate of the size of the NACA presence through the middle of 1948. And he achieved a real sense of personal satisfaction upon learning that Langley approved his appeals for a secretary.[26]

---

[26]Walter Williams to Melvin Gough, 28 October 1946, AFFTC/HO Historical Reference Collection (first quoted passage); Walter Williams to Melvin Gough, 6 November 1946, AFFTC/HO Historical Reference Collection; Walter Williams to Melvin Gough, 9 November 1946, AFFTC/HO Historical Reference Collection (second and third quoted passages); Walter Williams to Melvin Gough, 15 November 1946, AFFTC/HO Historical Reference Collection; Hartley Soulé to Walter Williams, 11 February 1946, AFFTC/HO Historical Reference

# A MAN IN A ROCKET PLANE


At the end of November 1946, all considerations but the XS-1 flight test program assumed secondary importance. The dynamo Stanley appeared again at Muroc the morning of the 27th and with his coming the sparks flew once more. "[I]n a stew to get a flight [of the XS-1] since he arrived," he canceled the Friday after Thanksgiving holiday for his subordinates and planned a flight that day even though the Bell factory itself closed for the long weekend. But this time his own staff seethed with mutiny, referring to him as the "Great White Father. You would expect to find him floating face down in the lake any morning if there was water in the lake. He treated all the people up to and including Dick Frost in a manner...you would expect under the serf system. When he saw that it would not be possible to get the flight Friday he really got in a foul mood and possibly cut corners too closely." Indeed, Frost felt uneasy about the perfunctory preparations for the final engine pressure tests and said so. Still, Stanley raced ahead. Williams, at ease with the completeness of the plane's instrumentation suite, offered no objections and hoped to collect worthwhile data on the loss of stability during turns at high Mach numbers. The flights on December 2nd aboard the XS-1 number two turned out to be less than satisfactory. Remarkably, Stanley allowed the loaded B-29 to take-off with the XS-1's nose gear *unable to lock in the up position*, taking the gamble that after being dropped, the little plane would release its ballast (added to simulate the handling qualities of a fully fueled aircraft), and glide safely to the runway. But Frost proved to be a prophet. The technicians could not obtain pressure in the liquid oxygen tank which meant the fuel could not be jettisoned and the XS-1 could not be released. After landing, the ground crew struggled with the malfunction for about

---

Collection.

two hours, finally succeeding in raising the nose gear. Goodlin then flew the plane--with its fuel

tank filled with a water-alcohol mixture--but because of the time spent dumping the load and the

plane's low altitude when the weight was gone, the NACA collected only a little data.

Goodlin felt from the start that these graduated ballast glides wasted time and Stanley,

unchastened by the day's close call, decided to cancel the rest of them and attempt the first

powered flight in a few days. The next day he ordered ground tests of the Reaction Motors

engine. But when the pressure-fed powerplant was ignited, only one chamber fired due to low

nitrogen pressure at the propellant valves. On December 5 more ground tests revealed the chill

of the liquid oxygen caused the plane's hydraulic brake lines to freeze. Bob Stanley refused to be

deterred by this development and announced a powered flight on the 6th. Even though it rained

early in the day a clear sky at noon persuaded him to fuel and launch the vehicles. But once the

B-29 was airborne the cloud cover deepened, forcing a postponement until Monday the 9th.

That morning, in perfect weather, Slick Goodlin lowered himself into the XS-1 at 9,000 feet. On

the way to 27,000 feet and release, however, he noticed declines both in the bleed pressure of the

rocket engine and in the pressure in the liquid oxygen tank. Despite these danger signs, the XS-1

separated from the bomber just before noon, after which the pilot felt it drop quickly and become

somewhat tail-heavy under the full load of fuel. Ten seconds later Goodlin ignited the first

chamber, detected no noise or vibration, but felt it start to accelerate. Climbing to 35,000 feet he

fired the second chamber and brought the machine almost to Mach 0.80. Then, as he descended

without power to 15,000 feet to begin a second set of tests, the plane started to oscillate and the

fuel tank pressures started to build. Nonetheless, at the desired altitude he adhered to the flight

plan, tripped all four chambers, and found himself propelled at a tremendous rate of acceleration.

But a howling noise forced him to close down the rockets and a light indicating engine fire

prompted him to radio Dick Frost in the P-51 chase plane to verify signs of smoke. Closing on

the XS-1, Frost did detect a plume streaming from the horizontal stabilizer fairing. Although he

smelled nothing out of the ordinary, Goodlin dumped fuel and liquid oxygen and nineteen

minutes after dropping out of the B-29's belly, touched down on the Muroc runway. Subsequent investigations by Bell Aircraft and by Reaction Motors of the damaged powerplant suggested the

fire occurred as a result of two factors loose nuts on one of the engine igniters caused a fuel

leak; and the engine igniters themselves overheated after the pilot lit all four cylinders almost at

once. To avoid the resulting combustion in future firings, Bell and Reaction Motors

recommended that technicians take special care to tighten the nuts and that the pilots light the

four chambers in slower succession. But John Gardner, one of Williams' engineers, discovered a

more workable and fundamental solution to the conditions that nearly resulted in disaster.

Because the automatic igniter delay cut-off circuit evidently malfunctioned as the igniters

reached high heat, he suggested shortening the interval of time before the igniter cut-off switch

activated itself, thus preventing the igniters from overheating in the first place.[27]

Indeed, in the crucible of this intense project, the engineers, mechanics, and aviators of

the Muroc Flight Test Unit developed and perfected many flight research techniques during the

first weeks after arriving in the desert. A delegation from Ames watching the flight of an XS-1

on January 17, 1947, could not fail to be impressed by the sophistication of the process.

> Our party observed the tests from the location of the NACA radar and telemetering
> stations, which seemed to be the best location. The NACA radar and telemetering set-up
> was...quite elaborate. The...equipment consisted of about five trucks, three of which were
> radar trucks and one of which was a telemetering truck, and another apparently a power
> supply truck. At this location loudspeakers were set up to broadcast all radio
> conversations taking place in regard to the tests and we could hear the pilots of the B-29
> and the XS-1 and the chase plane, as well as the engineer directing the tests from the
> ground, and any comments of the NACA personnel stationed at the radar equipment. The
> radar itself was directed by two NACA men operating an optical direction finder. If they
> should at any time lose the airplane from view in the optical apparatus, they could

---

[27]Walter Williams to Hartley Soulé, 27 November 1946, AFFTC/HO Historical Reference
Collection; Walter Williams to Melvin Gough, 1 December 1946, AFFTC/HO Historical
Reference Collection (quoted passages); Walter Williams to Hartley Soulé, 3 December 1946,
AFFTC/HO Historical Reference Collection; Walter Williams to Melvin Gough, 8 December
1946, AFFTC/HO Historical Reference Collection; Joel Baker to Herbert Hoover, 9 December
1946, AFFTC/HO Historical Reference Collection; Rotundo, *Into the Unknown*, 143-149; John
Gardner to Langley Chief of Research, 28 March 1947, AFFTC/HO Historical Reference
Collection; Frank H. Winter, "Black Betsy: The 6000C-4 Rocket Engine, 1945-1989," Part 1
(paper presented at the 23rd Symposium on the History of Astronautics, 40th International
Astronautical Congress, Malaga, Spain, 7-13 October, 1989), 18, DFRC Library.

immediately switch the radar to automatic direction finding so that they could continue to take radar readings if this should occur. It appears that about six or seven men were needed during the test runs to operate this apparatus.[28]

This well-tried system, honed during eleven good flights after the powered inaugural, finally yielded Williams and his associates high quality data during the winter of 1946 and 1947. The success led the Army Air Forces to prepare for the conclusion of the contractor acceptance trials. The Air Materiel Command and the NACA principals met at Langley on February 6, 1947, to negotiate their respective roles after the transfer from Bell to the NACA, culminating in an agreement which expanded the research opportunities of Williams' team. Once the NACA took possession of one of the two XS-1s, it agreed to furnish the flight crew, the fuel, and the maintenance for the research aircraft. The AAF, in turn, pledged to supply the same for the B-29 and to support the D-558 flight research program with the necessary base infrastructure. Air Materiel Command then invited the NACA to present a list of the housing, office space, and equipment required to conduct the two high speed programs. A week later Colonels G.F. Smith and Donald Putt visited Muroc to solidify the new relationship and to plan for the phase-out of Bell. Meantime, a group of reinforcements prepared to embark from Langley to augment the existing Muroc workforce with a full maintenance complement for the NACA's XS-1, consisting of a project engineer, a foreman, a crew chief, a mechanic, an electrician, a nitrogen evaporator operator, and an instrument technician.

Yet, a lingering problem still remained: what constituted completion of Bell's contractual obligations? The accumulated bad feeling between the company and the NACA manifested itself again when this question was raised. Early in 1947, Hartley Soulé (who replaced Mel Gough as Williams' boss a few months later) listed the conditions under which the NACA would accept the XS-1s. He asked for a total of 20 powered flights to prove the machine's mechanical elements, its control and stability, its structural integrity, and the efficacy of contractor modifications designed to eliminate any deficiencies. The NACA flight research pilot Joel

---

[28]Lawrence A. Clousing to Engineer in Charge, 29 January 1947, AFFTC/HO Historical Reference Collection.

Baker, who had observed 12 XS-1 flights with and fourteen without rocket power, identified a number of these "relatively minor" corrections. Slick Goodlin and several NACA engineers and mechanics also discovered some problems worth solving. These points surfaced at another Army-NACA meeting, this one on March 5, 1947, at Wright Field. Bob Stanley also attended the conference. The main complaints involved poor placement of the pilot's instruments and controls, failure to label the cockpit devices fully or at all, a non-adjustable rudder pedal designed for the tall Jack Woolams, wheel brakes which required too much pre-flight attention, and the need for a removable panel on the left forward portion of the windshield to combat fogging or frosting on approach. Impatient as always, Stanley pressed the question of whether his firm had or had not met its contractual obligations. Soulé admitted it had, but Robert Gilruth dodged, acting "as timorous as an old maid [who] didn't want to say yes and didn't want to say no...," according to Stanley. The consensus of those assembled, which included Mel Gough and Walt Williams, found that Bell had indeed delivered as promised and should be released pending the 20 flights. Meantime, the NACA would dispatch its newly formed maintenance crew to Buffalo to be trained in servicing the XS-1 and Bell would send a senior representative (like Dick Frost) to Muroc to act as an advisor during the NACA flights of the XS-1. The contractor also agreed to consider some of the modifications proposed by the NACA. In a surprise development, during the meeting Stanley advocated a two-pronged approach to further XS-1 testing: Bell would operate an accelerated flight test program while the NACA concurrently conducted a more data-oriented series of experiments (which, in a confidential memorandum, Stanley referred to as "slow and tedious and fruitless"). "This suggestion," wrote Stanley, "was not well received by the NACA" and the Army politely declined with the comment, "We don't have the funds." [29]

---

[29]John Crowley to Commanding General of the Army Air Forces, 19 February 1947, AFFTC/HO Historical Reference Collection; Walter Williams to Hartley Soulé, 14 February 1947, AFFTC/HO Historical Reference Collection; Hartley Soulé to Langley Chief of Research, 24 February 1947, AFFTC/HO Historical Reference Collection; Joel Baker to Langley Chief of Research, 25 February 1946, AFFTC/HO Historical Reference Collection; Henry Reid to Air Materiel Command Liaison Officer, 21 March 1947, AFFTC/HO Historical Reference

But Stanley's basic idea took root and despite the muted reaction at Wright Field, it appeared he would win the follow-on contract for Bell. Early in April Lawrence Bell, George Lewis, and General Laurence Craigie (Chief of Research and Engineering at Headquarters Army Air Forces) conferred and designated Bell to overcome the sound barrier. Walt Williams took this news hard and saw ahead a nightmare in which the combats of the last months would be extended into the foreseeable future. Stanley talked of a short series of tests leading to Mach 1 but Williams suspected (and feared) the XS-1 manufacturer might try to pad its test program to last as long as 60 weeks and include up to 60 flights. If Stanley got away with this rumored objective, what mission did it leave for the NACA Muroc Test Unit? More important, Williams wanted to know "who had primary control of the program." The AAF referred to the Bell flights as part of the NACA investigation, but "[d]oes this mean that NACA will be able to hold [postpone] flights in order to have all instrumentation working?" Just as the Army readied itself to offer Bell Aircraft a contract, events took a sharp about-face. When the service offered the Buffalo firm a fixed-price contract to stay on the project, Bell withdrew from the negotiations, arguing such a "highly experimental" project should be better rewarded. Apparently, faced with a severe post-war contraction of funds, Air Materiel Command only had so much to allocate and refused to sweeten the offer. By April 1947 the Materiel Command apparently decided to assign its own Flight Test Division the mission of flying the XS-1 past the sound barrier. By May 1 Bell excused itself from further consideration.[30]

The Air Materiel Command Flight Test Division quickly received instructions about its

---

Collection (first quoted passage); Robert Stanley to Messrs. Bell, Whitman, Strickler, Elggren, 6 March 1947, AFFTC/HO Historical Reference Collection (second third, and fourth quoted passages); S.R. Brentnall to John Crowley, 10 March 1947, AFFTC/HO Historical Reference Collection.
[30]Walter Williams to Hartley Soulé, 9 April 1947, AFFTC/HO Historical Reference Collection (first and second quoted passages); Floyd Thompson to Memorandum for Files, 11 April 1947, AFFTC/HO Historical Reference Collection; unknown to Commanding General of the Army Air Forces, 1 May 1947, AFFTC/HO Historical Reference Collection (third quoted passage); Memorandum, "XS-1," author unknown, 20 April 1947, AFFTC/HO Historical Reference Collection; Hartley Soulé to Walter Williams, 2 May 1947, AFFTC/HO Historical Reference Collection (fourth and fifth quoted passages).

role in the XS-1 project. Much like Bob Stanley, the AAF leadership wanted a flight program which led to Mach 1 "in the shortest possible time" with the "minimum instrumentation...required to adequately measure the speeds and altitudes obtained during the tests." It called for about five glide and powered familiarization flights at speeds up to Mach 0.80. Then in a series of flights the aircraft would be flown to altitudes as high as 100,000 feet, achieving the highest speeds during climbs. The climax of these tests would occur when the pilot, attaining an altitude of 70,000 feet, attempted to reach a speed of about 800 miles per hour. On other occasions, at 60,000, 50,000, 40,000, and 30,000 feet the XS-1 would be leveled off and accelerated "to the highest practical speed." Despite the service's emphasis on this accelerated program, the Army recognized the importance of the NACA's complementary transonic research and promised that "all work would be done in full cooperation with the Committee's organizations at Langley Field and Muroc." Moreover, Materiel Command pledged to instruct its flight test team to "work directly with NACA personnel at Muroc." A first sign of cooperation occurred when Colonel George Smith, Chief of the Materiel Command's Aircraft Projects Section, offered to make available to the NACA both XS-1 number 2 and (in the intervening period between Bell's completion of the acceptance tests and the time when the AAF began its accelerated flight program) also the XS-1 number 1. Originally, the NACA had asked for XS-1 number one. But the military brass satisfied Williams' more recent desire for the number two aircraft, more useful to the NACA because it had experienced most of the flight tests to date. Moreover, its 10 percent wing--in contrast to the number one's eight percent--offered better handling at low speeds. In addition, although both aircraft had been instrumented by May 1947, only the XS-1 number 2 was outfitted with a more comprehensive suite which included sensor capability. In recognition of the Army Air Force's flexibility, Hartley Soulé acceded to Colonel Smith's request to delay the modifications requested of Bell until the command could better afford them and until the NACA began its flights above Mach 0.80.[31]

---

[31]See the written comments on a draft version of this chapter by Ed Saltzman, DFRC Historical Reference Collection; Memorandum, "XS-1," author unknown, 20 April 1947, AFFTC/HO

During June 1947 a series of conferences between the AAF and the NACA clarified their working roles. As Williams congratulated himself on this blossoming relationship between his colleagues and the Army Air Forces personnel, still more disputes broke out with Bell during the contractor's final weeks in the program. He accused the company of inattentive work habits after one of the XS-1s was damaged and dire consequences almost ensued. Bell's technicians removed the bleed pressure from the system after loading the oxygen tanks but failed to make sure the propellant valves had been tightened completely. As a result, alcohol seeped into the liquid oxygen head and an explosion occurred during ignition of the cylinder. Williams fumed at the carelessness. "It was all a matter," he said, "of having a good procedure which had worked successfully and then they get in a hurry and throw procedure aside with what could have been disastrous results." Indeed, Williams so mistrusted Bell's apparent tendency toward haste in fulfilling its contractual obligations that he sent Donald Borchers, one of his mechanics, back to Buffalo to observe the repairs on the aircraft. "I felt," he explained to Mel Gough with typical candor, "we should have a man there full-time because there is so much that can be covered over with a can of paint. It is just the fact that I am afraid of their expediting which has always gotten them into hot water." In contrast, relations with the Army seemed workable and almost routine. First, the service held its own conference on June 25 to clarify its objectives in the transonic program and to identify the wherewithal to achieve them.. Among the Flight Test Division attendees, Colonel Albert Boyd, Chief of the Flight Test Division, introduced a 24 year old Army Air Forces Captain whom he had selected to fly the XS-1 past Mach 1. Although young, he possessed a notable war record. Flying for the 8th Air Force he downed one enemy aircraft in eight missions before being downed over France. He evaded capture, scaled the Pyrenees Mountains, and trekked the length of Spain to Gibraltar. There the Royal Air Force returned him

---

Historical Reference Collection (first and second quoted passages); Commanding General, Air Materiel Command to George Lewis, 6 May 1947, AFFTC/HO Historical Reference Collection (third and fourth quoted passages); George Smith to Hartley Soulé, 8 May 1947, AFFTC/HO Historical Reference Collection; "Notes on the XS-1," author unknown, 16 May 1947, AFFTC/HO Historical Reference Collection.

to England. He rejoined his squadron and flew 56 more missions, shooting down 12 more aircraft and earning a double ace, two Silver Stars, three Distinguished Flying Crosses, a Bronze Star, and a Purple Heart. It surprised no one that the tough West Virginian went by Chuck, rather than Charles E. Yeager.[32]

Captain Yeager and most of the Army conferees met again on June 30th and July 1 at Wright Field with NACA representatives including Clotaire Wood from Headquarters, Hartley Soulé and pilot Herb Hoover from Langley, and Walt Williams. Both sides seemed eager to end the many months of bickering that afflicted the program since the Pinecastle flights. The Army members expressed a willingness to be guided by the NACA and to cooperate fully; the NACA contingent wanted to be as exacting as possible regarding equipment, facilities, and personnel in order to avoid conflicts in the future. With that, some rules of engagement were discussed and agreed upon: [33]

1. The NACA would offer technical supervision as needed.

2. The Air Materiel Command's Flight Test Division would control the XS-1 number one phase of the program, but promised to coordinate "all activities" with the NACA.

3. The Muroc base commander would supply all required facilities.

4. Richard Frost of Bell Aircraft would be resident at Muroc for technical assistance.

5. Air Materiel Command would be kept informed through channels of the project's progress.

6. The Flight Test Division would assume overall responsibility for the B-29.

7. Either the Flight Test Division or Muroc would be responsible for the P-80 chase plane.

---

[32]For the details of Chuck Yeager's life as told by Yeager himself, see Charles E. Yeager, *Yeager: An Autobiography* (Toronto and New York: Bantam, 1985); Flint O. Dupre, compiler, *U.S. Air Force Biographical Dictionary* (New York: Franklin Watts, 1965), s.v., "Yeager, Charles Elwood"; Walter Williams to Melvin Gough, 11 June 1947, AFFTC/HO Historical Reference Collection (first and second quoted passages); P.B. Klein to George F. Smith, 25 June 1947, AFFTC/HO Historical Reference Collection; Handwritten Notes, "Conference, 25 June 1947," author unknown, AFFTC/HO Historical Reference Collection.

[33]Herbert Hoover to George Lewis, 8 July 1947, AFFTC/HO Historical Reference Collection. Fred Dent to Hartley Soulé, n.d., AFFTC/HO Historical Reference Collection.

8. XS-1 number one would be furnished, at a minimum, with the NACA six channel telemeter equipment and direct recording equipment.

9. XS-1 number two would be equipped with full NACA instrumentation.

10. The Army Air Forces would supply oxygen, nitrogen, and alcohol for the project.

11. During the early stages the NACA agreed to maintain both XS-1s, but the Flight Test Division crews would assume an increased role as it became acquainted with the planes.

12. The B-29 would be maintained by the Flight Test Division during the XS-1 number one flights, by Muroc base operations during the NACA flights.

13. Muroc agreed to maintain the P-80.

14. Muroc enlisted men would continue to maintain the project's radar equipment.

15. The NACA would assume responsibility for the installation and maintenance of the telemetering and data recording equipment of both aircraft, with service assistance as needed.

16. A Flight Test Division crew would operate the B-29 during the AAF part of the program, the base during the NACA part.

17. The P-80 would be flown either by Muroc fliers or by XS-1 pilots.

18. Walt Williams remained the Engineer in Charge for the NACA at Muroc, Captain Jack Ridley (a graduate of the Caltech school of aeronautics) assumed a parallel role for the Flight Test Division.

19. The NACA and the service representatives agreed that Captain Yeager would make the demonstration flights on aircraft number two before the NACA received it.

Despite the atmosphere of cordiality, one unpicked bone of contention remained. Both sides hinted at it. Colonel Albert Boyd ended the proceedings by expressing the expectation that "the AAF flight test program is to be *fairly progressive and brief* [author's italics] to attain the maximum speed considered safe on each flight." Soulé spoke last for the NACA side and made it a point to discuss the NACA instruments essential to the project. When pressed, he expressed the opinion that "it is better to plan initially for all equipment, then delete it at the very end if necessary, than to leave it out and then try to put [it] in." Thus, the historic NACA predisposition for full and systematic data collection manifested itself even as the Army Air Forces declared its role in the program to be short and accelerated. Walt Williams probably saw confrontation coming when he read Soulé's description of Captain Yeager as "an enthusiastic young man," who knew a great deal about conventional aircraft but little about high speed flying. Indeed, a good deal of tension did develop between Williams and Yeager over this very question, especially at the start of their relationship. The combat pilot made no secret of his unhappiness when flights were delayed "because some instrument wouldn't work." But Williams would not yield; he "was very intent on not flying the airplane unless the data could be recorded properly." The difference between these competing styles of flight research did not escape Williams.

> We were enthusiastic, there is little question. The Air Force group--Yeager, Ridley--were very, very enthusiastic. We were just beginning to know each other, just beginning to work together. There had to be a balance between complete enthusiasm and the hard, cold facts. We knew that if this program should fail the whole research airplane program would fail, the whole aeronautical effort would be set back. So, our problem became one of maintaining the necessary balance between enthusiasm and eagerness to get the job completed with a scientific approach that would assure success of the program. That was accomplished.[34]

While Yeager and Williams at first fought over the specific applications of instrumentation in the daily decisions about scheduling and data collection, there were no arguments about the actual equipment. Using a six channel telemeter the NACA sought to obtain airspeed, altitude, elevator position, normal acceleration, stabilizer position, aileron position, and elevator stick force. In addition, the NACA team outfitted XS-1 number one with four strain gauges to capture information on air loads and vibration. But caution in pursuit of this data and in the conquest of Mach 1 seemed only sensible to Soulé and other NACA figures in light of the proven dangers of the experimental aircraft. The rocket engine already proved capable both of fire and of explosion. Altitudes of 60,000 would not sustain life should cockpit pressurization fail or the pilot be forced to abandon the plane. Compressibility forces caused radical changes in

---

[34]Fred Dent to Hartley Soulé, n.d., AFFTC/HO Historical Reference Collection (first quoted passage); Herbert Hoover to George Lewis, 8 July 1947, AFFTC/HO Historical Reference Collection (second quoted passage); John W. Russell, interview with Dr. J.D. Hunley, 7 and 11 March 1997, 16, DFRC Historical Reference Collection (third and fourth quoted passages); Walter Williams, "The X-1 Story: The Background," The NACA High-Speed Flight Station X-Press, 14 October 1957, 3, Hallion Papers, DFRC Historical Reference Collection (block quote).

aerodynamic characteristics. But the struggle to maintain vigilance in the face of Yeager's "damn the torpedoes" attitude paled in comparison to the Muroc team's battle to obtain aircraft parts and supplies. Vital tools and fittings ordered from Wright Field simply failed to materialize as the paper trail extended from Muroc to Dayton to the Sacramento Depot and back again to Muroc. Under these conditions, suggestions that the NACA crew failed to move quickly enough infuriated Williams. But at least now, when he vented his frustrations he could do so in private; the NACA team moved into the more spacious Bell offices during early August. More important, on the 6th of August the Army Air Forces completed its first glide flight of the XS-1. Thrilled with its light and easy performance, Captain Yeager called it the "best damn airplane I ever flew." A little more than three weeks later (August 29th) he completed his first powered flight in which he surprised himself by piloting the rocket ship through a 90 degree climb at Mach 0.85.[35]

The NACA team did not celebrate this long step toward Mach 1. Because Yeager exceeded the pre-arranged 0.80, no telemetering data was recorded and Williams scheduled a new test in the 0.80 to 0.85 range. Never having attended college himself, Yeager bridled at these fine points and resented direction from men with more formal learning. But Colonel Boyd, who admired the pilot's skill and determination, also admonished him to follow the flight plan. Thus, the careful pre-flight briefings, painfully tedious to Yeager and Ridley, went on as before, with Williams and De Beeler reminding the two captains of the lessons from the last flight and the objectives of the upcoming maneuvers. At this point, however, events in Washington, D.C., conspired to strengthen Williams' hand in guiding the course of XS-1 research. On September 1 the respected and familiar George Lewis, who did more than any other person to mold the character and mentality of the NACA, resigned as Director of Research due to ill-health after a 38 year association. Associate Director of the National Bureau of Standards Dr. Hugh L. Dryden

---

[35]Hartley Soulé to Langley Chief of Research, 21 July 1947, AFFTC/HO Historical Reference Collection; Walter Williams to Hartley Soulé, about July 1947, AFFTC/HO Historical Reference Collection; Walter Williams to Hartley Soulé, 15 August 1947, AFFTC/HO Historical Reference Collection; Rotundo, *Into the Unknown*, 248 (quoted passage), 250-254.

stepped into his role the following day. One of the world's leading scientists in the field of transonic flight and thus sympathetic to the Muroc mission, Dryden not only named the Flight Test Unit a permanent NACA facility (reporting still to Langley and Soulé) but visited the research oasis before the end of his first month on the job. For its part, the air power branch of the Army experienced an even greater transformation. On September 18 Congress reconstituted the Army Air Forces as the United States Air Force, an independent military service.[36]

After the initial powered flight mix-up, the telemetered data flowed in consistently and well. But the run-up to Mach 1 failed to occur without incident. A flight in early September attained altitudes of 30,000 and 35,000 feet but yielded no data from the airplane's internal instruments because the pilot neglected to throw the switch. It would have to be flown a second time. Nonetheless, the maneuvers proved to be highly instructive. At both altitudes, turns caused heavy buffeting at 2g but appeared to be accomplished with a high degree of stability. Level flight induced mild buffeting. Yeager also experienced the first nose-down trim change, yet at the flight's maximum speed of Mach 0.88 he felt a tendency for the nose to rise. At mid-month, Yeager pushed the speed envelope in powered flight #4 to between Mach 0.91 and 0.92, at which velocities "[d]efinite tuck-under tendencies are shown in the records." On October 6 Yeager and Ridley encountered Colonel Boyd at Wright Field and received a sobering lecture, designed to channel their youthful spirits. If they thought the path to Mach 1 was theirs for the taking, the senior officer warned them to think again. The recent data showed mild buffeting at one speed, severe buffeting at another; nose up at Mach 0.87, nose down at 0.90. "That aeroplane," Boyd concluded, "is liable to go in any direction, or all of them at once." Properly reminded, they faced the big flight on October 14. The day before, the NACA and Air Force participants reviewed a phenomenon of growing concern. Previous flights suggested that elevator effectiveness on the XS-1 declined between the shock wave's first appearance on the

---

[36]Rotundo, *Into the Unknown*, 255; Hallion, *On the Frontier*, 14-15; Anon., "Dr. George W. Lewis: Past Director of Aeronautical Research," *Thirty-Third Annual Report of the National Advisory Committee for Aeronautics*, (Washington, D.C.: GPO, 1950), ix.

55

wing at Mach 0.88 and its rearward progression approaching Mach 0.94 indicated airspeed.

Fortunately, the XS-1 design staff at Langley had insisted on an adjustable horizontal stabilizer

for just such an eventuality, allowing Williams to proceed with the Mach 1 flight with

confidence that this NACA innovation would compensate for the brief lapse in elevator control.

Ultimately, the realization of this new speed regime proved surprisingly attainable. In an

otherwise uneventful flight, Yeager crossed Mach 0.94 at 42,000 feet, noticed diminished

elevator effectiveness, but found the stabilizer compensated for the loss. At Mach 0.96 elevator

control returned. As he rose to Mach 0.98 a sudden surge of acceleration occurred, and as the

shock wave passed over the aircraft the Machmeter needle froze, then disappeared from view. A

three line cable from Muroc Base Commander Colonel Signa Gilkey to Colonel George Smith in

Dayton told the results. "XS-1 BROKE MACH NO ONE AT 42,000 FT ALT P[ERIO]D FLT

CONDITIONS IMPROVED WITH INCREASE OF AIRSPEED P[ERIO]D DATA BEING

REDUCED AND WILL BE FORWARDED WHEN COMPLETED P[ERIO]D END."[37]


## A DISCIPLINE TRANSFORMED


The pursuit of Mach 1 positioned the NACA to share in one of the great technical

achievements in aviation history. Of course, the NACA owed a large debt to the unquestioned

courage and piloting acumen of Chuck Yeager, and to the engineering contributions of Jack

---

[37]Walter Williams to Hartley Soulé, 10 September 1947, AFFTC/HO Historical Reference
Collection; Joseph Vensel to Cleveland Chief of Research, 15 September 1947, AFFTC/HO
Historical Reference Collection (first quoted passage); Rotundo, *Into the Unknown*, 268 (second
quoted passage), 274-279; See the written comments (relating to the XS-1's movable horizontal
stabilizer) on a draft of this chapter by Ed Saltzman, DFRC Historical Reference Collection; S.A.
Gilkey, telegram to George F. Smith, 14 October 1947, AFFTC/HO Historical Reference
Collection (third quoted passage).

Ridley. As a result of the collaboration with the military service, the NACA succeeded in collecting data about the transonic and supersonic flight regimes that proved absolutely essential to the future design of vehicles traveling in those ranges. Moreover, the movable horizontal stabilizer suggested by the NACA to obviate loss of elevator control at transonic speeds remains a major aeronautical innovation. Yet, the surmounting of the speed of sound represented far more than a technological triumph. As a result of the NACA's participation, Walt Williams and his colleagues established the ground rules of modern flight research. The tools, the techniques, and the personnel all underwent a transformation in order to cope with the immense technical difficulties encountered. Yet, the Muroc Flight Test Unit not only discovered new approaches to flight research; its men and women worked and lived in an unfamiliar environment which imposed some hardships, but also induced group cohesion and camaraderie. The demands of transonic and supersonic experimentation required a complete redefinition of what constituted adequate physical conditions for flight research. As a consequence, during the year-long collaboration with Bell Aircraft and with the Air Force, Langley and the NACA Headquarters in Washington conceded that in its new embodiment flight testing required not merely a division in a multidisciplinary laboratory, but a home of its own. Yet, for all the astounding changes wrought in such a short time, the research undertaken at Muroc from October 1946 to October of the following year still remained squarely in the traditions evolved at Langley since the end of World War I. The insistence on carefully designed and graduated experiments; on the full, safe, and precise collection of data; and on close collaboration between engineers, pilots, technicians, and mechanics continued to characterize the NACA approach to flight research. Although a continent distant from "mother Langley," Muroc perpetuated the style of flight research developed in Hampton, despite all of the surprises offered by the desert.

# CHAPTER 5

## A Leap Out of Water:
## The Research Airplane Program

### BENEFICIARIES OF SUCCESS

Despite his extraordinary tenacity, Walt Williams finally conceded defeat. Try as he might to adhere strictly to the NACA traditions of flight research during the early phases of the XS-1 program, he saw a portion of these time-honored practices transfigured in the wake of Chuck Yeager's success. Typically, Langley flight test programs received little or no public notice, focused on a set of conservative experimental objectives, and operated with ingenious frugality. The Research Airplane Program swept away each of these conventions except cost-consciousness. Attempts by the U.S. Air Force to disguise or deny the conquest of Mach 1 only intensified press and public speculation about the rocket planes thundering above the California desert and rendered concealment impossible. Along with anonymity, modest research expectations also disappeared. The surprising ease with which *Glamorous Glennis* finally crossed the supersonic threshold emboldened many at Muroc, at Langley, and at NACA Headquarters to envision new experiments and new vehicles capable of transforming both military and civil aeronautics. Finally, Williams and his staff found after September 27, 1947, that his desperate early appeals for housing and facilities now received due attention and funding.

Indeed, during 1948 the Muroc Flight Test Unit assumed a number of the characteristics associated with well-rooted bureaucracies. Its staff grew from 27 to 60 and with it, a fully

1

realized organizational structure, devised at Langley and imposed by Henry Reid himself, went into effect. Although clearly the man in charge, Walt Williams shared control of daily operations with three others. De Elroy Beeler, formerly a Langley loads engineer, became Head of Engineering, responsible for six project offices (the XS-1-1, the XS-1-2, the D-558-I-2, the D-558-I-3, the D-558-II-2, and the XS-4), each directed by an aeronautical engineer. Beeler also supervised a group of women known as computers. Their specialized function involved reducing to plotted or numerical form the raw flight data recorded on film. On the other hand, Head of Operations Joseph Vensel, a one-time Langley test pilot more recently employed at the Lewis Laboratory, assumed the management of pilots Herbert Hoover (from Langley) and Howard Lilly (from Lewis), four crew chiefs, eight mechanics, and the maintenance staff. Finally, Gerald Truszynski assumed the position of Chief of Instrumentation, overseeing the work of technicians involved in internal instruments, telemetry, radar, and calibration. Williams retained overall authority under Hartley Soulé's oversight.[1]

But if the NACA's desert oasis progressed toward normality on paper, in reality it remained austere. Earlier promises by the Air Materiel Command to rectify the stark living and working conditions proved inadequate. With Muroc's new stature, however, Williams and those who worked for him no longer found themselves voices in the wilderness. Now reports of the situation not only reached the desk of Soulé, but of Henry Reid as well. Edmund Buckley, Langley's Chief of the Instrument Research Division who had recently returned from Muroc, sent Soulé a blistering report. Buckley's observations--often strident and probably exaggerated--nevertheless explained the essence of Williams' dilemma: how to recruit and retain the most able people to participate in programs of great technical and, indeed, national importance when their workplace provided few personal comforts. He painted a Dickensian portrait of a workforce

---

[1]Written comments on a draft of this chapter by Gerald Truszynski, 27 January 1999, DFRC Historical Reference Collection; Chronology, Muroc Flight Test Unit, Spring 1946 to January 1954, 1, DFRC Historical Reference Collection; Henry Reid to the NACA (Hugh L. Dryden), 26 February 1948, with Organization Chart, 1 February to 1 July 1948, DFRC Historical Reference Collection; Hallion, *On the Frontier*, 13.

worn down by overtime, lacking recreational opportunities, and lodged in Spartan circumstances. In quarters, Buckley described cell-sized rooms outfitted with community toilets. Barracks D, a step down, featured big, unheated communal bays, no common areas, and open lavatories. In Barracks A and B Buckley witnessed some high-spirited partygoing during the small hours of the morning. At the bottom of the housing chain, unfinished prefabricated buildings containing no furniture and no toilets were occupied by those who took the housing shortage into their own hands. Yet, Buckley reserved his most critical comments for the dining facilities. Quite unfairly, he called the Officers' Mess inferior to the Langley cafeteria and the Post Exchange (PX) chow line less clean than many pool halls in Hampton. The soldier's mess on the north base, where many NACA employees drove for their meals, astonished the Langley engineer. "When I was there," he told Soulé, "the concrete floor had recently been hosed and although covered with water was not dirty except as the desert dust was tracked in. Here on a greasy metal tray, without dishes but with sticky and rusty utensils, was deposited some sort of undetermined greasy mess in two or three colors but of remarkably similar taste. Somehow the...European [Displaced Persons] Camps came to mind."[2]

Soulé routed this acid correspondence to Henry Reid, a man of long administrative experience and recognized discernment. Reid at first reacted with disbelief to the assertions, considering them "perhaps facetious and overstated...." But after a personal tour of Muroc, the Engineer in Charge--perhaps by then somewhat predisposed by Buckley's harsh portrait--agreed with the substance of his comments. Reid felt compelled to inform Headquarters that Buckley's assessment (which he enclosed) "has not painted too bleak a picture of the situation...." During his short stay on base Reid slept in Barracks A, ate lunch in the PX, and toured the "fire trap...best...described as a barn" where pilot Herb Hoover lived. Reid conferred with base commander Colonel Signa A. Gilkey about the situation. The Langley leader tried to be

---

[2]Edmund C. Buckley to Hartley Soulé, 22 January 1948, adjacent to File number 001446, Headquarters NASA Historical Reference Collection.

conciliatory, emphasizing the importance of cooperation between the USAF and the NACA in the successes already achieved, and about to be achieved, in the XS-1 program. But to maintain this level of efficiency, Reid insisted on better living conditions for the Langley contingent and asked Gilkey how and when he intended to make improvements. Gilkey really had no answer, only offering a long harangue about the dangers to Air Force morale if the NACA built housing superior to that of the military. Before leaving Muroc Reid examined a large, well-constructed, and empty new structure (Building T-83) which Walt Williams recently requested from the Air Force to alleviate the housing pinch. Although eight miles from the main base, Reid nonetheless saw its potential for conversion into excellent NACA quarters. He warned NACA Headquarters that if Williams failed to win Building T-83 from Gilkey, Langley would not accept the decision quietly.

> It is definitely desirable...that we be permitted to make the living conditions of our personnel as satisfactory as possible. Unless we can show our employees at Muroc that something is being done for their personal comfort and they have a brighter future to look forward to, we can expect to have operations at a very low efficiency and a damaging turnover. Crowded conditions, inconvenience, and even dirty and unsanitary living conditions can be put up with as a temporary measure for a short period of time, but this work has already assumed a permanent status and our people are right in expecting better conditions under which to live and work.[3]

Simultaneous to the pursuit of this vital objective, Williams and his co-workers capitalized on the recent technical achievements of the Muroc Flight Test Unit to win expanded facilities for the NACA team. In addition to the NACA's existing East Main Hangar, he requested space from the base commander in the East Butler Hangar for offices, shops, and for a sealed room in which to calibrate instruments without the contamination of the desert dust. Ames Director Smith J. DeFrance--perhaps thinking that as it grew in stature the Muroc facility might be drawn into his laboratory's orbit--offered to free some of his model makers and

---

[3]Henry Reid to the NACA, 18 March 1948, Hallion Collection, DFRC Historical Reference Collection (first quoted passage and block quote); Henry Reid to Memorandum for the Files, 18 March 1948, adjacent to File number 001446, Headquarters NASA Historical Reference Collection (third and fourth quoted passages).

carpenters to construct the desired modifications. Colonel Gilkey rejected the NACA's incursion into another building, but permitted the Ames craftsmen to widen the sides of the NACA hangar to add a total of 6,400 square feet of aircraft maintenance bays, instrumentation "clean" rooms, offices for the increasing number of female "computers," and lavatories. Gus Crowley at Headquarters sent funds to pay for materials. By November 1948 Williams and his staff not only occupied these new surroundings, but by spring of the following year took possession of the converted dormitory (Building T-83) coveted by Henry Reid.[4]

While the conditions of the NACA employees improved, some of the familiar characteristics of Muroc Air Force Base as a whole underwent a transformation. A tragic crash precipitated one of the changes. Early in June 1948, Captain Glen Edwards, a 30 year old Air Force test pilot of great promise, lost control of a YB-49 Flying Wing at 40,000 feet over Muroc. He and four others perished after the large experimental jet stalled and then disintegrated as it plunged into the sand. Once the USAF declared its intention to rename Muroc in his honor, the NACA also felt obliged to redesignate its desert outpost. Accordingly, on November 14th, 1949, Williams and his staff , now numbering about 100, became known as the NACA High-Speed Flight Research Station (HSFRS). This announcement not only recognized the facility's mission and implied its permanence, but also suggested a distinctness from the military reservation surrounding it, which was itself re-named Edwards Air Force Base on December 8 of the same year. But for good or ill, as a tenant organization the NACA operation never escaped the impact of the base authorities. Once the fledgling Air Research and Development Command absorbed Edwards (as well as many other engineering and science installations) from Air Materiel Command, ambitious plans took effect. The new Air Force Flight Test Center assumed control

---

[4]Walter Williams to the Commanding Officer, Muroc Air Force Base, 9 March 1948, adjacent to File number 001446, Headquarters NASA Historical Reference Collection; Edward Betts to Smith DeFrance, 30 April 1948, Hallion Collection, DFRC Historical Reference Collection; Smith DeFrance to John W. Crowley, 5 May 1948, Hallion Collection, DFRC Historical Reference Collection; Smith DeFrance to Colonel S.A. Gilkey, 15 June 1948, Hallion Collection, DFRC Historical Reference Collection; Powers, *Women in Flight Research at NASA Dryden Flight Research Center*, 8; Walter Williams, interview by Richard P. Hallion, 13 June 1977, 13.

of all experimental flying activities in June 1951. A $120 million Master Plan won Air Force

approval at the start of 1952 and unleashed a metamorphosis at Edwards, eliminating its transient

World War II character and creating a permanent infrastructure. The appropriation paid for the

removal of the Atcheson, Topeka, and Santa Fe railroad running through the northern portion of

the Rogers lakebed; bought out mud mines (for silt) situated along the railway right-of-way;

provided for the relocation and reconstruction of the entire Main Base two miles west of the

original site on the Western shore of Rogers; furnished the wherewithal to acquire Rosamond

Dry Lake further to the West; financed the building of a 15,000 foot runway, as well as the

expansion of the Rocket Engine Test Facility on the Eastern side of the lake; and supplied

capital for new housing, schools, and a shopping center.[5]

The money which poured into Edwards improvements reflected much more than a desire

by a military service to improve one of its bases. When the December 22, 1947, extra edition of

the *Los Angeles Times* roared the two-tiered headline "U.S. Mystery Plane Tops Speed of

Sound," the Antelope Valley became a recognized crossroads in the Cold War landscape. Unlike

the more routine flight test projects, the vehicles built under the aegis of the Research Airplane

Program Committee represented the leading edge of national defense, as well as the leading edge

of aeronautical research. First convened in May 1944 at Langley and comprised of NACA,

Navy, and Army Air Forces members, the committee began by brokering a compromise between

factions desiring a supersonic rocket plane (the AAF) and others seeking a transonic research

vehicle (the NACA and the Navy). Langley's John Stack--as legitimate a claimant as anyone to

the title of father of the Research Airplane Program--assembled the committee with the narrow

intention of augmenting his research on high speed aerodynamics, stalled at the time by the

failure of the existing generation of wind tunnels to provide reliable data in the transonic range.

---

[5]Handwritten transcript of Charles V. Eppley and N.A. Frank, "History of the Air Force Flight Test Center, 1 July to 31 December 1965," vol. 1, 17-18, 22-23, Hallion Collection, DFRC Historical Reference Collection; "Dryden Historical Milestones," *NASA Facts*, Dryden Flight Research Center Public Affairs Office, n.d., 1; Flint O. DuPre, *U.S. Air Force Biographical Dictionary*, "Edwards, Glen Walter."

If the tunnels could not prevail, his wartime experiences with airplane dive tests convinced Stack that properly instrumented, piloted aircraft could serve as flying laboratories capable of solving the supersonic conundrum. Still, neither Stack nor anyone else involved in the initial meeting in Hampton could have envisioned the long-term influence of a committee assembled solely to sort out the parallel research roles of the XS-1 and the D-558. But Cold War necessity, as well as the internal dynamic of technological discovery, transformed the Research Airplane Program from a project of limited objectives and duration into a long-term American inquiry into the science of high performance aeronautics. (See chapter 4 for a related discussion of the origins of the XS-1 and D-558 programs).

Leaders of the NACA recognized the enduring importance of high speed flight as early as September 1948 when Associate Director of Aeronautical Research Gus Crowley named Hartley Soulé Chairman of the Interlaboratory Research Airplane Projects Panel "in recognition of the increasing complexity and difficulty of coordination in all stages" of the XS-1 and D-558 aircraft. Unlike the broader representation present in Stack's Research Airplane Program Committee, the Research Airplane Projects Panel only included NACA personnel. The reporting chain of the panel was unambiguous; Williams reported to Soulé, and Soulé not only answered to Crowley, but sat on the staff of the NACA Director. The regular attendance by Hugh Dryden at the Research Airplane Projects meetings further underscored the pivotal role ascribed to high speed flight research at the headquarters. The group met annually and its number consisted of one member from each laboratory, from Headquarters, and from Muroc. Although Williams always attempted to set the agenda, he did not escape the frank opinions of his colleagues as he presented his programs. Perhaps the greatest value lay in the network of scientific and engineering experience it opened for Williams and his staff. Each laboratory designated project engineers whose knowledge related to an aspect of the supersonic program. Hartley Soulé could tap any of them for technical coordination and Williams often availed himself of the service.[6]

---

[6]Powers, *Women in Flight Research at NASA Dryden Flight Research Center*, 34 (*LA Times*

# FIRST OVER THE TOP: THE X-1 RESEARCH

During the years in which the XS-1 and its successors streaked over Edwards, Williams and his colleagues needed all the assistance they could find. For twelve years (1946 to 1958) flight research at Muroc contributed to aeronautical knowledge to an extent inconceivable at the end of World War II. These flights yielded unparalleled engineering data. The XS-1, Number 1 (or more simply X-1 as it came to be known and as it will be called hereafter in this narrative) flew between 1946 and 1950 and not only surpassed Mach 1, but eventually reached 957 miles per hour and an altitude of nearly 80,000 feet. It also provided valuable flight data used to validate wind tunnel calculations. Its sister ship, the X-1 Number 2, possessed a different airfoil profile (a 10-percent thickness to chord ratio for the wing versus 8-percent in the Number 1) and the NACA employed it to investigate both the transonic and supersonic regimes ranging from Mach 0.70 to 1.20. Under contract to the U.S. Air Force, Bell Aircraft also fabricated a second generation of X-1s (the A, B, and D, but no C) five feet longer and about 2,500 pounds heavier than the originals and outfitted with the 8 percent wing. The A model was flown by the USAF from 1953 to 1955 for high altitude and Mach 2 research. It set records for speed (1,650 miles per hour) and altitude (90,440 feet). Just before the launch of its second flight for the NACA on

---

headline); Meeting Minutes, NACA Committee on Aerodynamics, 24 May 1944, Hallion Collection, DFRC Historical Reference Collection; Interview, Ira Abbott by Walter Bonney, Sandurst, New Hampshire, 28 October 1971, 18, Hallion Collection, DFRC Historical Reference Collection; Walter Williams, interview by Richard P. Hallion, 13 June 1977, 45-47, Hallion Collection, DFRC Historical Reference Collection; John Crowley to All [NACA] Laboratories, 9 August 1948, Hallion Collection, DFRC Historical Reference Collection (quoted passage); Hartley Soulé to the NACA, 30 August 1948, Hallion Collection, DFRC Historical Reference Collection; Ira Abbott to Hartley Soulé, 2 September 1948, Hallion Collection, DFRC Historical Reference Collection; Hartley Soulé to Ames and Cleveland Laboratories, 8 September 1948, Hallion Collection, DFRC Historical Reference Collection; Hartley Soulé to Walter Williams, 9 September 1948, Hallion Collection, DFRC Historical Reference Collection.

August 8, 1955, the X-1A's liquid oxygen tank detonated while the aircraft was being carried by a B-29 bomber. Pilot Joe Walker found safety by climbing back into the mothership, but the vehicle was lost. The X-1B flew over Edwards from 1954 to 1958 . During its early flight program, NACA pilots Jack McKay and Neil Armstrong tested the X-1B in entirely different aspects of flight: McKay obtained considerable data on high speed aerodynamic heating and Armstrong became the first pilot to experiment, however briefly, with reaction controls. The X-1D, the first of the elongated fuselage series, suffered an early demise; just before its second flight it exploded while being carried by a B-50A bomber. One other X-1 succumbed to disaster. In 1951 the X-1 number 3 blew up on the ground after only one glide flight. Its intended role involved the testing of a steam-powered turbopump designed to transfer propellants from the tanks to the motors.[7]

While discoveries resulting from the X-1s transcended any individual aircraft or any particular flight, researchers involved in the program faced two fundamental challenges: to render the aircraft and the pilots fit to perform the desired maneuvers and return safely; and to design and execute tests yielding the widest possible range of knowledge. Although marvels of engineering in many respects, the machines demanded careful handling. Conceived and constructed by Bell under intense time pressure and fabricated with highly combustible fuel systems, they might break down or blow up unexpectedly. Similarly, regardless of their skill in subsonic vehicles, the pilots who entered this new and unpredictable flight regime could never be fully prepared. Finally, the crews repairing and maintaining these delicate and often idiosyncratic ships found themselves improvising solutions for malfunctions large and small.

One launch of X-1 number 2 illustrated what might go wrong. During a late afternoon on

---

[7] In aeronautics parlance, the chord is a straight line connecting the leading to the trailing edge of an airfoil. Louis Rotundo with J.D. Hunley, "The X-1 Research Airplane," DFRC Office of External Affairs, n.d., 11-15, DFRC Historical Reference Collection; anon., "Flight Research Vehicle Resume," n.d., Milt Thompson Collection, DFRC Historical Reference Collection; anon., "50 Years of Dryden Research Aircraft," draft, 4 April 1996, 2-4, DFRC Historical Reference Collection; anon., "Experimental Research Aircraft," n.d., Milt Thompson Collection, DFRC Historical Reference Collection.

October 21, 1947, veteran Langley aviator Herbert Hoover attempted the first NACA flight on the rocket plane. Well past the first blush of youth at 35, Hoover had flown for the NACA for seven years and held a degree in mechanical engineering. He began his glide run, designed to provide stability and control data as well as to familiarize him with the vehicle, after being dropped by the B-29 bomber at 24,000 feet. Flying westerly for six to eight minutes he flew level, executed three left turns, and at 2.8 g experienced stall oscillations, preceded by mild buffeting. Hoover found he could control the airplane laterally only for the briefest periods due to the difficulty of finding the trim setting for the ailerons. Landing the little machine proved more difficult still. Over the east end of the railway line Hoover turned the craft to align with Runway 24. At 13,000 feet he lowered the landing gear and accelerated from 200 to 250 miles per hour; at 1,000 he decided to decelerate back to 200. By this time normal cockpit distortion combined with an approach directly into the setting sun rendered his vision poor. Hoover found himself unable to see the landing strip looking straight ahead, so he tried a yawing maneuver in order to look out the side panels. Now he could see, but unfortunately could not distinguish height. For five seconds before impact, during nine seconds of repeated strikes on the ground, and through a skid of about 2,500 feet, the pilot found himself in a situation of great potential danger.

> As contact was more closely approached and a gradual flaring attempted, a porpoising flight-path resulted. This porpoising was pilot induced and resulted from overcontrolling with an elevator having low stick forces and good response. Uncertainty of height with concern over stalling too high off the ground or striking the ground with too small vertical velocity complicated the picture. Several ground contacts were made, the last of which was very closely followed by collapse of the nose wheel. Following this, the airplane skidded fairly smoothly to a stop 1/2 mile, more or less, from the final contact point. On each contact, an effort was made to hold the airplane on the ground by use of down elevator. Each contact was made main gear first and at no time did the nose gear appear to be in contact except for the final one and then only after the main gear struck. The contacts did not seem excessively rough or out of the range of normally acceptable impacts.[8]

------

[8]Walter C. Williams and Hubert M. Drake, "The Research Airplane: Past, Present and Future," *Aeronautical Engineering Review* (January 1958): 39-40; Hansen, *Engineer in Charge*, 303, 418;

Understandably, Hoover sought to minimize the seriousness of the incident, although other pilots also collapsed the X-1's nose gear. He estimated only two weeks to correct the damage to the landing strut, not counting delays in acquiring parts. In fact, despite the presence of an able pilot, a ready vehicle, and a select flight crew, the X-1 number two did not fly again for seven weeks due to repairs and to uncooperative weather. While unavoidable, vagaries such as these beset the program's operations and added to the difficulty of the mission. Everyone involved shared a common sense of the uncertainties and risks associated with placing men and machines in this mysterious flight environment. It fell to Walt Williams and the Muroc team to transform this shared realization into purposeful activity. He did so by imposing a simple but rigorous standard on his staff: "He expected people to do the job they were there for," recalled an admiring research pilot, "and if they did it was really a great relationship." Conversely, employees who failed this test found themselves at the short end of Williams' patience.[9]

Williams may have lost the war to preserve the traditional atmosphere of research nurtured at mother Langley, but he won the battle to prevent ever increasing rates of speed from becoming the obsession of his research staff. Indeed, during his watch the pursuit of these records assumed an important, but not a predominant role. Although the USAF received the cooperation of the High-Speed Flight Research Station in obtaining supersonic data crucial to the design of military aircraft, the NACA participated in the Research Airplane Program not simply to serve defense needs, but to arrive at an understanding of the fundamental forces affecting aircraft flying through and over the speed of sound. To foster reliable travel in this regime the NACA conducted tests over a continuum ranging from subsonic through the highest Mach

---

Herbert Hoover, Flight Notes on the XS-1 number 2, 21 October 1947, DFRC Historical Reference Collection; Herbert Hoover to the Chief of Research, 22 October 1947, AFFTC/HO Historical Reference Collection (block quote).

[9] Herbert Hoover to Chief of Research, 22 October 1947, AFFTC/HO Historical Reference Collection; Hansen, *Engineer in Charge*, 303; Charles L. Hall, "Future Program," in *Air Force Supersonic Research Airplane XS-1*, Report Number 1, 9 January 1948, 45, DFRC Historical Reference Collection; John Griffith, (telephone) interview with Michael Gorn, 26 May 1998, transcript in DFRC Historical Reference Collection (quoted passage).

numbers safely attainable. Walt Williams' engineers and pilots concerned themselves primarily with four categories of X-1 research: overall loads and buffeting; drag measurements using the eight percent and the 10-percent-thick wings; stability and control characteristics; and pressure distribution. By the late 1940s they had much to report.

One of the most important investigations involved *buffeting*, a condition common to the X-1 and a good one to study because the rocket plane possessed the power to fly through the entire range in which it occurred. Installing six strain gauge stations on the NACA rocket plane, the team of investigators testing the 10 percent thick wing established for the first time the relationship between speed, lift, and intensity of buffeting, finding that it occurred most severely near the point of maximum lift at Mach 0.90. Although the engineers lacked as much data for the thinner-winged and more sparsely instrumented USAF X-1, they felt confident reporting that it encountered far less buffeting than its sister ship. Moreover, using accelerometer data obtained from these tests, the researchers arrived at conclusions about the properties of *aerodynamic drag* at transonic speeds. Once again, the eight percent wing demonstrated clear advantages over the ten percent. At Mach 1.1 the Air Force X-1 flew with *60 percent* less drag than the NACA aircraft. However, at the thin wing's highest speeds and altitudes, the plane's aerodynamics suggested that re-designing the tail-fuselage-wing combination might yield an aircraft capable not only of flying very fast, but for much longer duration.

Conclusions related to *stability and control* yielded some invaluable clues about supersonic handling but they proved difficult to obtain. The thrust of the rocket engines could be varied only in increments of 1,500 pounds, rendering steady flight difficult. Moreover, the high rate of fuel consumption caused rapid changes in the weight and center-of-gravity of the aircraft. Nonetheless, researchers found in both the number one and two aircraft a similar pattern of behavior: between Mach 0.78 and 0.99 a gradual nose-down change in trim occurred, followed by a pitch-up at Mach 1, and finally, a nose-down tendency above the speed of sound. Moreover, elevator control effectiveness diminished to such a low value between Mach 0.93 and

0.99 that stabilized trim became difficult to achieve, although elevator effectiveness gradually increased above .99 and returned during deceleration below it. (Of course, the moveable horizontal stabilizer allowed control in pitch when the elevator became ineffective). Similarly, rudder efficacy all but vanished at .99. Finally, measurements of *pressure distribution* on the wings and tail during supersonic flight not only indicated the degree of loading on these members, but also the migration patterns of the centers of pressure as speed increased. For example, between Mach 0.75 and 0.85 the center of pressure on the upper surface of the wing shifted to the rear, from 25 to 41 percent chord; at 0.88, it advanced forward again to 25 percent. At this point, the pressure on the upper surface remained nearly stationary and rearward movement occurred on the lower wing surface. At Mach .95 the upper surface shock wave pushed the center of pressure back to 48 percent of chord and it continued to proceed in this direction as speeds approached Mach 1. At Mach 1.25 the center of pressure positioned itself at 51 percent chord.[10]

By the early 1950s the NACA X-1 had finished most of its research program. During 1951 it flew thirteen times, completing its pressure distribution measurements and its lift and drag work. To extend the aircraft's usefulness and further explore the relationship between thinner airfoils and reduced buffeting, members of the Interlaboratory Research Airplane Projects Panel meeting at NACA Headquarters decided in February 1952 to ask authorities at Air Materiel Command's Wright Air Development Center to sponsor replacement of the aircraft's ten percent wings with ones only four percent thick. This attempt represented the second bid to

---

[10]Hall, "Future Program," in *AF Supersonic Research Airplane XS-1*," 45; Anon., "Transonic-Supersonic Research Tools," *The Pegasus*, (August 1949): 5; Walter Williams and Hubert Drake, "The Research Airplane," in *AF Supersonic Research Airplane XS-1*, 37-40; Harold R. Goodman, "Over-all Loads and Buffeting Measurements," in *AF Supersonic Research Airplane XS-1*, 47-57; John J. Gardner, "Drag Measurements in Flight on 10-Percent and 8-Percent Thick Wing XS-1 Airplanes," in *AF Supersonic Research Airplane XS-1*, 35-43; Hubert M. Drake, "Stability and Control Characteristics," in *AF Supersonic Research Airplane XS-1*, 21-31; De E. Beeler, "Pressure-Distribution Measurements," 61-71, in *Air Force-NACA Conference on the XS-1 Flight Research. A Compilation of Papers Presented at Muroc Air Force Base*, 19 May 1948, DFRC Historical Reference Collection.

transform the X-1-2. The Interlaboratory Committee had tried the year before but the generals declined due to the high cost estimated by Bell Aircraft . In the meantime, the four percent wing underwent wind tunnel tests at Langley and seemed to offer high promise. The flight research data comparing the USAF's thin wing X-1 to the NACA's thick wing tended to confirm the experimental results. As a consequence, Hugh Dryden not only attended the 1952 session, but gave the project his personal endorsement. Soon, Air Force headquarters expressed an interest in funding it. Then Hartley Soulé received instructions from Dryden to approach Dayton again and to determine whether Lockheed might be willing to undertake the modification. These steps caused Bell to reduce its original estimate and Air Materiel Command to reconsider. Eventually, Stanley Aircraft--headed by Williams' old nemesis Robert Stanley--won the contract and Wright Air Development Center paid the bills with $900,000 appropriated from no less a source than the Secretary of Defense's emergency fund. The company predicted completion of the retrofit during 1953.

In the interim the High-Speed Flight Research Station engineers devised a comprehensive test program for the reincarnated NACA aircraft. Renamed the X-1E because of its radical differences from the X-1-2, it rolled out with the new 4 percent thickness to chord wings, a canopy, an ejection seat, as well as modified XLR-11 engines improved by a low-pressure fuel system fed by turbine pump. Due to its expanded performance profile, this transfigured vehicle looked forward to a broader research program than its predecessor. The X1-E would be tested for longitudinal, lateral, and directional stability and control from the subsonic range to Mach 2.2. Its wings and horizontal tail loads would be measured through the same speed range in level flight, in turns, and in pull-ups. Finally, an aerodynamics program would analyze buffeting boundaries, lift-to-drag ratios, aerodynamic heating, and wing aeroelasticity. Unfortunately, long delays ensued. Williams and his staff waited until January 1955 for the wings, until spring for the improved powerplant, and until the following summer for the first powered flights. Once delivered, the NACA pilots flew the X-1E from 1955 to 1958 in a demonstration program much like that of a new aircraft, consisting of four ground tests of the rocket engine; several captive flights; a number of powered launches to determine handling and stability qualities; flights to Mach 0.80 to check rocket engine reliability and the aircraft's overall structural integrity in maneuvers; and symmetrical pull-ups at supersonic speeds to evaluate the structural integrity of the thin wings. Pilots also received familiarization training during these preparatory runs. In addition, the aircraft underwent structural testing and calibration at the HSFS. Ultimately, the X-1E flew 26 times and remained in service until November 1958. It demonstrated the slender airfoil at speeds below and above Mach 1 and added important knowledge about the aerodynamic forces likely to be encountered by the coming generation of hypersonic vehicles.[11]

## THE OTHER RESEARCH VEHICLE

If the X-1 program faced the daunting tasks of penetrating and then exploring an unknown flight regime, it also had the advantage of a clear and straightforward mission. The waters may have roiled when Bob Stanley and Walt Williams collided, but they fought more

[11]Anon., "Report for Research Airplane Projects Panel of Research Activities of NACA High-Speed Flight Research Station for the Year 1951," 1, 37, Hallion Collection, DFRC Historical Reference Collection; anon., 'Meeting Minutes of Interlaboratory Research Airplane Projects Panel," 4-5 February 1952, 4-5, Hallion Collection, DFRC Historical Reference Collection; anon, "Report for Research Airplane Projects Panel of Research Activities of NACA High-Speed Flight Research Station for the Year 1952," 2-3, 51-61, Hallion Collection, DFRC Historical Reference Collection; Hartley Soulé to NACA Headquarters, 5 January 1953, "Agenda Items for Research Airplane Projects Panel meeting on January 14 and 15, 1953," 5, Hallion Collection, DFRC Historical Reference Collection; anon., "Addendum, Minutes of Meeting of Interlaboratory Research Airplane Projects Panel," 4-5 February 1954, 3, Hallion Collection, DFRC Historical Reference Collection; anon., "Report for Research Airplane Projects Panel of Research Activities of NACA High-Speed Flight Station for the Year 1954," 2, 57, Hallion Collection, DFRC Historical Reference Collection; anon., "Flight Research Center Vehicle Resume, n.d., Milt Thompson Collection, DFRC Historical Reference Collection; Wallace, *Flights of Discovery*, 53.

about timing than objectives. The D-558 program experienced a more complicated life cycle. The Douglas engineers coped with designing an experimental aircraft first with straight and later with swept wings; with jet, with rocket power, and with a combination of both; and, at the mutual instigation of the Navy and the NACA, with the capacity for combat service. Thus, Douglas Chief Designer Ed Heinemann instructed his staff to fabricate the most conventional machines possible, consistent with their exotic missions. But the complexities inherent in their performance rendered both the D-558-I Skystreak and the D-558-II Skyrocket far from commonplace. For example, the Skystreak's main landing gear rolled on special thin wheels capable of being stored in the plane's uncommonly thin wings; the forward portions of its wings were sealed to act as 230-gallon kerosene fuel tanks; its thick magnesium alloy skin fastened to aluminum-alloy frames allowed designers to dispense with the customary stiffeners, thus reducing weight, increasing internal fuselage capacity, and permitting a smooth exterior due to countersunk rivets. Unlike the X-1, the Skystreak flew off the runway on its own power rather than being air launched. The D-558-I also flew longer missions than the X-1 and actually collected more data. Still, the Douglas machine lacked the comparative performance of Bell's creation. Although bigger than the X-1, the D-558-I's 35 foot fuselage (more than four feet longer) and 12-foot high tail (four feet taller) were powered by a General Electric TG-180 turbojet which produced 4,000 pounds of thrust. In contrast, the X-1's liquid oxygen and alcohol rocket engine developed 6,000 pounds. Although there appeared to be a significant weight differential between the two machines when empty (nearly 7,711 pounds for the Skystreak versus 4,900 for the X-1), Bell aircraft required more than 5,000 pounds of fuel compared to a mere 1,400 for the turbojet airplane.

Douglas ultimately delivered three D-558-Is to the Navy Bureau of Aeronautics. The first of these vehicles with the straight, stubby, 10-percent thick wings arrived for testing at Muroc early in 1947. The NACA crew found the new aircraft a sight to behold. Scarlet colored, highly polished, with a slender fuselage and a long, elegant canopy, D-558-I-1 seemed to breathe speed and modernity. The Douglas test team, with NACA assistance on calibrating the instruments, readied it for the initial flights and project pilot Gene May first took the controls on April 15, 1947. Less than auspicious, the journey ended abruptly when a partial power loss occurred. Another such incident happened a week later. Then the landing gear refused to lock on the following six runs. By mid-July the difficulties seemed to abate; during the first week in August the red line in the sky reached Mach 0.85. Later that month the second D-558-I, destined for NACA testing, arrived at Muroc and installation began on a full NACA instrumentation package much like that in the X-1s: a 12-channel oscillograph for the strain gauges; a manometer to record pressure distribution; wheel and pedal force transmitters; aileron, elevator, and rudder position recorders; a three- component accelerometer; a four-channel telemeter to signal airspeed, altitude, acceleration, elevator, and aileron positions; an airspeed-altitude recorder; a sideslip angle transmitter; and a camera to photograph the readings on the control panel. Finally, on November 25, 1947, NACA pilot Howard Lilly, formerly of the Lewis Laboratory, made the first NACA flight aboard Skystreak number two, a familiarization run which ended with an instrumentation malfunction.

On the ground, Bureau of Aeronautics representatives established a clear delegation of authority calculated to avoid the bickerings in the X-1 program. Under contract to the Bureau, Douglas agreed to undertake the flight program of number 1 and perform major maintenance and modifications on all three aircraft. The Navy would support engine overhauls and replacement. The NACA committed itself to fly the programs for aircraft two and three, conduct routine maintenance and inspections, and procure fuels and lubricants from the USAF.[12]

---

[12]Richard P. Hallion, "The Douglas D-558-I Skystreak," draft manuscript, 20 December 1971, 10-20, Hallion Collection, DFRC Historical Reference Collection; Hallion, *On the Frontier*, 13, 300; anon., "50 Years of Dryden Research Aircraft," draft, 4 April 1996, 4-6, DFRC Historical Reference Collection; Anon., "Experimental Research Aircraft," n.d., 3, 5, Milt Thompson Collection, DFRC Historical Reference Collection; Charles V. Eppley, *The Rocket Research Aircraft Program, 1946-1962* (Edwards Air Force Base, California: Air Force Flight Test Center, 1963), 3, 6-9, Hallion Collection, DFRC Historical Reference Collection; anon., "D-558-I," *NASA Facts On Line*, Dryden Flight Research Center, 2 February 1998, NASA Dryden Web Site

The pilots who tested its handling qualities and performed the flight experiments described it as a plane easy to love but whose eccentricities commanded respect. The whole flying corps at Muroc admired its sleek appearance and found it "easy to become very comfortable on take-offs and climb outs," a "fun" aircraft with excellent response and control in the subsonic range, and one capable of attaining altitude at a then unheard of rate of 10,000 feet per minute. The Skystreak's aviators also felt a reassuring sense that its airframe could withstand whatever pressures the flight plan subjected it to. Its design limit of 18gs resulted in an aircraft remembered for its strength, "built so strong, that they were--aerodynamically...virtually rigid. And the aeroelastic effects hardly ever showed up...." But from the pilot's viewpoint at least, these positive features coexisted with some decided liabilities. One remarked that in the absence of an ejection apparatus, "when they bolted the canopy over your head you became an airplane part number." This remark also applied to the extraordinary configuration of the cockpit, designed by the Douglas engineers for minimum aerodynamic drag. Walt Williams apparently hired pilot Stanley Butchart after asking just one question: "Will you fit in the [Skystreak]?" Eager to please, the young flier replied, "Yes, sir." "Okay," said Williams. "You're on." Butchart, a World War II naval aviator and graduate of the University of Washington's Guggenheim Aeronautical School, arrived at the High-Speed Flight Research Station in 1951. The tightly-corseted interior of the D-558-I astonished him. He found it impossible to read his instruments when he sat up straight and looked out the glass, and unable to see ahead as he craned his neck downward to read the gauges. No wonder Butchart felt constricted; the Skystreak measured

> only 22 inches wide, straight down the sides. You flew it with your elbows in, and the wheel between your knees, and crunched down. Your helmet was up into a tight canopy. We had a chamois skin on our helmets to keep from scratching the inside of the plexiglass. There was a double layer--glass and then plexiglass with air in between to keep the frost off. And if you turned your head a little bit to try to see out to a chase

(http://www.dfrc.nasa.gov/PAO/PAIS/HTML/FS-036-DFRC.html), filed in DFRC Historical Reference Collection.

[plane] or wing tip, your head would get stuck, and you'd have to suck it back down to see forward again. If you ever had claustrophobia, that was the airplane to get it in.[13]

Worse still, the D-558-I-1 assumed an altogether different character above Mach .75 than the easygoing machine found at lower speeds. Suddenly, the pilots got the "feeling that it just wasn't going to go any faster." They experienced a phenomenon called "wing dropping" in which shock waves eddied across the wings and the control surfaces causing the instruments to shake and the aircraft to oscillate. It became impossible to level out despite recourse to the controls. Buffeting and vibration increased toward Mach 1 and when the flight plan called for steeper and steeper dives, the plane grew increasingly difficult to control and it shook violently. Under such adverse conditions the de-briefings of the research pilots added an important augmentation to the instrumentation data. Clearly, the unsteadiness they experienced at increasing speeds eliminated the prospect of mounting guns in a combat role and, conversely, the ingredients necessary for good handling properties at very high velocities needed to be factored into the design equation. Pilot observations like "it really didn't roll very good, or there was a terrible amount of buffeting after I deflected the control, or when I did the pull-up...there was pitch-up and it was difficult to control," while qualitative, formed a significant part of the overall evaluation of the vehicle. In the pursuit of such knowledge, Howard Lilly mounted D-558-I number 2 at noon on May 3, 1948, for its 19th flight. Problems with the landing gear door failing to lock recurred consecutively on flights four to seven and surfaced again on this date, forcing the outgoing and popular West Virginian aviator to return to the hangar for repairs. Late

---

[13]W. G. Williams, "Machbuster: A Test Pilot Recalls the Early Days of Supersonic Flying, Where You Either Broke the Sound Barrier or it Broke You!" *Wings* (February 1991), reprinted as "Testing the First Supersonic Aircraft: Memoirs of NACA Pilot Bob Champine," *NASA Facts*, Langley Research Center, January 1992, 10, DFRC Historical Reference Collection (first quoted passage); J.D. Hunley, ed., *Toward Mach 2: The Douglas D-558 Program. Featuring Comments by Stanley P. Butchart, Robert A. Champine, A. Scott Crossfield, John Griffith, Richard P. Hallion, and Edward T. Schneider* (Washington, D.C.: NASA SP-4222, 1999), 36, 38 (fourth quoted passage and block quote), 58 (second quoted passage); A. Scott Crossfield and Walt Williams, "When Flight Test Was the Only Way," in the *Twenty-second Symposium Proceedings of the Society of Experimental Test Pilots*, 1978 Report, (September 27-30, 1978): 165, DFRC Historical Reference Collection (third quoted passage).

in the afternoon he strapped in and tried to complete the day's assignment. But shortly after taking to the air a component in the engine compressor disintegrated and hurled metal shards into both the fuel and the control lines. Flying close to the ground, the five year NACA pilot lost control of the aircraft, the tail caught fire, and the machine dove toward the lakebed and exploded on impact. Langley's Mel Gough chaired an accident investigation of this first NACA research pilot fatality. Its final report urged all of the laboratories to equip their aircraft with the latest engine models (D-558-I number 2 flew with an older TG-180 powerplant) and to armor-plate propulsion parts in proximity to fuel and control conduits. Still, the death of Howard Lilly numbed Walt Williams and his co-workers, sobering everyone with the reality that the Research Airplane Program would result not only in successes, but on occasion, in the loss of lives.[14]

For nearly a year after Howard Lilly's death, the NACA D-558 flight research program quieted down. Then, in Spring 1949 it returned with renewed force. First, Douglas delivered the Skystreak number 3 to Muroc and on 22 April Bob Champine began a series of dive and pressure distribution flights, joined by former Lewis icing pilot John Griffith. Then, little more than a month later, Champine and Griffith transferred to the newly minted Skyrocket number 2 and starting on May 24 flew hazardous longitudinal stability and control, as well as stall missions. The two planes, which vied for the Antelope Valley airspace, seemed almost as different from one another as either did from the X-1. Both possessed horizontal stabilizers, but unlike the Skystreak's straight wings and vertical tail, the Skyrocket featured 35-degree swept-back wings and a 49-degree swept-back tail. The D-558-II also measured a full seven feet longer than its predecessor. Fully loaded, the heaviest Skyrocket weighed nearly twice as much (roughly 16,000 pounds) as the D-558-I at take-off. Finally, the Skystreak always flew as a turbojet while the D-558-II powerplants varied widely and changed over time. The contractor Skyrocket (number 1) began its career with a Westinghouse J-34 turbojet engine capable of 3,000 pounds of

---

[14]Williams, "Champine Memoirs," 10, DFRC Historical Reference Collection; Hallion, *On the Frontier*, 27-29, 300.

thrust. The NACA's D-558-II number 2 featured the same propulsion system until November 1950 when Douglas retrofitted it for air launch with a 6,000 pound thrust LR-8-RM-6 rocket motor, essentially the Navy version of the LR-11 used on the X-1. The final Skyrocket (number 3), outfitted with jet (J-34) and rocket (LR-8-RM-5) engines eventually flew programs combining both types of propulsion.[15]

Like the Skystreak, the Skyrocket exhibited some temperamental handling qualities. If, as one pilot remarked, one of the principal objectives the D-558-II test program involved "develop[ing] the savvy to practically resolve transonic and supersonic handling problems,' these machines certainly provided the necessary range of flying experiences. Below the speed of sound the D-558-II flew reasonably well, although not without peculiarities. During the early flights with Skyrocket number 2, Jet Assisted Take-Off (JATO) rocket canisters (early versions of which were developed during the 1940s at Caltech and elsewhere) were required to compensate for the inadequate Westinghouse powerplant. Pilots gunned the engine to achieve maximum ground speed, fired the JATOs, and found, "just enough speed to take-off and retract the landing gear." Reversing the process could be more hazardous. Robert Champine, an experienced naval aviator who transferred to Muroc after Howard Lilly's death, thought his first landing might be his last. He experienced "a terrible Dutch roll" in which the aircraft swung 15 to 20 degrees in two second intervals. Using the ailerons at the end of each oscillation seemed to worsen the problem, so he "punched it a couple of times with the ailerons" while the plane rocked back and forth. This cured the malady. "I briefed every guy who flew after me and said, 'you're not going to crash. You'll control it...in the end...right before landing. But you'll have serious doubts until that point.' We got used to it but it was never very comfortable."[16]

---

[15]Contractor flights of the D-558-II-1 occurred from 1948 to 1951, after which Douglas transferred it to the NACA. Hallion, *On the Frontier*, 300-314; "Research Airplane Characteristics Summary," 43, filed in "Flight Research Background and History," DFRC Historical Reference Collection; Richard P. Hallion, *Supersonic Flight: The Story of the Bell X-1 and the Douglas D-558* (New York and London: Macmillan, 1972), 66-77; anon., "50 Years of Dryden Research Aircraft," draft, 4 April 1996, 6, DFRC Historical Reference Collection.
[16]Crossfield and Williams, "When Flight Test Was the Only Way," 165, DFRC Historical

The Skyrocket also offered ample opportunity to evaluate the handling qualities of swept wing vehicles flying at high speed. The big surprise occurred at high altitudes and at high angles of attack. As shock waves traveled over the wings the tips stalled before the roots. When this phenomenon happened aft of the center of gravity, the aircraft pitched up. Before the HSFRS undertook a series of experiments with wing "fences," slats, and chord extensions, pilots like Robert Champine faced sudden, catastrophic encounters over Rogers lakebed.

> If you pulled up and got to 4 or 5gs, it would suddenly stall in such a manner that the lift distribution on the wing would cause it to pitch up violently. It would go to extremely high angles of attack, between 45 and 60 degrees, and then it would start to roll violently, so the aircraft became completely and totally out of control--just spinning around in the sky. Once you fell into it, you had no way of controlling it. You just had to ride it out until you eventually were falling nose down in a spin. Once you were able to unstall the wing with nose-down elevator you just used opposite rudder and it would recover in a vertical dive.[17]

Cantankerous at times to fly, the D-558s did not fulfill a pilot's every wish; but for the HSFRS engineers they held a place of high importance. The two flight test programs ran simultaneously between 1948 and 1953 and during these five years the researchers gathered and interpreted data about the fundamental character of flight below, at, and well over the speed of sound. The aircraft industry and military leaders swiftly incorporated these findings into high performance machines. Indeed, the knowledge gleaned from the NACA research helped decode the behavior of the Korean War's front-line F-86 fighter, another swept wing aircraft prone to pitch up but assisted (in later models) by the moveable horizontal stabilizer common to the X-1 and the D-558s. The so-called Century Series fighters (the F-100, 101, 102, 104, 105, and 107) also owed a tremendous debt to the aerodynamic data collected during the X-1, Skystreak, and Skyrocket trials. Finally, at the dawn of commercial jet travel the results of subsonic turbojet flight research assumed great significance to the manufacturers of the nation's airliners.

---

Reference Collection (first quoted passage); Williams, "Champine Memoirs," 10, DFRC Historical Reference Collection (second to fifth quoted passages).
[17]Hunley, *Toward Mach 2*, 24-25; Williams, "Champine Memoirs," 10, DFRC Historical Reference Collection (block quote).

But specific applications such as these reflect only the obvious by-products of research. By the early 1950s fundamental data from both programs flooded in. The Skystreak yielded important aerodynamic knowledge through speeds approaching Mach 1. In 1951, for instance, the number three aircraft flew 28 times and concentrated on buffeting phenomena. Researchers discovered no relationship between altitude and buffeting up to Mach 0.88 but did succeed in mapping other operative factors such as tail loads and wing pressures, leading to the conclusion that above the range of maximum lift rising angles of attack resulted in sharp increases in buffeting. The HSFRS engineers also measured and defined the mechanics involved in the loss of aileron effectiveness encountered between Mach 0.88 and 0.90. Finally, Skystreak number 3 went aloft 15 times in 1953 before its retirement in June of that year. Seven of these missions investigated longitudinal, lateral, and directional dynamics over a broad band of velocities. The pilots' flight plans concentrated on elevator, aileron, and rudder controls. For the most part the flights took place at 50,000 feet and steady speed, although longitudinal stability and control received additional attention at a variety of altitudes and loads. After August 1, 1953 the NACA technicians removed the aircraft's instrumentation preparatory to its transfer to the Navy.[18]

The Skyrocket research concluded two years after that of its sister program, but not before accomplishing even more far-reaching objectives than the D-558-I. Most of the structural members of the D-558-II underwent detailed loads evaluations. Pressure measurements transmitted from five span stations on the Skyrocket wing yielded the aerodynamic characteristics of airfoil sections from Mach 0.65 to 1.2. Perhaps most important of all, strain gage measurements of wing loading (up to the limiting Mach number of the aircraft in level flight and in turns) revealed span and chord centers of pressure, degree of pitching, lift, and the

---

[18]Hunley, *Toward Mach 2*, 25; "Report for Research Airplane Projects Panel of Research Activities of NACA High-Speed Flight Research Station For the Year 1951," 4-5, 22, 59, Hallion Collection, DFRC Historical Reference Collection; "Report for Research Airplane Projects Panel of Research Activities...for...1952," 3-4, 66, Hallion Collection, DFRC Historical Reference Collection; "Report for Research Airplane Projects Panel of Research Activities...for...1953," 6, Hallion Collection, DFRC Historical Reference Collection; Hallion, *On the Frontier*, 300-314.

aerodynamic center of the wing. Complementary assessments of horizontal tail loads (recorded in pull ups over the lift coefficient range and in level flight to Mach 1.6) led D-558 investigators to calculate the loads during balanced and maneuvering conditions and to determine the wing-fuselage aerodynamic center. Furthermore, by combining the wing with the horizontal tail load data, HSFRS engineers arrived at vital generalizations about the ratio of load carried by the Skyrocket's wings, fuselage, and horizontal tail; and also the role of each part in overall aircraft stability.[19]

Invaluable as such conclusions may have been, the life-threatening problem of pitch up received even more attention. Between September 1951 and summer 1953 Skyrocket number 3 delved into its mysteries. Aerodynamicists at Langley undertook wind-tunnel analyses and HSFRS engineers pored over the data from Bob Champine's hair-raising flight in August 1949. The airworthiness and air safety not just of swept-wing, but also of similarly afflicted delta-wing aircraft hung in the balance. Initially, the Langley researchers suggested placing an outboard "fence" on the wings to alleviate pitch up. The flight program consisted of piloting the Skyrocket number 3 to its maximum capabilities, collecting data relative to the points of instability, and then employing the fences in various configurations--sometimes singly, sometimes in parallel pairs, both inboard and outboard. A. Scott Crossfield, a World War II Naval aviator and aeronautical engineer who reported to the Flight Station in 1950, flew most of the missions. He collected data which tested long, narrow auxiliary wing slats by themselves and with the fences, both in fixed and in free floating positions. Following these trials, Crossfield tried a series of variously shaped leading edge chord extensions to determine whether they alleviated the problem. By 1953 research ceased on the fences and concentrated on the more promising slats, at times in full extension and other times retracted. Ultimately, the most useful of all measures proved to be locking the slats in the open position, effective over the entire speed

[19]"Report for Research Airplane Projects Panel of Research Activities...for...1951," 75-80, Hallion Collection, DFRC Historical Reference Collection.

range except that of Mach 0.83 to 0.87. The fences also showed some value in curbing pitch up. Despite encouraging wind-tunnel analyses, however, extending the chords seemed to have no beneficial effect. Despite all of these worthwhile results--particularly applicable to delta wing aircraft whose tail configurations could not be modified--in the end the Research Airplane Projects Panel members admitted that the Skyrocket's problem stemmed from its "high tail location...[which] practically prohibits curing its pitch up tendency." As a result, the committee transferred the last remaining portion of the flight schedule (chord extension) to the X-5 program. In doing so, the NACA informed aircraft manufacturers of one simple solution to pitch up, at least on swept-wing aircraft: avoid positioning the tail far above the fuselage, as the Douglas engineers had done on the D-558-II.[20]

But the Skyrocket research program did not merely wither away. To celebrate the 50th anniversary of powered flight (and to thwart an Air Force claim on the next great speed mark) Scott Crossfield, with the connivance of Walt Williams and the support of the Navy, quietly planned an attempt on Mach 2 in the D-558-II-2. The engineers and technicians extended the plane's rocket nozzles for added thrust and made careful trajectory calculations to squeeze out the last bit of speed. Then the NACA prepared to bask in a rare moment of celebrity. On November 20, 1953, Crossfield and his aircraft were released from the B-29 bomber and climbed to 72,178 feet, at which point he pushed over into level flight. His instruments revealed the ascent left him "with a full minute of [power]....So, after leveling out...I blasted along on all four rockets for a full 45 seconds, faster and faster. I suddenly heard the rockets begin to misfire and

[20]"Report for Research Airplane Projects Panel of Research Activities...for...1951," 72, Hallion Collection, DFRC Historical Reference Collection; "Report for Research Airplane Projects Panel of Research Activities...for...1952," 71, Hallion Collection, DFRC Historical Reference Collection; "Report for Research Airplane Projects Panel of Research Activities...for...1953," 8, 44, Hallion Collection, DFRC Historical Reference Collection; Hartley Soulé to NACA Headquarters, 5 January 1953, "Agenda Items for Research Airplane Projects Panel Meeting on January 14 and 15, 1953," 6-7, Hallion Collection, DFRC Historical Reference Collection (quoted passage); Hallion, *On the Frontier*, 36-37, 50-51, 300-314; Milton O. Thompson, *At the Edge of Space: The X-15 Flight Program* (Washington and London: Smithsonian Institution Press, 1992), 2; Hallion, *Supersonic Flight*, 166-173.

knew this was the end of the line. I glanced quickly at the Machometer--it read 2.05! There could be an instrument error, but it still startled me for a moment. Had we really flown more than twice the speed of sound?"[21]

## BANK AND TURN

Just as the successful X-1 program won improved physical conditions, a growing staff, and sharper organizational focus for the Muroc Flight Test Unit, the D-558 flights further defined the NACA's flight research mission. By the end of the D-558 program the essential value of flight research had been proven beyond a doubt. Indeed, its well recognized contributions prompted many in the NACA to support the HSFRS becoming the master of its own house. Accordingly, on St. Patrick's Day 1954 good fortune smiled on the High-Speed Flight Research Station with the publication of NACA General Directive Number 2, authorizing the desert facility complete separation from Langley effective July 1 of that year. Henceforth, new employees of the HSFRS traveling East for the Langley indoctrination also received instructions to "plan a few days" at NACA Headquarters in nearby Washington, D.C. To "define and clarify" the role of individuals and of the Station as a whole, Hartley Soulé suggested to Headquarters the need for a procedures manual for the new entity. Walt Williams released an Operations Manual late in May 1954, one which reflected a virtual revolution in the institutional structure which had evolved since summer 1946. Originally, each program office represented a single major research project and six or eight of them dominated the top line of the organization

---

[21]A. Scott Crossfield (as told to Don Dwiggins), "Flight Through the Heat Barrier," n.d., 6, File number 000403, Headquarters NASA Historical Reference Collection (quoted passage); Hallion, *On the Frontier*, 67-69.

chart. By mid-1954, however, four division chiefs (Research, Flight Operations, Instrumentation, and Administration) each supervised three or four functional branches. Once the airplane-based project offices fell victim to the size and complexity of the Research Airplane Program, each of the three branches of the Research Division--stability and control, loads, and performance--operated in its own, discrete organizational niche. This fractured arrangement raised obvious questions about project coherence and command, answered in the persons of Project Coordinators. The responsibility vested in these figures suggests the institutional stratification occurring even in the early years of NACA flight research.

> It is the project coordinator's responsibility to see that the airplane which is being used as a test vehicle is used to complete the program in an orderly and logical sequence. This includes coordination and scheduling of the various investigations being run, and coordination of scheduling of flight operations and instrumentation with the Operations and Instrumentation Branches. The project coordinator is responsible for and must approve all instrument or airplane modifications, changes or additions that are made on *his* [author's italics] airplane by the Instrumentation or Operations Branches. He has the responsibility of making all direct contacts with the Instrumentation and Operations Branches in order to accomplish the work necessary for various investigations being carried out on the airplane by the project engineers. All contacts and arrangements concerning programs and instrumentation on the airplane between the HSFRS and outside companies that have had prior approval of the Station Head will be the responsibility of the assigned project coordinator.[22]

The stratification manifested itself not only in the need to coordinate and assemble the necessary labor and equipment to undertake flight research projects, but in the subtle transformation of on-the-job relationships. Inadequate as the early conditions on the South Base may have been, nearly everyone worked under one roof, on one floor. But with a surge in the

---

[22]E.H. Chamberlin to Langley, 30 March 1954, "Establishment of the NACA High-Speed Flight Station as an Autonomous Station, Effective 1 July 1954," filed behind File number 001446, Headquarters NASA Historical Reference Collection; E.H. Chamberlin to Langley, 5 April 1954, filed behind File number 001446, Headquarters NASA Historical Reference Collection (first quoted passage); Hartley Soulé to NACA Headquarters, 19 April 1954, filed behind File number 001446, Headquarters NASA Historical Reference Collection; Organization Chart of the High-Speed Flight Station, July 1954, DFRC Historical Reference Collection; Walter Williams to HSFS Staff, 27 May 1954, with attached Operations Manual, Milt Thompson Collection, DFRC Historical Reference Collection (block quote).

number of aircraft awaiting flight testing, Williams' staff multiplied six fold from 1948 to 1954. Attuned to this trend, in 1951 NACA Headquarters won from the Congress an appropriation of $4 million to construct new NACA offices and laboratories on Edwards Air Force Base. Shovels first turned the sand in February 1953 on 1.5 square kilometers leased by the USAF to the NACA. When the HSFRS headquarters building opened in June 1954, office assignments reflected a differentiation among personnel functions; walls and distinctions began to appear among the workforce. Engineers and white collar employees sat at desks on the second floor, technicians and mechanics spent their days on the first floor. While the spirit of cooperation and friendliness remained, opportunities for completely free association present in the first years diminished. With a more formal organizational structure and changes in working patterns also came a new and simplified name. On July 1, 1954 the NACA High-Speed Flight Research Station dropped the word "Research" from its title. But for the High-Speed Flight Station (HSFS), research remained the essential ingredient of the mission.[23]

In fact, 1954 brought more than bureaucratic maturity to the High-Speed Flight Station. A remarkable confluence of events--including institutional independence, new facilities, and a string of technical achievements culminating in the Mach 2 Skyrocket flight-- prepared Walt Williams and his engineers to lead a program of unprecedented size and importance. At this juncture, aircraft configuration research failed to stir their imaginations; the HSFS's hangars already bulged with machines of all different shapes and types. Rather, three words loomed large: higher and faster. While Williams felt the traditional NACA flight research agenda needed to be preserved, he also recognized that the expansion of the supersonic envelope demanded intensive investigation. Indeed, the flights to date suggested no serious impediments existed to speeds and altitudes far in excess of those achieved by the X-1 and the D-558. As a consequence, *hypersonic* human flight became the touchstone of the High-Speed Flight

[23]Betty Love, interview with Michael Gorn, 10 April 1997, 12, DFRC Historical Reference Collection; "Dryden Designation Chronology," n.d., DFRC Historical Reference Collection; Hallion, *On the Frontier*, 43; Powers, *Women in Flight Research*, 17.

Station.[24]

The origins of this concept may be traced to the Second World War. During the 1940s Americans thought of very high speed flight only as the domain of missiles. German scientists Eugen Sänger and Irene Bredt, on the other hand, wrote with persuasive detail about the technical feasibility of propelling individuals over long distances at incredible velocities. Published in 1944, their paper appeared in the open scholarly literature after the end of the war. The article gave substance to the idea, which germinated for a few years. Then it began to appear in several places at once. At Edwards during 1950 and 1951, Robert Carmen and Hubert Drake pondered ways to attain speeds of Mach 3 and altitudes over 100,000 feet. They drew plans to modify the Bell X-2, an aircraft constructed of K-Monel nickel alloy and stainless steel and capable of withstanding the rigors of hypersonic speed. Their ideas were transmitted to Langley for further analysis. Meanwhile, a scientific discovery made at the Ames laboratory added momentum to the hypersonic project. H. Julian Allen found a way to mitigate the effects of extremely high temperatures encountered as missiles (and, presumably, aircraft) re-entered the earth's atmosphere. By designing blunt, rather than pointed noses for these vehicles Allen predicted a strong bow-shaped shock wave would safely deflect the high heat. Yet another voice entered the growing chorus when Robert J. Woods of Bell Aircraft, designer of the X-1, X-2, and X-5 airplanes, wrote to the NACA suggesting hypersonic flight and space travel be added to the Committee's list of research projects. But Woods' enthusiasm did not materialize out of thin air. It had been kindled by Walter Dornberger, a colleague at Bell and wartime director of Germany's Peenemünde rocket test facility. Dornberger knew Sänger and Bredt and he introduced their theories to Woods. This realization prompted Woods' correspondence with the NACA.[25]

---

[24]Walter Williams, "X-15 Concept Evolution," in *Proceedings of the X-15 First Flight 30th Anniversary Celebration* (Washington, D.C.: NASA Conference Publication 3105, 1991), 11.
[25]John V. Becker, "The X-15 in Retrospect," Third Eugen Sänger Memorial Lecture presented at the first annual meeting of the Deutsch Gesselschaft für Luft und Raumfahrt, Bonn Germany, 4-5 December 1968, 1, DFRC Historical Reference Collection; Williams, "X-15 Concept

In the face of these scattered but significant signs of interest, the NACA addressed itself officially to hypersonic flight. During its Spring 1952 meeting the Aerodynamics Committee recommended that the NACA undertake studies relating to hypersonic flight. That June the Executive Committee followed through, instructing all laboratories and stations to investigate flight at speeds *beyond* Mach 10 and into the realm of spaceflight. Early leadership emerged from three Langley engineers: Charles Brown (compressibility research division), Charles Zimmerman (stability and control division), and William J. O'Sullivan (pilotless aircraft research division, or PARD). After reading Sänger and Bredt, these men concluded that hypersonic travel should be pursued by a piloted aircraft flown to the limits of the atmosphere, then propelled by rockets into space, and finally returned to earth by control glide. The panel also received the X-2 proposal, transformed by PARD's David Stone into a Mach 4.5 vehicle capable of achieving earth orbit by using two expendable solid rocket boosters and reaction controls. Brown, Zimmerman, and O'Sullivan reviewed Stone's proposal in Summer 1953 and found it worthy of additional evaluation. The Air Force Scientific Advisory Board added further impetus to the project later that year when it declared its support for a piloted hypersonic aircraft.[26]

Events leading toward a hypersonic program quickened in 1954. On the 4th and 5th of February the NACA Interlaboratory Research Airplane Projects Committee held a regular meeting at Headquarters and reviewed all of the High-Speed Flight Research Station's pertinent activities. When a member raised the question of a new thin wing for the D-558-II, a general discussion ensued about the recent hypersonic proposals and whether any of the existing research

---

Evolution," in *Proceedings of X-15 First Flight 30th Anniversary Celebration*," 11-12; Richard P. Hallion, "Toward New Horizons: The Rocket Research Aircraft, 1956-1976" 1-2, DFRC Historical Reference Collection; Hansen, *Engineer in Charge*, 350.

[26]Robert S. Houston, "Development of the X-15 Research Aircraft," Published as Supplemental Volume III to *History of Wright Air Development Center, 1958*, June 1959, 1, DFRC Historical Reference Collection; Williams, "X-15 Evolution Concept," in *Proceedings of X-15 First Flight*, 12; Hansen, *Engineer in Charge*, 351-355; James R. Hansen, "Transition to Space: A History of 'Space Plane' Concepts at Langley Aeronautical Laboratory, 1952-1957," 70-71, DFRC Historical Reference Collection.

airplanes should be re-directed for this purpose . Chairman Hartley Soulé's group reached a consensus: rather than modify the X-2 or even the Skyrocket, an all new aircraft should be designed. It advised Headquarters authorities to canvas the laboratories for requirements for this new research airplane. Langley took the lead again, acting almost immediately to assemble a second hypersonic task group, led this time by John V. Becker, chief of the compressibility research division and designer of the laboratory's hypersonic wind tunnel. He had the assistance of Maxime Faget, a rocket propulsion expert, and Norris Dow, a aerodynamic heating researcher. Becker assumed the job with the essential insight that this project represented a perishable opportunity which needed to be grasped while favorable conditions prevailed.

> By 1954 we had reached a definite conclusion: the existing potentialities of these rocket-boosted aircraft could not be realized without major advances in all areas of aircraft design. In particular, the unprecedented problems of aerodynamic heating and high-temperature structures appeared to be so formidable that they were viewed as "barriers" to hypersonic flight. Thus no definite requirements for hypersonic vehicles could be established or justified. In today's environment [1968] this inability to prove "cost-effectiveness" would be in some quarters a major obstacle to any flight vehicle proposal. But in 1954 nearly everyone believed intuitively in the continuing rapid increase in flight speeds of aeronautical vehicles. The powerful new propulsion systems needed for aircraft flight beyond Mach 3 were identifiable in the large rocket engines being developed in the long-range missile program. There was virtually unanimous support for hypersonic technology development. Fortunately, also, there was no competition in 1954 from other glamorous and expensive manned space projects. And thus [the hypersonic proposal] was born at what appears in retrospect at the most propitious of all possible times for its promotion and approval.[27]

The Becker panel produced its findings in April. Its drawings and specifications largely presaged the aircraft which eventually materialized, both in weight and in dimensions. The design adhered as closely as possible to conventional patterns, thus reducing the chances of aerodynamic problems in the low and transonic ranges. It featured a cruciform tail configuration, a wedge-shaped vertical fin for high stability, and a suite of three or four small

---

[27]James A. Martin, "History of NACA-Proposed High-Mach-Number, High-Altitude Research Airplane, 1950 to 1955," rough draft, 16 December 1954, DFRC Historical Reference Collection; Hansen, "Transition to Space," 71; Becker, "X-15 Program in Retrospect," 1-2 (block quote).

rocket motors to be fired in the air-launch method employed at the HSFS. Re-entry heating concerned Becker and his team, so they insisted on constructing the machine using heat-sink techniques and fashioning it from International Nickel Corporation's Inconel-X chrome-nickel alloy. All this proved persuasive to NACA audiences, so plans were laid (in the customary Research Airplanes tradition) to present the concept to a joint meeting of Air Force, Navy, and NACA representatives. Before the briefing, Hartley Soulé won Headquarters approval to better define the practical aspects of the project by parceling out the necessary research preparations. Walt Williams received instructions to begin mapping out the operational objectives, while the powerplant work fell to Lewis, the aerodynamics studies to Ames, and the hypersonic wind tunnel tests and structures experiments to Langley.

The first session--one of many, as it turned out--occurred in Washington, D.C. on July 9, 1954. Hugh L. Dryden, one of the first scientists to study the aerodynamics of high-speed flight, opened the proceedings with a quiet summary of the arguments in favor of hypersonic research. John Becker then informed the listeners about what had become known as the Mach 7 aircraft. Since the military side included members of the Air Force Scientific Advisory Board, Becker appreciated the strong support his presentation received from the NACA leadership. "Fortunately," he later reflected, "it was not proposed as a prototype of any of the particular concepts in vogue in 1954, which have since largely fallen by the wayside. It was conceived rather as a general tool for manned hypersonic flight research, able to penetrate the new regime briefly, safely, and without the burdens, restrictions, and delays imposed by operational requirements." The Research Airplane Committee agreed in the end to let the NACA disseminate the details of the project to the services and to industry. Finally, the moment of decision arrived; on October 5, 1954, the NACA Aerodynamics Committee convened in Executive session at the High-Speed Flight Station to render a verdict on the project. Famed Lockheed Skunk Works Director Clarence "Kelly" Johnson dissented, arguing the previous research airplanes contributed little of practical value to the design of military aircraft. Walt

Williams countered that military aircraft designers absorbed the lessons of transonic and supersonic flight slowly, but surely, reminding the meeting that it took about six years for combat aircraft to equal the X-1's *1947 performance* of Mach 1.5. The rest of the room sided with Williams. Perhaps in an effort to strengthen Hugh Dryden's hand in negotiating an agreement with the uniformed services, the Aerodynamics Committee passed a resolution which began with a rhetorical nod to the Cold War, using language reminiscent of Theodore von Kármán's seminal report for General Hap Arnold entitled *Toward New Horizons.* "The necessity of maintaining supremacy in the air" intoned the first sentence, "continues to place great urgency on solving the problems of flight with man-carrying aircraft at greater speeds and extreme altitudes...." After this prologue, the committee stated the essential objectives: "the immediate initiation of a project to design and construct a research airplane capable of achieving speeds of the order of Mach 7 and altitudes of several hundred thousand feet...."[28]

Hugh Dryden then convened the Research Airplane Committee in his office on October 22, 1954. Gus Crowley joined him, as did Rear Admirals Lloyd Harrison and Robert S. Hatcher, Air Force Brigadier General Benjamin Kelsey, and USAF science advisor Albert Lombard. In Dryden, the NACA had a leader of extraordinary talents. Born in 1898 to a Baltimore streetcar conductor, he attended the Johns Hopkins University on a full scholarship, eventually apprenticing himself to Joseph S. Ames, a physicist known for his encyclopedic command of the many branches of his discipline. Ames called him, "the brightest young man [I] ever had, without exception." Dryden received his doctorate at age 20, the youngest to earn one from Hopkins. He accepted a position in the National Bureau of Standards' new Aerodynamics Section, became its chief in his early twenties, and eventually rose to the position of Associate Director of the NBS. Meantime, he earned an international scientific reputation in high speed

---

[28]Hallion, "American Rocket Aircraft," 23; Martin, "History of NACA ...High-Mach...Airplane," 4; Becker, "The X-15 Program in Retrospect," 2 (first quoted passage); Hansen, 'Transition to Space," 72-73; "Resolution Adopted by NACA Committee on Aerodynamics, 5 October 1954," "Flight Research Background and History" DFRC Historical Reference Collection, (second quoted passage).

aerodynamics, becoming one of the first to describe the physics of compressibility. As George Lewis' successor, Dryden became known as a patient but highly skilled administrator, a man capable of the most sophisticated scientific counsel, and a person of unquestioned integrity. Less hearty and affable than Lewis, the mild and unassuming Dryden nonetheless acquired considerable recognition outside the NACA, especially among Washington's military establishment.[29]

His well-known qualities and personal connections stood him in good stead on the day of the hypersonic meeting. Without attempts at argumentation or even at overt persuasion, he convinced his colleagues in the armed forces to support a risky endeavor fraught with expense, technical difficulty, and questionable operational utility. Dryden may have won their support by raising the specter of *"national urgency"* (distinct from the less compelling phrase *"great urgency"* employed by the Aerodynamics Committee), a term later used by military figures to denote the project's fundamental security implications. The stronger words appeared in the last point of the Memorandum of Understanding (MOU) agreed upon that date, stating simply that "Accomplishment of this project is a matter of national urgency." The parties also decided to adopt without alteration the Aerodynamics Committees' technical objectives. On the procedural level, the MOU ceded to Dryden the technical chairmanship, but not direct program control, of the Research Airplane Committee. The NACA Director acted with the concurrence of the other two members of the committee: General Kelsey (Air Force Deputy Chief of Staff for Research and Development), and Rear Admiral Hatcher (Assistant Chief of Research and Development in the Navy's Bureau of Aeronautics). In addition, the USAF agreed to let competitive contracts, to administer the design and construction phases, and to split the costs with the Navy. In contrast to their activities during the X-1 and D-558 investigations, the services chose not to conduct

---

[29]James McCormack to AFDDC, AFOCS, OSAF, 28 October 1954, "Flight Research Background and History," DFRC Historical Reference Collection; Michael H. Gorn, *Hugh L. Dryden's Career in Aviation and Space*, Monographs in Aerospace History no. 5 (Washington, D.C.: National Aeronautics and Space Administration, 1996), 1-10 (quoted passage, 2).

separate flight research on the hypersonic vehicle. Instead, once accepted by the Air Force and Navy, the completed vehicles would be transferred to the NACA for tests at the HSFS, the results of which would be shared by all. Due to delays in coordination and in obtaining signatures, final approval languished some seven weeks. But two days before Christmas 1954 Hugh Dryden, Assistant Secretary of the Navy for Air J. H. Smith, and Special Assistant for Air Force Research and Development Trevor Gardner became signatories to the MOU. On the same day, Dryden informed Langley, Ames, Lewis, and the High-Speed Flight Station of the birth of Project 1226, more commonly known as the X-15 research airplane program.[30]

Apparently, "national urgency" meant just that. Only one week after the three signatures inaugurated the X-15, Air Materiel Command's Aircraft Division mailed invitations-to-bid to the leading American aircraft manufacturers, including Bell, Boeing, Chance-Vought, Convair, Douglas, Grumman, Lockheed, Martin, McDonnell, North American, Northrop, and Republic. Of the twelve, few possessed the experience necessary to compete in the high performance field. Moreover, since no production contract would follow the prototype phase the industries realized profits would be slender. Further winnowing down the interested candidates, the Air Force decided to allow just two and one-half years development time. After studying the USAF's preliminary program outline, cost analysis, and the NACA's design study the contractors sent representatives to a briefing at Wright Field on January 18, 1955. Not surprisingly, the May 9 deadline to submit proposals passed with only four companies--Bell, Douglas, North American,

---

[30]James McCormack to AFDDC, AFOCS, OSAF, 28 October 1958, "Flight Research Background and History," DFRC Historical Reference Collection; "Memorandum of Understanding: Principles for the Conduct by the NACA, Navy and Air Force of a Joint Project for a New High Speed Research Airplane," 23 December 1954, DFRC Historical Reference Collection (quoted passage); Trevor Gardner to Hugh L. Dryden, 9 November 1954, DFRC Historical Reference Collection; J.H. Smith to Hugh L. Dryden, n.d., DFRC Historical Reference Collection; Hugh L. Dryden to Trevor Gardner, 23 December 1954, DFRC Historical Reference Collection; Hugh L. Dryden to NACA Langley, Ames, Lewis, and HSFS, 23 December 1954, DFRC Historical Reference Collection; Richard P. Hallion, "American Rocket Aircraft: Precursors to Manned Flight Beyond the Atmosphere," International Astronautical Federation 25th Congress, Amsterdam, 30 September to 5 October 1974, 24.

and Republic--vying for the contract. Due to the cumbersome process of coordination among the Bureau of Aeronautics, the NACA, the Wright Air Development Center, and the Research Airplane Committee in Washington, evaluations wore on for more than three months. At last, North American Aviation won the competition but, to the surprise of everyone, declined the award. The company informed the USAF that its existing back orders prevented delivery of the aircraft in less than three years. Douglas submitted the next most attractive proposal, but because it offered to construct the airframe from magnesium (rather than the Inconel-X agreed to by North American) a laborious process of design modification would be required. To avoid delay, Hugh Dryden and Air Force General Howell Estes of the Air Research and Development Command brokered the terms of this important agreement. They persuaded Department of Defense officials to fund the start-up costs, and acceded to North American president J. L. Atwood's demand that his company receive 38 months (rather than the allotted 24) to complete the project. Early in December 1955 North American returned the letter contract, which entailed $2.6 million for initial work and $39 million for the design, development, and delivery of three aircraft, in addition to a flight demonstration program. Meanwhile, Reaction Motors agreed to collaborate once again on the development of a research airplane, signing a $9 million letter contract in February 1956 to plan, design, and fabricate the first X-15 engine.[31]

The project did not begin auspiciously in North American's Los Angeles hangars and offices. The company's manager of research and development, Harrison A. Storms, received his fundamental technical instructions from Hartley Soulé: "You have a little airplane and a big engine with large thrust margin. We want to go 250,000 feet altitude and Mach 6. We want to study aerodynamic heating. We don't want to worry about aerodynamic stability and control, or the airplane breaking up. So if you make errors, make them on the strong side. You should have enough thrust to do the job." Storms found a similar clarity in his relationship to North

---

[31]Houston, "Development of the X-15 Research Aircraft," 11-22, DFRC Historical Reference Collection.

American management. He learned right after the company won the competition that the top bosses lacked any enthusiasm for the project. Its relatively low pay-off rendered it less profitable than other work and because of its inherent complexities, the leadership worried that it might become a drain on the firm's limited pool of talent, occupying the most able engineering minds while the more lucrative projects got second best. Indeed, Chief Engineer Raymond Rice told Storms the company would go ahead with the X-15 only if Storms agreed to be the sole technical decisionmaker and, in fact, to be the sole North American representative on *all* X-15 matters. Astonishingly, Storms also found himself compelled to promise never to refer a single X-15 problem to the Chief Engineer's office. Orphaned from its first days, his program could operate with virtual autonomy. "This was fine with me," and with his engineers, he later remarked. "I felt that the X-15 was vital to the future of aerospace and I wanted to be intimately involved with the future of this industry and would have no hesitation in agreeing to do most anything in order to be associated with [it]."[32]

Storms quickly assembled a team of 35 persons led by chief project engineer Charles Feltz and initiated the preparation of detailed specifications. Feltz also gave top attention to the problem of adequate tankage for the aircraft. Meanwhile, the NACA assumed an important advisory role from the inception of the undertaking. Scott Crossfield, hired by North American as a test pilot in 1955, played a crucial part as a staff consultant who applied his experience flying rocket planes to the actual X-15 design process. Walt Williams assigned members of his staff to work with their technical counterparts at North American. Other NACA organizations also made contributions. Starting in February 1956 engineers associated with the Langley 9-inch blowdown tunnels ran tests on an X-15 model to determine the aerodynamics of the side portion of its fuselage and to learn the effects of extending the speed brakes. At the same time, the Lewis Laboratory raised serious questions about the use of ammonia as part of the fuel mixture

---

[32]Harrison A. Storms, "X-15 Hardware Design Challenges," in *Proceedings of the X-15 First Flight 30th Anniversary Celebration*, 27, 33 (first and second quoted passages).

in the Reaction Motors engine. Tests at Lewis revealed the proposed propellant combination of ammonia and oxygen seemed inadequate to the operational requirement for many re-ignitions; the mixture also proved to be corrosive. Indeed, some worried more about the technical development of the powerplant than of the airframe. Despite the many hurdles to be overcome (a very high standard of safety imposed on a powerful, reusable engine equipped with variable thrust), Reaction Motors received its letter contract only four months after North American. On the other hand, the airplane mockup easily passed the development engineering inspection in winter 1956, although by that time several changes in configuration already had occurred. North American engineers concerned about the vehicle's longitudinal stability decided to re-locate the side fairings aft 100 inches, and to improve directional stability by replacing the original twin vertical tail wedge design with a single tail wedge. The contractor also elected to deviate from the initial plans by re-positioning the landing gear skids rearward to a point almost under the aircraft's tail. At the same time, the auxiliary power units were shifted forward to achieve a workable center of gravity.[33]

The entire technical picture came into clearer focus when Hugh Dryden, acting in accord with his duties as Research Airplane Committee chairman, convened a conference on the status of the X-15 at the Langley Laboratory in late October 1956. It attracted over 300 attendees from inside and outside the program and included those affiliated with the NACA, all the major aircraft manufacturers, the Navy Bureau of Aeronautics, and the Air Force research and development facilities. Charles Feltz described the characteristics of the new research plane in full detail. Measuring roughly 49 feet in length, 22 feet in wingspan, and 14 feet in height at the vertical tail, its 31,275 pound launching weight almost doubled that of the heaviest D-558. Its stubby wings, swept 25 degrees in front, were thin--just five percent thick. Its long slender

---

[33]"Minutes of Meeting of Interlaboratory Research Airplane Projects Panel...30-31 January 1956," 3-8, Hallion Collection, DFRC Historical Reference Collection; "Report for Research Airplane Projects Panel of Research Activities...for...1956," 3, Hallion Collection, DFRC Historical Reference Collection; Hallion, "American Rocket Aircraft," 29; Thompson, *Edge of Space*, 2-3.

fuselage concealed four compartments: in the forward section, the reaction control rockets embedded in the nose, the cockpit, and the equipment bay; behind it, the liquid oxygen, liquid nitrogen, and helium tanks; third in line, the anhydrous ammonia and hydrogen peroxide tanks (paralleled on the exterior by the other reaction controls mounted on the wings); and farthest aft, the rocket engine. The greatest departure from past research aircraft occurred in the powerplant. Capable of propelling the airframe up to 6,600 feet per second, this single engine, if completed successfully, would develop 57,000 pounds of thrust, many times that of any aircraft yet flown. On the other hand, Hubert Drake of the HSFS reminded the audience of the lessons afforded the X-15 project by the rocket planes which preceded it. He told them to expect serious problems in such flight fundamentals as longitudinal control effectiveness, high altitude dynamic stability, thrust misalignment, control at low dynamic pressure, roll coupling, and supersonic directional stability. If anything, Drake predicted even greater difficulties in a plane of such commanding performance as the X-15.[34]

During 1957, the NACA stepped up its direct participation in X-15 development. The Lewis Laboratory conferred with Reaction Motors about diminished thrust calculations. It seems the rocket company erred in the figures submitted with its bid and admitted in February that the planned powerplant would operate "well below the design value." The Lewis engineers suggested bell-shaped nozzles or extensions for greater power and also recommended fuel additives and better injector systems. While North American geared up for fabrication of the airframe and started assembly in September, the Ames Laboratory mounted X-15 models in four different wind tunnels to test longitudinal stability at various angles of attack, oscillation at Mach 2.5 to 3.5, and pitching at hypersonic speeds. The Moffett Field facility also compared Inconel-

---

[34]Sources differ about the thrust of the XLR-99 powerplant; some say 57,000 pounds, others (such as Milt Thompson in *At the Edge of Space,* p. 46) claim 60,000. Hubert M. Drake, "Flight Experiences with Present Research Airplanes," 15-22, and Charles H. Feltz, "Description of the X-15 Airplane, Performance, and Design Missions," 23-35, both in *Research Airplane Committee Report on Conference on the Progress of the X-15 Project: A Compilation of the Papers Presented*, Langley Aeronautical Laboratory, 25-26 October 1956, DFRC Historical Reference Collection.

X, beryllium, and copper as leading edge materials under temperatures up to 1200 degrees.

At this juncture, important X-15 work fell to the High-Speed Flight Station. Under the terms of the X-15 MOU, the NACA enjoyed sovereignty over the entire test program. But on reflection, Hugh Dryden thought the USAF ought to play some part in the process and asked Walt Williams "to work out arrangements with the...[Air Force Flight Test] Center for active participation by AFFTC personnel in the X-15 flight program." Williams interpreted Dryden's instructions liberally and in June 1957 created an X-15 Flight Test Steering Committee. The center director not only chaired the committee, but held a deciding vote in all deliberations. Its members consisted of paired representatives from the NACA and the USAF: a project officer, a pilot, and an engineer from each. Williams endowed the group with broad powers to enact its own policies and procedures, to be responsible for the flight test program as a whole, and to oversee flights of the X-15, its mother ship, and their maintenance; to reply to all inquiries related to the program; to supervise instrumentation, data reduction, and chase plane operations; and to manage high range activities, ground support, and overall range support. Despite the appearance of shared responsibility with the Air Force, Williams assured Dryden he had "no intention whatsoever of relinquishing the technical direction of the program" grounded in the MOU. Clearly provoked by Williams' assumption of authority, the ordinarily mild Dryden issued a stinging rebuke, warning him to not to assume greater powers than appropriate to his position nor to forget his obligations to headquarters:

> *For the future protection of your position* [author's italics], it is suggested that you make certain that the Flight Test Center personnel are aware that the apparent scope of the authority to be exercised by the committee, as contemplated in the enclosure to your reference letter goes beyond matters over which the local level has jurisdiction. You should consult with the Flight Test Center about the instrumentation of the airplane and the planning of the research flights to achieve so far as possible the objectives of both groups. At the same time, you and the High-Speed Flight Station are responsible in the final analysis for the instrumentation and for the planning of the flights, because they are research problems. Any major changes in the scope or intent of the program have to be cleared with the NACA Headquarters. It is presumed there are similar restraints on the Flight Test Center. It should be understood at the outset, therefore, that the steering committee would have jurisdiction only in regard to matters that would normally come

under jurisdiction of the Flight Test Center or the High-Speed Flight Station. Control and dissemination of NACA data should remain with the High-Speed Flight Station. It would be best if the committee could work without charter, at least until some experience with its operation and interest was obtained.[35]

Two days after Dryden wrote this magisterial letter to Williams, the X-15 assumed far greater prominence when the Soviet Union launched Sputnik I, the world's first earth satellite. Now the Research Airplane Program represented not merely an opportunity to widen the aeronautical flight envelope, but a national imperative to leap into space. If achieved, its success would restore some of America's tarnished technical prestige; but if it failed, it would result in intense embarrassment. Perhaps to foreclose future acts of independence by Williams or others in the extraordinarily high-stakes X-15 program, at the end of 1957 NACA headquarters abolished the Interlaboratory Research Airplane Panel, confining oversight to Dryden and his Air Force and Navy counterparts on the Research Airplane Committee. But a far bigger surprise lay ahead. As a consequence of the American public's outrage and panic in the wake of Sputnik, some political figures accused the nation's scientific elite of failing to stay abreast of Soviet technological advances. The Congress initiated high-publicity hearings concerning the "Space Race." Few remembered amid all the controversy that before Sputnik there existed virtually no public demand or political expression for travel outside the atmosphere. Nevertheless, Hugh Dryden, the NACA, and the Research Airplane Program had pursued the realm of spaceflight for years. Now that it was within the NACA's grasp in the X-15 program, the Congress decided to merge the venerable committee, its laboratories, and its 8,000 employees into a new organization created by the National Aeronautics and Space Act of 1958. The resulting National Aeronautics

---

[35]"Minutes of Meeting of Interlaboratory Research Airplane Projects Panel...February 4-5, 1957." 4, Hallion Collection, DFRC Historical Reference Collection (first quoted passage); "Quarterly Report on Status of Ames Projects Relating to Research Airplanes, January 1, 1957 to March 31, 1957," 4-7, Hallion Collection, DFRC Historical Reference Collection; Hugh L. Dryden to Donald L. Putt, 2 October 1957; DFRC Historical Reference Collection; Walter Williams to Hugh L. Dryden, 5 June 1957; DFRC Historical Reference Collection; Walter Williams to Hugh L. Dryden, 21 June 1957; DFRC Historical Reference Collection; Donald L. Putt to Hugh L. Dryden, n.d.; DFRC Historical Reference Collection; Hugh L. Dryden to Walter Williams, 2 October 1957; DFRC Historical Reference Collection (block quote).

and Space Administration (NASA) came into being on 1 October 1958. While Hugh Dryden retained his duties as chairman of the X-15 Research Airplane Committee, he found himself second in command and Deputy Administrator under NASA's first administrator, T. Keith Glennan of Case Western Reserve University. Similarly, while aeronautics held pride of position in the name of the new agency, space would assume the lead functional role.[36]

Nonetheless, North American raced ahead with construction of the three airframes during 1958. But as the year progressed it became clear that Reaction Motors' accompanying XLR-99 powerplant would not be ready for the first flight scheduled in 1959. Instead, two interim XLR-11s would serve the engine requirements of X-15 number 1 during the entire first year of powered launches. Meantime, at its plant in the Los Angeles suburb of Inglewood, North American rolled out the first X-15 in public ceremonies held on October 15, 1958. Interestingly, the highest ranking official representing the company happened to be Raymond Rice, the same man who gladly ridded himself of the X-15 albatross by giving it root and branch to Harrison Storms. Now he praised the "team spirit" which informed its management. Walt Williams then spoke briefly and likened the upcoming flight research program to that of the X-1. Just as the earlier plane vanquished the misconception that man could not penetrate beyond Mach 1, the X-15 "will give a good portion of the answers to the problem of man's place in a space mission." The two projects also shared instrumentation techniques, the X-15 benefiting from the data collection procedures and devices developed for the X-1, the D-558, and other high performance aircraft flown previously at the HSFS. During the upcoming hypersonic tests, instrumentation would be arrayed to learn the limits of such factors as aerodynamic heating in flight, the effects of weightlessness on pilots, the phenomena encountered approaching the edge of space and re-entering earth's atmosphere, and the reliability of navigational equipment. To gather the data, Williams' staff prepared external and internal equipment for all three aircraft. Outside the

---

[36]J.W. Crowley to Charles Donlan, 19 December 1957, Hallion Collection, DFRC Historical Reference Collection; Roger D. Launius, *NASA: A History of the U.S. Civil Space Program* (Malabar, Florida: Krieger Publishing, 1994), 24-33.

aircraft, precision radar monitoring recorded all range activities between Wendover, Utah, and Edwards, California. Telemetering devices gathered system function and crucial airplane parameters. The bulk of the devices, embedded on-board the X-15, measured airspeed, altitude, angle of attack, sideslip, structural temperatures, surface loads and surface pressures, as well as the skin temperature and heart rate of the pilots. The X-15 instrumentation did depart from earlier practices in one important respect. Rather than the FM/FM telemetry installed in all three X-15s (and backed up by on-board oscillographs), a new system called Pulse Code Modulation (PCM) went into service late in the program on aircraft number 3. The PCM converted signals from the sensing devices into binary numbers with values proportionate to the strength of the incoming impulses. On the programmatic level, Hugh Dryden described X-15 flight research as a two part inquiry. During the initial phase the Number 1 aircraft, equipped with the XLR-11 engines, would begin to investigate the aircraft's performance and expand the flight envelope to the powerplants' maximum capacity. Beginning in summer 1960 its work would be continued by the number 3 aircraft powered by the XLR-99. The second phase, designated the NASA X-15 Research Program, would be flown on aircraft Number 2 and concentrate on detailed scientific experiments conceived by the Flight Station and by the NASA laboratories. The HSFS pilots and crew would fly these missions until mid-1961, at which time an entirely new round of investigations would be planned based on the existing research results and the proven capabilities and limitations of the vehicles.[37]

The anticipation finally ended on June 8, 1959, when the long, dark rocket plane took to the air. This event initiated the 14 contractor demonstration flights, all flown by North American's Scott Crossfield. Crossfield contributed greatly to the final design of the aircraft not only because of his intimate knowledge of high performance flight, but also because he

---

[37]Hallion, "American Rocket Aircraft," 29-30; Raymond Rice, "The X-15 Rollout Symposium," 15 October 1958, AFFTC/HO Historical Reference Collection (quoted passage); Walter C. Williams, "The X-15 Rollout Symposium," 15 October 1958, AFFTC/HO Historical Reference Collection; Hugh L. Dryden to Chief, Bureau of Naval Weapons, n.d., X-15 Files, DFRC Historical Reference Collection; Powers, *Women in Flight Research*, 14.

understood thoroughly the dual role of research pilot, combining the engineer's skills with the aviator's finesse. One of his admiring HSFS superiors said that in the D-558 program "he got intimately involved with the analysis of the data...from the wind tunnel before we ever flew it; he got intimately involved with the flight planning and the reasons for it; he was intimately involved with results after that. And when we attempted to go to a Mach Number of 2 for the first time in a D-558, I would say he was the project engineer on it." No one knew the X-15 better than Crossfield, which proved to be an immense advantage in the first flight. While attached to the B-52 mothership, he discovered an inoperable pitch damper. Even though the flight rules called for aborting the mission, Crossfield had the final say and decided to proceed since the flight path was simple and the runway consisted of the immense lakebed. But as he attempted to reduce the steep glide path in preparation for landing, longitudinal oscillation began, a pilot-induced situation which worsened as he approached the ground and the speed decreased. In danger of a crash, he succeeded in putting the plane down at the bottom of an oscillation, minimizing damage to a broken landing gear. Investigators later realized the fault lay in the settings of the horizontal stabilizer actuators whose rate needed to be increased from 15 degrees per second to 25. All subsequent flights on all three aircraft followed this guideline. Aircraft number 1, however, had to be trucked back to North American where it underwent six months of repair.[38]

Just as the consequential first powered X-15 flight neared, two landmark changes occurred at the High-Speed Flight Station. First and far more noteworthy, in September 1959 Hugh Dryden drafted Walt Williams to be Associate Director of the Space Task Group, created to oversee Project Mercury. Williams later served as Mercury's Operations Director. After thirteen years on the bridge at the HSFS--hard years fraught with all of the struggles of the formative period--Williams' hard-driving, hands-on style cast a long shadow over every corner

---

[38]Thompson, *Edge of Space*, 91-92; De Elroy Beeler, interview with Richard P. Hallion, December 1976, 30-31, Hallion Collection, DFRC Historical Reference Collection; "X-15 Program Summary," Milt Thompson, n.d., 1, Milt Thompson Collection, DFRC Historical Reference Collection.

of the facility. In his place, Glennan and Dryden selected a man just as demanding and equally determined, but less confrontational in style. Paul F. Bikle, the 43-year-old Technical Director of the Air Force Flight Test Center, simply drove across Edwards to assume his new duties. Bikle inaugurated his career at Wright Field in 1940 as an aeronautical engineer and by 1944 became chief of the Aerodynamics Branch in the Flight Test Division. His manual entitled "Flight Test Methods," was a landmark attempt to codify flight research procedure and became a classic in the field.

The HSFS experienced one other change at this time. On the 27th of September 1959 it received the new designation of NASA Flight Research Center (FRC). The elimination of "high speed" from the name recognized that NASA's space vehicles achieved not merely supersonic, but hypersonic velocities; at the same time, elevating the HSFS to center status gave the desert facility organizational parity with the venerable Langley, as well as with Ames and Lewis.[39]

Bikle endured two stern tests almost immediately after replacing Williams. In order to complete the X-15 test plan he needed firm dates when the Air Force planned to transfer the three X-15s to the FRC. He proposed to the AFFTC Commander, Brigadier General J.W. Carpenter, November 1959 for aircraft number 1 and June 30, 1960, for numbers 2 and 3. This occasioned a pointed discussion between the two men in which the question of pilot precedence arose. Perhaps testing the rookie's resolve, Carpenter expressed the bald fact that the $100 million the USAF had already sunk into the program entitled Major Robert White to have the first non-contractor flight. Bikle held firm to the NASA choice, pilot Joseph Walker, arguing the NACA conceived the program in 1952 and had itself devoted considerable resources to it. In the end, Bikle won out with Walker flying first and White sharing most of the phase I flight demonstration of the X-15 design objectives. Bikle's second test involved X-15 number 2, flown by Scott Crossfield on Thursday, November 5, 1959, on its third powered flight. The first two

---

[39]Hallion, *On the Frontier*, 103; NASA Biographical Data, "Paul F. Bikle," 7 October 1964, DFRC Historical Reference Collection.

proved uneventful, as did this one until the pilot fired the upper chamber on the lower engine at 45,000 feet and Mach .82. There followed a blast and fire which tore off the last few inches of the chamber and the nozzle, blew off the explosion doors, and caused extensive damage inside the engine compartment. When it occurred, Crossfield knew only what his chase planes and his instrument panel told him, so he immediately shut off the engines, jettisoned the fuel, and attempted a glide landing. He touched down safely but when the nose gear hit the Rosamond Dry Lakebed, the weight of the propellants caused the fuselage to fail, severely buckling on top, just forward of the liquid oxygen tank. At the same time, the joint at the bottom of the fuselage opened and sheared many of the bolts. Momentum dragged the aircraft on the ground for 1,500 feet. Like X-15 number 1, the broken remains of number 2 were hauled by truck to Inglewood where North American technicians and designers worked for three months to mend the wreckage. At the same time, the Reaction Motors' engineering staff puzzled over the causes. Meanwhile, to avoid a repetition of the threat to man, machine, and program, Bikle ordered the XLR-11 motors removed after every other flight and shipped to the Air Force for major maintenance and repair. On their return, the NASA technicians subjected them to short ground tests on the propulsion system test stand before re-installing them on the aircraft and conducting flight qualifying ground runs. This intensive engine work persisted for the first 100 flights.[40]

By 1960 the infrastructure of the X-15 program reached the stage of full maturity. Paradoxically, the buildup at first made little impact on the FRC's organizational structure. But it nonetheless unleashed a revolution in the corridors of the Flight Research Center. To stay abreast of the workload, an increasing number of staff members found themselves drawn into the X-15 orbit until, by the early 1960s, nearly everyone worked in some capacity on the project. At the same time, contract personnel swelled the ranks of the center. Gradually, as other projects

---

[40]Paul F. Bikle to AFFTC Commander, 23 September 1959; DFRC Historical Reference Collection; Paul F. Bikle to NASA Headquarters, 2 October 1959; DRFC Historical Reference Collection; X-15 Flight Records, Flight No. 2-3-9, November 5, 1959, DFRC Historical Reference Collection; Milt Thompson to Walt Williams, 31 May 1978, Milt Thompson Collection, DFRC Historical Reference Collection.

completed their normal lifespans, the X-15 became *the* leading FRC activity with no challengers in sight. But the informality and the direct human contact possible in a small, isolated operation--the kind of access achieved during the earlier research airplane projects--vanished with the complex, high-stakes work being pursued on the flightline. Walt Williams thought nothing of issuing spoken instructions to a team of engineers who, in turn, were free to ask technicians, flight crews, and carpenters for support as needed. But to be able to re-trace the trail of labor and money in a climate of intense national scrutiny, Paul Bikle *needed* paper--paper to monitor the intense contractor involvement, paper to track his staff's time, and paper to account for the progress of the flight program itself. In this sense, the institutional stratification begun during the earlier phases of the research airplane program reached full realization during the X-15 project.

The importance of record-keeping materialized again when the third X-15 followed models one and two to the disabled list. During preparations on June 8, 1960, for the first XLR-99 flight, a ground run on aircraft number 3's engine resulted in an explosion in the hydrogen peroxide tank. The blast wrecked the entire machine aft of the wings, necessitating yet another trip by truck to Inglewood. Because the whole rear portion of the plane required complete reconstruction, it remained in the North American hangar for more than a year. Fortunately, by mounting an XLR-99 on aircraft number 2 the program suffered no lag in activity. But the extent of the damage resulted in a full Air Force accident investigation by a 19-member board comprised of only one NASA representative and 18 others from North American, Reaction Motors, and the Air Force. Forced to explain how an aircraft costing the USAF millions blew up on the ground, the FRC could be thankful for its paperwork. [41]

---

[41]Flight Research Center Organization Chart, July 1960, DFRC Historical Reference Collection; Thomas Finch, interview with Richard P. Hallion, 8 December 1976, 79, Hallion Collection, DFRC Historical Reference Collection; Betty Love, interview with Michael Gorn, 10 April 1997, 13-14, DFRC Historical Reference Collection; Milt Thompson, "X-15 Program Summary," n.d., 2-3, Milt Thompson Collection, DFRC Historical Reference Collection; Arthur Murray to AFFTC, ? June 1960, X-15 Files, DFRC Historical Reference Collection.

While the ultimate success of the program seemed unclear after more than a year of flight research, by the start of 1961 the X-15s had flown 31 times at speeds up to Mach 3.3 and altitudes up to 136,000 feet. Of greatest importance to date, during the 26th flight on November 15, 1960, the XLR-99 engine underwent its first flight test and, without incident, powered Scott Crossfield and aircraft number 2 to a speed of Mach 2.97. If this powerplant proved a long-term success, the X-15 might well fulfill its promise. But in an assessment of the project made the first day of 1961, Paul Bikle seemed cautious. He described the data on performance, flight dynamics, control, and structural loads as "fairly complete" within the envelope flown. At the same time, he conceded that the X-15 structural temperature research suffered from the "short duration and highly transient nature of each flight [which] have generally precluded extensive and systematic measurements." Because of lingering uncertainties about both the XLR-99 and the evolving flight expansion tests, Hugh Dryden felt the moment right to organize a second X-15 conference sponsored by the Research Airplane Committee. His present Air Force partner on the panel, Major General Marvin Demler, agreed that the program stood "at the threshold of payoff" but also reminded Dryden that not just the acquisition of scientific data, but "the expenditure of more than $150 million of public [more specifically, Air Force] funds" weighed in the balance.[42]

The conference opened at the Flight Research Center on November 20th, 1961. Paul Bikle assumed the thankless job of describing to the plenary session the future course of this as yet unpredictable program. For the immediate period, he promised further exploration of such important areas as flight characteristics at high angle of attack, the effects of aerodynamic heating, the operation of the reaction controls, the adequacy of the adaptive control system, the

---

[42]Paul Bikle to NASA Headquarters, "X-15 Status Report, January 1, 1961," 30 December 1960, DFRC Historical Reference Collection (first and second quoted passages); Milt Thompson, "X-15 Program Summary," 3, Milt Thompson Collection, DFRC Historical Reference Collection; Hugh L. Dryden to Marvin C. Demler, 9 February 1961, X-15 Files, DFRC Historical Reference Collection; Marvin C. Demler to Hugh L. Dryden, 15 March 1961, X-15 files, DFRC Historical Reference Collection (third quoted passage).

aircraft's overall performance, and the efficacy of its of displays and its energy management systems. During these flights Bikle's engineers also planned to define the lift and drag characteristics of the aircraft. In addition, greater knowledge of behavior at angles of attack ranging from 15 to 25 degrees needed to be ascertained before achieving altitudes approaching 250,000 feet. Should these tests and other hypersonic tools (such as a redundant stability augmentation system) warrant, Bikle foresaw altitudes as high as 400,000 feet at speeds between 2,000 and 5,500 feet per second, at which point experiments with displays, guidance, precision control, and bioastronautics might be attempted. Toward the end stages of the program, Bikle envisioned a new generation of instrumentation not only capable of gleaning new data in the atmosphere, but of gathering flight data in space. Using pilots as on-the-spot investigators, the FRC Director proposed a final series of sophisticated research projects covering such subjects as ultraviolet stellar photography, infrared exhaust signature, computer-guided landings, detachable high temperature leading edges, horizon definition, and hypersonic propulsion. The exact nature of these missions, scheduled to begin after about 30 more phase I flights, still awaited decisions by the Research Airplane Committee. In the meantime, Bikle urged his listeners to regard the X-15 flight program as progressive rather than static, as evolutionary rather than fixed in its objectives.

> When the X-15 was first approved, the objectives were clearly stated in terms of aerodynamic heating , speed, altitude, reaction control research, and bioastronautics. As the program has progressed, it appears that, while these worthwhile objectives have been or will shortly be achieved, many important benefits have been of a different sort. The X-15 program has kept in proper perspective the role of the pilot in future programs of this nature. It has pointed the way to simplified operational concepts which should provide a high degree of redundancy and increased chance of success in future space missions. And, perhaps most important, is the fact that all of those in industry and in the government who have had to face up to the problems of design, building the hardware, and making it work have gained experience of great value to the future aeronautical and space endeavors of this country.[43]

---

[43]Paul Bikle, "Future Plans for the X-15," "Research Airplane Committee Report on Conference on the Progress of the X-15 Project," NASA Flight Research Center, November 20-21, 1961, 329-333, Milt Thompson Collection, DFRC Historical Reference Collection (block quote, 331).

In the short term, persistent XLR-99 engine problems clouded the fulfillment of these loftier visions. Not that the flight program failed to make progress. During spring 1962 the Mach numbers inched up to and over 5 and altitudes topped 200,000 feet, tremendous feats in themselves. But the question of reliability and safety kept arising. At a meeting of the Joint Operating Committee of the X-15 program in March 1962 the greatest difficulty involved the frequency of engine maintenance, inspections, and repairs. One crew chief admitted the XLR-99 was not just new and unfamiliar but more complex and demanding than any his men had ever seen, requiring more training, more sophisticated checkout procedures, more ground equipment, and more ground runs. Everyone on the Operating Committee believed the USAF should pursue an XLR-99 improvement contract with Reaction Motors to solve the problems. But neither Paul Bikle nor his counterpart on the Operating Committee, Colonel Chuck Yeager, had the funds to underwrite it. As a consequence, yet another costly and dangerous engine failure occurred. During flight number 74 on November 9, 1962, NASA pilot Jack McKay, at the controls of X-15 number 2, found it impossible to attain more than 30 percent thrust after launch from the B-52. Apparently, the engine's governor actuator failed. Ground control instructed McKay--a former Navy fighter pilot who not only flew D-558s-I and -II, but two of the X-1s--to shut off the XLR-99, jettison the liquid oxygen and the anhydrous ammonia, and attempt an emergency landing at Mud Lake. There followed a chain of events almost culminating in disaster. On approach, the wing flaps failed to operate, forcing McKay to come in faster and harder on the nosegear than normal. Unfortunately, the automatic flight control system imposed additional heavy airloads on the landing gear, causing it to fail. As the plane slammed down, both the wing and the horizontal stabilizer buried themselves in the lakebed, in turn flipping the aircraft onto its back, where it came to rest. Rescuers pulled McKay from the wreckage without great difficulty, but he sustained cracked vertebrae, and although he again flew the X-15, the crash ultimately resulted in his retirement. Aircraft number 2 also fared poorly; grounded for 19 months, it reduced the X-15 "fleet" to two until June 1964 and taxed the limits of numbers 1 and 3.[44]

While aircraft number 3 underwent repairs, numbers 1 and 3 continued to expand the flight envelope. The August 22, 1963, flight of the FRC's Chief Pilot Joe Walker proved that when the XLR-99 worked, it worked very well, indeed. On that date, Walker and number 3 dropped from the B-52 over Smith Ranch, Nevada, with the mission to fly the highest altitude ever achieved by an aircraft, an attempt at *360,000 feet*. Almost everything worked according to plan. Climbing under power for 86 seconds at an average pitch angle of approximately 51 degrees, the aircraft then coasted to an indicated 362,000 feet after engine burnout. Walker, outside the earth's atmosphere, experienced no difficulty maneuvering with the reaction controls. Like the other pilots before and after him, he experienced the sensation of "coast[ing] over the top ballistically" achieving apogee while lingering in space. He finally used the reaction controls to drop the X-15's pitch angle to -39 degrees and begin his descent. This flight marked the high point of the X-15 altitude program: 354,200 feet (just over 67 miles) according to the final measurement.

Just as the altitude milestone won recognition, the phase II experimental program assumed concrete form. The Research Airplane Committee sifted many suggestions before arriving at viable candidates to fly on the X-15 platform, but pressures from various sponsors could also be brought to bear. For instance, Secretary of Defense Robert McNamara wrote to NASA Administrator James Webb in July 1963 saying that he found the X-15 "an eminently successful example of a joint NASA-DOD research endeavor." At the same time, due to the "substantial additional funding" entailed by the extended program of basic research, McNamara ordered "very careful consideration" of NASA's request for continued Department of Defense

---

[44]"Minutes of Meeting: X-15 Joint Operating Committee...March 6, 1962," 2-3, DFRC Historical Reference Collection; John W. Russell, interview with Dr. J.D. Hunley, 7 and 11 March 1997, 27-28, DFRC Historical Reference Collection; X-15 Flight Records, Flight No. 2-31-52, November 9, 1962, DFRC Historical Reference Collection; Milt Thompson, "X-15 Program Summary," 4-5, Milt Thompson Collection, DFRC Historical Reference Collection; Hallion, *On the Frontier*, 330-333.

support to assure "the additional costs...are warranted by...significant research results...." In all likelihood, the Defense Secretary pressed for such assurances because by the time Webb received his letter, the 88th X-15 flight (November 1961) surpassed Mach 6 and the record altitude mark lay just a few weeks in the future. With the flight envelope stretched almost to its limit, an increasing number of the remaining missions would employ the X-15 as a research platform, not as a hypersonic flight demonstrator whose data often held much military utility. Still, many of the later flights involved expanded investigations of aspects of high-speed aerodynamics and aeronautics undertaken since the start of the X-15 program. Indeed, some of them dated to the start of the Research Airplane Program itself. These subjects included handling qualities, stability augmentation, guidance, display, flow fields, heat transfer, drag derivatives, air loads, structural heating, landing gear loads, and so forth. On the other hand, entirely new avenues of research--many related to space--had been approved by Dryden and his military colleagues: high altitude sky brightness, micrometeorite collection, atmospheric density measurements, ultraviolet stellar photography, horizon definition experiment, advanced integrated data systems, and others. Once underway, these investigations rendered the X-15 a ship of dual uses; at once the world's supreme hypersonic research vehicle and an unparalleled flying laboratory that helped prepare the way for the Apollo program.[45]

Some of the platform experiments merely used the X-15 as a passive carrier. The collection of miremeteorites, studied in order to determine the risks to space vehicles posed by these tiny flying objects, simply exposed a box poised on a wing-tip pods to the upper

[45]X-15 Flight Reports, Flight No. 3-22-36, 22 August 1963, DFRC Historical Reference Collection; William Dana, interview with Peter Merlin, 14 November 1997, 5-6, DFRC Historical Reference Collection (first quoted passage); Hugh L. Dryden to James Ferguson, 15 April 1963, DFRC Historical Reference Collection; James Ferguson to Hugh L. Dryden, 29 July 1963, DFRC Historical Reference Collection; NASA Headquarters (Aeronautical Division) to NASA FRC, 8 August 1963, DFRC Historical Reference Collection; Robert S. McNamara to James E. Webb, 12 July 1963, DFRC Historical Reference Collection (second through fifth quoted passages); Milt Thompson, "X-15 Program Summary," 6, Milt Thompson Collection, DFRC Historical Reference Collection; "Semi-Annual Status Report of X-15 Program, Report No. 1," October 1963, DFRC Historical Reference Collection.

atmosphere. The infrared scanning radiometer, on the other hand, designed to record the earth's radiation, compelled the pilot to conform his flight plan to accommodate the collection of data (three flights, 70,000 to 100,000 feet, Mach 3 to 5). On the other end of the spectrum of difficulty, a device known as a zero-gravity heat exchanger (to test heat-transfer designs for space ship cooling systems) required the X-15 pilots to conduct four flights penetrating space for the maximum duration possible. For those experiments related to flight research, the X-15 revealed some important new findings on high speed aerodynamics. The theoretical assumptions and ground testing techniques applied to stability and control at hypervelocities could now be challenged (or bolstered) by actual flight data. Likewise, the heat transfer measurements contributed to a field with little experimental data and suggested that while intense temperatures might result in *localized* structural failures, they need not result in generalized, catastrophic events. Finally, the standard methods of predicting hypersonic aerodynamic characteristics--theoretical mathematics and wind tunnel tests--were augmented and corrected by the full-scale flights which revealed the actual flying conditions.

The X-15 continued to collect such data until NASA pilot Bill Dana flew the 199th and final mission on October 24, 1968. During these last few years, the earlier problems of landing-gear durability and engine reliability did not diminish so much as the flight crews and pilots learned to cope with the inherent weaknesses of these delicate components. Zealous preventive maintenance and exacting pilot practice permitted the X-15 to soldier on despite the infirmities. But one fatality did occur. Air Force pilot Michael Adams died when X-15 number 3 crashed north of Edwards Air Force Base on November 15, 1967. Accident investigators discovered a long list of contributing factors, some related to human physiology, others to aircraft system failures. On the human side, scientists at the time had little understanding of the effect of intense stresses (such as high g forces) on the nervous system. Vertigo represented one symptom of sensory overload experienced by many X-15 pilots, including Adams. On a previous flight it affected him so badly that he became disoriented. During the fatal incident, extreme dizziness

53

occurred during the climb to altitude, probably causing Adams to mistake his roll indicator for a heading indicator. He then used the reaction controls to unwittingly turn the aircraft 90 degrees from the correct flight path, causing aerodynamic loading and an eventual spin. Yet the machine also contributed its share to the calamity: an electrical disturbance shortly after launch reduced the effectiveness of the control system and added to the pilot's workload; and the adaptive control system tore apart the X-15 upon re-entry into the atmosphere. Adams' death represented a bizarre reversal of fortune for the X-15; only six weeks earlier pilot Pete Knight pushed the number 2 aircraft to the program's maximum speed record, attaining Mach 6.7, or 4,520 miles per hour. But this great achievement almost presaged Michael Adams' fate. This X-15 was modified with a dummy version of a device conceived by Langley called the Hypersonic Ramjet Experiment (HRE), designed to propel the X-15 to Mach 8 by adding two immense fuel drop tanks, a thermal protection system, and a powerful ramjet engine. The initial opportunity to test the aerodynamics of the system presented itself when NASA and Air Force representatives agreed to pay for the installation of a non-operating ramjet on X-15 number two, already undergoing repairs in the North American hangars after the landing accident in November 1962. During Knight's successful attempt at the speed record, localized aerodynamic forces raised the aircraft's skin temperature to 3,000 degrees Fahrenheit, searing the ramjet off of its pylon and burning a hole in the ventral fin. Although less boundary layer heating occurred than predicted, any more meltdown and the plane's hydraulics would have been threatened. Thus, even in a moment of triumph, the X-15 and its pilot barely escaped disaster.[46]

---

[46]"X-15 Joint Program Coordinating Committee, September 16, 1964, Summary of Committee Action," DFRC Historical Reference Collection; "Contributions to Technology by the X-15 Program," author unknown, n.d., Milt Thompson Collection, DFRC Historical Reference Collection; Thompson, *At the Edge of Space*, 24-25, 252-264; "X-15 Flight Log," *NASA Facts*, Dryden Flight Research Facility, n.d., DFRC Historical Reference Collection; Hallion, *On the Frontier*, 119-125.

# THE LAST OF ITS KIND

Under the influence of the Research Airplane Project, flight research underwent a metamorphosis. Not only did the techniques adapt to the demand for higher and higher speed and altitude, but the Flight Research Center, the principal institution dedicated to civilian flight research, changed completely. Because of the national importance attached to the work and because of the courage involved in much of the flying, flight research won the greater notice, as well as the greater admiration from the public at large. Finally, the flights that propelled human beings into space--like fish flying out of water, Hugh Dryden said--accustomed the world to human spaceflight well before it became a routine accomplishment. Hypervelocity aeronautics also supplied much of the scientific and engineering know-how necessary to sustain these leaps for periods of longer and longer duration. The knowledge gleaned from X-15 flight research lent itself especially to the Space Shuttle by shedding light on many of the mysteries of the hypersonic flight regime. For example, although engineers discovered boundary layer turbulence at these extreme speeds, they also found less boundary layer heating than predicted. Those who analyzed the X-15 data likewise concluded that the degree of skin friction turned out to be lower than surmised. Important insights about aerodynamic heating taught spaceplane designers that irregularities on the surface of aircraft resulted in local hot spots and that excessive temperatures induced by severe shock-interaction inhibited flight at the top velocities.[47] Despite these and all of the other contributions of the X-15 program to aeronautics, once the space program took off, flight research found itself relegated to the background. In the post-X-15 world, after nearly 25 years pursuing supersonic and hypersonic speeds, what new role would flight research carve out for itself?

---

[47]Wallace, *Flights of Discovery*, 182-183.

# CHAPTER 6

## Slower and Cheaper:
## Lifting Bodies Flight Research

### REVERSE COURSE

Even during its halcyon days, the X-15 worried many at the Flight Research Center. At first, the worry stemmed from technical uncertainties. But once its place in the annals of aeronautics became obvious, another set of anxieties arose, this time involving the aircraft's total effect on the Flight Research Center. In time, the program's achievements and notoriety almost overwhelmed the center, engulfing the staff and diverting it from other projects. The FRC found itself taxed as never before to account for the unprecedented flow of money, hire new staff, handle the crush of public inquiries, monitor the contractors and their subcontractors, maintain and repair three aircraft of unprecedented complexity, conduct and analyze a long series of experiments, and present the results in published form. No wonder paperwork escalated, procedures mounted, and old hands mourned the passing of direct and personal working relationships. By the time he left the Flight Research Center in 1959 Walt Williams directed a staff of about 340 (88 in the Operations Division alone) who, in turn, managed a greatly expanding contractor workforce. The signs of the X-15's bureaucratic influence and programmatic supremacy were plain to see. An organization manual released at the end of 1962 did not just list every unit at the center; a one paragraph description of *every* functional responsibility followed, each identified by three- and four-digit code numbers. Moreover, during the three short years since the first X-15 powered flights began, the boxes on the FRC organization chart multiplied, doubling in both the Data Systems and Research Divisions and reflecting the deepening influence of the one big project. Indicators such as these raised troubling questions about the future well-being of the center. What would happen to the FRC and to its employees when the X-15 fulfilled its mission? One young engineer saw "nothing on the horizon" after it. "We knew," he said, "we weren't going to get to go faster and faster and faster...." Two senior men in the Research Division expressed their apprehensions more forcefully. Hubert Drake, the Assistant Chief of the Division, and Donald Bellman, Chief of the Performance Branch, although intimately involved in the X-15 since its origins, "were pretty much against this total involvement, and we were trying to propose some continuing work" which would endure beyond the hypervelocity project. Drake and Bellman thought ramjets might open a future for the Flight Research Center beyond the earlier research airplanes. But to the surprise of many, an almost whimsical flight vehicle proposed by Robert Dale Reed, a 32 year old FRC engineer from Idaho , evolved gradually into the X-15's stablemate and, in some respects, its research successor. Unassuming though this new flying machine and its descendants may have seemed, they proved in the end to be as worthy and as daring in their own right as the black rocketplane itself.[1]

The champion of this project conceived of it as a solution to one of the most vexing questions associated with the early American space program: how to return the astronauts to earth safely and efficiently once they completed their missions outside the atmosphere. Since arriving at the High-Speed Flight Research Station fresh out of college in 1953, he had witnessed

[1]Betty Love, interview with Michael Gorn, 10 April 1997, 13-14, DFRC Historical Reference Collection; HSFS Telephone Book, ca. 1959, Milt Thompson Collection, DFRC Historical Reference Collection; "Organizational Manual," 31 December 1962, DFRC Historical Reference Collection; Flight Research Center Organization Chart, December 1963, DFRC Historical Reference Collection; Ronald "Joe" Wilson, interview with Michael Gorn, 11 April 1997, 20-21, DFRC Historical Reference Collection (first quoted passage); Donald Bellman, interview with Richard P. Hallion, 4 March 1977, 29, DFRC Historical Reference Collection (second quoted passage); "Robert Dale Reed," *NASA Facts*, Hugh L. Dryden Flight Research Center, April 1976; Hallion, *On the Frontier*, 102-103, 273.

2

a succession of ideas geared toward mastering the re-entry conundrum. The first answer involved the parachute which proved to be safe and reasonably effective. But this method of breaking the fall of incoming spacecraft had significant deficiencies; not only was it impossible to predict the exact point of impact, but it consigned retrieval to the vastness of the ocean and entailed great expense due to the need for a recovery vessel or vessels. Moreover, this method exposed the astronauts to some undeniable risks. A failure to pinpoint the location of a splashdown opened the possibility of crew deaths due to drowning, and to the loss of capsules due to immersion. An invention by a Langley researcher offered the promise of eliminating parachute descents. Francis Rogallo, an aeronautical engineer who managed two of the laboratory's wind tunnels, fabricated a simple kite-like structure capable being flown by a pilot who achieved pitch and roll control by moving wires which shifted the center of gravity. Immensely lightweight and portable because it required no supports, this double arched, rectangular parawing for which Rogallo received a patent in 1951, seemed a promising alternative for spacecraft re-entry and recovery. The project first came to light at the Flight Research Center in August 1960 when pilots Milt Thompson and Neil Armstrong proposed towed flight testing of a simple paraglider research vehicle, or Paresev, to be designed and constructed at the Flight Research Center. Paul Bikle encouraged his staff to make research suggestions, but he also insisted the work be unique to the tools and talents of the FRC and that it be supported by budget and manpower estimates. He denied their request. But less than a year later (July 1961) he relented when a contract funded by the NASA Space Task Group changed his mind. The same North American Aviation then building the world's most complicated aircraft also won the right to construct this humble glider. If the prototype trials were a success, the Paresev would become a candidate for use in the Gemini space program. In summer 1961 the company's scale models underwent tests at the FRC. These experiments and the wind tunnel research on the Paresev seemed promising. But by mid-1962 the sad truth revealed itself; full-scale research on the ground and in towed flights (piloted by Thompson and Bruce Peterson) proved the glider to be too complex, not too reliable, incapable of supplying the lift-to-drag ratio

3

required to master the Gemini capsule, and lacking in even satisfactory control responsiveness. Cancellation befell the Paresev.[2]

But as the Rogallo wing fell out of favor, a second plan to achieve controlled re-entry and ground landings from space started to germinate. The kernel of the concept originated at the Ames Research Center. There, in 1950, H. Julian Allen announced his theory of blunt bodies re-entering the atmosphere from space. Allen calculated that as it plunged to earth, an object with a rounded, compact shape would protect itself from incineration by heating only the air surrounding it, through a mechanism known as *pressure* drag. Conversely, spacecraft with pointed noses or protuberances tended to become intensely hot themselves due to the effects of *frictional* drag. Later in the decade a colleague at Ames named Alfred J. Eggers asked himself what ideal shapes might best embody the blunt body proposed by Allen? To satisfy Allen's conditions, it needed to be free of wings or other appendages. In collaboration with C.A. Syvertson, G.C. Kenyon, and G.G. Edwards, in 1957 Eggers announced that a cone shape-- actually, since asymmetry imparted lift, a cone sliced in half lengthwise--offered the closest incarnation of Allen's blunt body. Hypersonic wind tunnel tests in 1958 and 1959 showed the

---

[2]While the Paresev failed to succeed the parachute as a means of spacecraft re-entry, the sport of hang gliding evolved from Francis Rogallo's creation, as did a parafoil used to slow the descent of the X-38 vehicle. Milton O. Thompson and Victor W. Horton, "Exploratory Flight Test of Advanced Piloted Aircraft--Circa 1963," *25th Symposium Proceedings of the Society of Experimental Test Pilots*, Beverly Hills, California, 23-26 September 1981, 230; Robert Zimmerman, "How to Fly Without a Plane: A Would-be Aviator Who Couldn't Get a Pilot's License Invented Hang-Gliding Instead," *American Heritage of Invention and Technology*, Spring 1998, 22-30; M.O. Thompson and N.A. Armstrong to Paul Bikle, 9 August 1960, Milt Thompson Collection, DFRC Historical Reference Collection; Paul Bikle to M.O. Thompson and N.A. Armstrong, 31 August 1960, Milt Thompson Collection, DFRC; "Paresev Flight Test Program," rough draft, n.d., Milt Thompson Collection, DFRC Historical Reference Collection; Paul Bikle to Commanding General, Aircraft Fleet Marine Force Pacific, 28 July 1961, Milt Thompson Collection, DFRC Historical Reference Collection; Victor W. Horton to Memorandum for Files, 22 August 1961, Milt Thompson Collection, DFRC Historical Reference Collection; Paul Bikle to Commander, Air Force Flight Test Center, 28 February 1962, Milt Thompson Collection, DFRC Historical Reference Collection; Paul Bikle to George Mellinger, 12 June 1962, Milt Thompson Collection, DFRC Historical Reference Collection; Robert Champine to Langley Associate Director, 19 December 1962, Milt Thompson Collection, DFRC Historical Reference Collection.

lift-to-drag ratios for this configuration to be almost right for stable high speed flight, but at subsonic speeds it exhibited gross instability. A series of intuitive modifications remedied this deficiency, resulting in a flat, upwardly inclined slope on the lower part of the tail and a downwardly tapered slope on the upper tail surface. The Ames scientists added rear vertical fins for directional stability, triangular elevon controls, and a cockpit canopy. Eggers and his associates refined the design and continued to run high- and low-speed wind tunnel experiments until 1964. They referred to it as the M(odification) 2 lifting body, a wingless flying machine held aloft by lift sustained only by the shape of its fuselage.[3]

At the beginning of 1962, Dale Reed--the FRC engineer whose almost fanciful wingless aircraft concept ultimately recast the center's research agenda--found himself, like most of his colleagues, in the throes of the X-15 project. For all of its engineering complexity, perhaps the greater complications arose from its role as a national testbed, influenced by many and controlled collectively. In contrast, the independent Idahoan initiated a project virtually indigenous to the Flight Research Center. It germinated in Reed's mind after he read the papers presented at a conference on high speed aerodynamics at Ames in 1958. In particular, "Preliminary Studies of Manned Satellites--Wingless Configurations: Lifting Body" announced the practicality of aerodynamic controls over lifting reentry and the probable flight paths. A practiced modelmaker, Reed decided to flight test the concept by building and flying a half meter-sized radio-controlled machine faithful to Alfred Eggers' designs. Before doing so, he embarked on an incremental flight test "program" undertaken entirely on his own initiative. His amused and puzzled co-workers watched as he flew countless paper airplane variants fashioned in the characteristic half-cone shape. Then he applied his results to a scale model fabricated from thin balsa sheets and stringers, to which he added adjustable outboard elevons and vertical rudders for flight control. First, he launched the little plane by hand into soft, tall grass to test its gliding qualities. After

---

[3]Reed contributed to the X-15 program by predicting the aircraft's structural integrity during overheating, a cause of concern in the planned speed build-up flights. C.A. Syvertson, "Aircraft without Wings," *Science Journal* (December 1968): 46-48; R. Dale Reed with Darlene Lister, *Wingless Flight: The Lifting Body Story* (Washington, D.C.: NASA SP-4220, 1997), vii.

experimenting with different control surface positions, Reed climbed onto the roof of the NACA hangar and onto other FRC buildings, released the aircraft, and watched time and again as it made a steep descent but landed upright on its tricycle landing gear. He tried flying the little plane like a kite, running as it trailed behind and above him on a string. Encouraged by its stability in flight, Reed attached the glider to his gas-powered model plane and observed it being towed in free flight. At sufficient altitude a timer released the balsa lifting body, which again demonstrated superior stability, a steep glide, and a safe landing. Finally, in February 1962 he flew the pair under radio control with similar encouraging results. By this time, the initial puzzlement of Reed's co-workers had vanished and a few became motivated to enlist in this home-grown project. His first recruit possessed perfect qualifications; an aeronautical engineering degree, experience in constructing his own gliders, and close friendships with the many wood and metal craftsmen at the FRC who shared his weekend hobby. Dick Eldredge and Reed acted as an ad hoc design team, working out the problems of control systems, fuselage skins, and structural bracing.[4]

Eventually, Reed needed Paul Bikle's imprimatur to give the project legitimacy. But Bikle's sympathies sometimes proved difficult to predict. He had at first turned down Milt Thompson and Neil Armstrong on their initial bid to launch another shoestring program, the Paresev. Moreover, he imposed on the two pilots high standards of planning and accountability before he finally gave his go-ahead. On the other hand, Bikle was an accomplished sailplane flier and designer who had set records for altitude and duration. He might well have a soft spot for the humble lifting bodies project. Reed and Eldredge sought to win an affirmative response from Bikle by first contacting Alfred Eggers at Ames. Eggers carried weight as the concept's originator and could also be very helpful in scheduling wind tunnel time to test the M2's flight characteristics. The Ames researcher agreed enthusiastically to support Reed's project, both in

[4]Reed, *Wingless Flight*, 9-15; Syvertson, "Aircraft Without Wings," 48; Milt Thompson, handwritten essay about the early lifting body program, n.d., 1, Milt Thompson Collection, DFRC Historical Reference Collection.

the attempt to win over Bikle and in obtaining access to the test facilities at the Sunnyvale laboratory. But Reed still lacked an ingredient necessary for him to win a victory for his project. Without the support of at least one research pilot Reed knew Bikle would probably say no. Indeed, in conversations the FRC Director already seemed to have closed the door on piloted lifting bodies flights. Eager to launch a full flight research program, Reed prevailed upon Milt Thompson, one of the center's most likable and popular figures. Thompson arrived at the center in 1956, having served in World War II as a naval aviator. He then earned a B.S.degree in engineering from the University of Washington and worked for three years as a flight test engineer with Boeing Aircraft. At the Flight Research Center he participated in many and varied projects. Between 1959 and 1963 he served (with Neil Armstrong and Bill Dana) as a pilot-consultant to the USAF's Dyna-Soar program. Thompson also initiated and assumed pilot duties in the Paresev test program. Most importantly, he flew the X-15 a total of 14 times between October 1963 and August 1965. An exuberant personality, Thompson nonetheless earned a reputation as a calm, precise, cerebral aviator who possessed an engineer's instincts and a will to solve problems. Reed hoped the backing of a man so highly regarded personally and professionally might persuade Bikle to relent and to allow the FRC pilots to fly his wingless machines.[5]

Accordingly, Reed and Eldredge asked Thompson if he would fly their strange little contraption, should it ever actually be completed. Thompson liked the technical kinship between the lifting bodies and his earlier experiences with Dyna-Soar and the Paresev. "I also had some free time," he nonchalantly remarked, and he accepted the proposal. Reed then arranged a

---

[5]Reed, *Wingless Flight*, 16-17, 31; Milton O. Thompson and Curtis Peebles, *Flying Without Wings: NASA Lifting Bodies and The Birth of the Space Shuttle* (Washington and London: Smithsonian Institution Press, 1999), 13, 18-29; Syvertson, "Aircraft Without Wings," 48; Thompson, handwritten essay about the early lifting body program, n.d., 1, Milt Thompson. DFRC Historical Reference Collection; Ronald "Joe" Wilson, interview with Michael Gorn, 11 April 1997, 15, 61, DFRC Historical Reference Collection; "Milton O. Thompson," *NASA Facts*, April 1976, Dryden Flight Research Center, Milt Thompson Collection, DFRC Historical Reference Collection.

meeting at Edwards with Eggers, Bikle, and the FRC principals.. All went as Reed hoped. Eggers promised to make available whatever equipment the project required; Reed offered to oversee the construction of the vehicle now known as the M2-F1 (for Modification 2, Flight Version 1), the product of his labors with Dick Eldredge. Once assembled, the machine would be trucked to Ames for tests in the 40 X 80 foot wind tunnel. But Bikle still balked at making a commitment to a piloted flight research program. After all, he had no authorization from headquarters and the strange looking aircraft did not inspire great confidence among the uninitiated. Bikle preferred a more cautious approach to the lifting bodies, choosing to defer decisions about flying them until the wind tunnel tests offered some proof of their true worth. But this did not satisfy Thompson. He wrote a memorandum to Bikle in May 1962 arguing that even a *minimal* initial flight test schedule should be undertaken. "The value of even a limited flight program utilizing a vehicle of this configuration," said Thompson, "is worth any amount of support which can be made available." He reminded the director of the rewards reaped by Paresev "with relatively insignificant expenditures and minimum personnel support." It demonstrated human control of the vehicle in flight and in so doing stimulated interest in the military and in NASA. In case budgetary or manning considerations prevented Bikle from planning for lifting bodies flight research, Thompson presented him a list of expedients by which to economize and simplify and thus preserve lifting bodies flight research: reduce instrumentation by up to 60 pounds; rather than fashioning it from metal, fabricate the M2-F1 structure and hull from materials like those of light aircraft; contract the design and construction instead of building it in-house; and even *eliminate the pilot's parachute*. Thompson, by now an ardent believer in the project, made a final appeal to his boss.

> The[se] suggestions are offered to reduce the amount of effort and money required to obtain a manned...vehicle. If one flare and landing can be demonstrated, the expenditures incurred would have been repaid. Even if funding for additional testing did not result, we could speak with some assurance to others interested in promoting this configuration and again create interest within the NASA and [the] military. Until the demonstration of piloted flare and landing capability of this vehicle has been

accomplished and some investigation of manned control attempted to correlate and support model and tunnel testing, this configuration will not be given proper consideration during competitive selection of a configuration for space missions.[6]

Bikle conceded Thompson's point and during spring and summer 1962 the principals immersed themselves in building a piloted lifting body. Reed assembled a staff of about 13 full- and part-time engineers and fabricators to design all of the parts and fittings, the tricycle landing gear (borrowed wheels and nose gear from a Cessna 150), the tubular steel carriage, and the aluminum sheet metal tail fins and controls. Encouraging a sense of cohesion, Reed delineated his team's workspace by cloistering them behind a canvas curtain. Developing an almost paternal interest in the project, Paul Bikle visited Reed's offices nearly every day. By the time construction of the first vehicle started in October 1962, Bikle had reduced the burden on the lifting bodies group by contracting out the assembly of the fuselage and the canopy; leaving the FRC engineers and mechanics to fabricate the machine's internal structure and components. Bikle further assisted the cause by persuading Gus Briegleb, a friend in the sailplane community, to construct the lightweight wooden shell whose complex contours would cover the interior framework being erected by Reed and his staff. An artisan in cloth and wood, Briegleb soon found his hangar in nearby El Mirage, California, teeming with the whine of saws slicing mahogany plywood, mahogany ribs, and spruce supports. Yet, because Briegleb was more a craftsman than a businessman, his $5,000 bid proved to be woefully inadequate. Bikle padded his reward to $10,000, which still constituted a bargain for a hand-crafted fuselage weighing less than 300 pounds and capable of withstanding the rigors of flight. Briegleb's twin-keeled, cross-braced structure satisfied every demand. Another of Bikle's sailplane comrades, Ed Mingele of Palmdale, fashioned a plexiglas canopy. In February 1963 the FRC technicians required only four bolts to attach their carriage to the Briegleb hull. Before them stood the 1,000-pound M2-F1 in its wood and metal incarnation, sheathed in Dacron, doped for durability, and measuring 22

---

[6]Thompson, handwritten essay about the early lifting body program, n.d., 1, Milt Thompson Collection, DFRC Historical Reference Collection; Syvertson, "Aircraft Without Wings," 48; Reed, *Wingless Flight*, 16-17; Milt Thompson to Paul Bikle, 11 May 1962, Milt Thompson Collection, DFRC Historical Reference Collection (all quoted passages).

feet long, 14 feet wide at the elevons, and 9.5 feet tall at the rudder.[7]

No mere wind tunnel dummy, this aircraft contained all of the structures and supporting equipment necessary for flight. Milt Thompson asked for, and Reed supplied, a simulator to practice on before actually taking to the skies. Devised by two junior engineers named Bertha Ryan and Harriet Smith, the machine duplicated the feel of pilot input on the aircraft's stick and on its rudder pedals. The women programmed their full-cockpit simulator with aerodynamic characteristics derived from Ames wind tunnel tests on M2-F1 models. Meanwhile, Reed and Eldredge designed the flight controls with Thompson's help, together deciding the best way to regulate roll and yaw in the cockpit. Like Bertha Ryan and Harriet Smith, a keen young engineer named Ken Iliff also volunteered to serve on the project. Reed put him to work on several complex mathematical problems designed to predict the aircraft's controllability, its optimum lift-off speeds, and the amount of power required to tow the wingless aircraft. Since the lifting body had not yet been proven in flight, Reed and his associates decided to use an automobile, rather than an aircraft, for the initial trials. Iliff calculated that the half-ton M2-F1 needed to be accelerated to a speed of at least 100 miles an hour before it would take-off, thus requiring a car with greater horsepower than any on the center. Bikle answered the problem with money from his discretionary account, the same source which paid Briegleb the $10,000. With it, his staff purchased a new, 1963 Pontiac Catalina softtop convertible with the biggest production engine available. One of Joe Vensel's assistants then drove the length and breadth of Southern California, stopping at one specialist mechanic after another until the Pontiac was converted into a virtual dragster, complete with racing tires, heavy duty shock absorbers, transmission, radiator, and rear end, dual exhaust, and roll bars. In the end it became a 150-mile per-hour speedster more than capable of the task outlined by Iliff. But, so no one might think NASA spent federal

---

[7]Reed, *Wingless Flight*, 19-23, 32; Thompson, handwritten essay on the early lifting bodies, n.d. 1-2, Milt Thompson Collection, DFRC Historical Reference Collection; Syvertson, "Aircraft Without Wings," 48-49; Ronald "Joe" Wilson, interview with Michael Gorn, 11 April 1997, 17, DFRC Historical Reference Collection; Thompson and Horton, "Exploratory Flight Test," 232.

money on exotic race cars, it appeared on the center's procurement rolls as separate parts.[8]

Milt Thompson practiced on the simulator much of February 1963 in anticipation of the first flight. At the end of the month Reed's team finished its work and on the first of March Thompson mounted the little vehicle and rode in tow behind the speeding Pontiac. At 75 knots he gradually pulled up the nose of the M2-F1 and it rose, only to stay aloft for a moment at a time as it bounced from one main wheel to the other. The machine was uncontrollable. After two attempts, everyone realized the time had arrived to call in Alfred Eggers' offer to test the M2-F1 in the Ames full scale wind tunnel. Eggers managed to free the machine for two full weeks so the lifting body could undergo all of the necessary tests. Dale Reed decided to pack his small trailer and drive to Mountain View with his wife and small children. Milt Thompson also prepared to stay the whole time. The small aircraft traveled the 350 miles between Edwards and Moffett Field uncovered on a flat-bed truck. Its strange appearance actually attracted a crowd of curious onlookers when the drivers stopped to eat lunch. Upon its arrival, the M2-F1's technicians mounted it on flexible poles inside the cavernous tunnel. During the entire two weeks Milt Thompson, still the sole research pilot associated with the project, alternated with Dick Eldredge in "flying" the aircraft as the screaming artificial winds flowed by. Both men worked the controls while instruments recorded force and moment information. The sensation of flight suggested some minor aerodynamic clean up work For example, the tests determined that a pulsating sensation in the stick arose when Kármán vortices formed on the base of the plane and flowed backwards, pounding its lower aft section. Two aluminum scoops placed on the base alleviated the problem. Perhaps more important than immediate fixes such as these, the tunnel work allowed the FRC visitors to return home with full sets of new and more accurate data plottings. Overall, the tests suggested the aircraft should be airworthy.[9]

---

[8]Reed, *Wingless Flight*, 23-35; Thompson, handwritten essay on the early lifting bodies, 2, Milt Thompson Collection, 2, DFRC Historical Reference Collection; Syvertson, "Aircraft Without Wings," 48; Walter Whiteside, interview with Betty Love, n.d., 8-9, DFRC Historical Reference Collection.
[9]Thompson and Peebles, *Flying Without Wings*, 25, 69-70; Thompson, handwritten essay on the early lifting bodies, n.d., 3, Milt Thompson Collection, DFRC Historical Reference Collection;

The lifting bodies team arrived back at Edwards in late March and prepared to resume the flight research program. First they programmed the new aerodynamics information into the simulator and found the results differed from those generated by the smaller tunnel tests of M2-F1 models, leading Reed and his associates to reconfigure the lifting body's flight controls. In the meantime, Paul Bikle received an internal FRC inspection report of the project which praised the preparations for the impending flight research program but advised close scrutiny before embarking on the more advanced stages of envelope expansion. This report was forwarded to the Director on April 4; on the very next day Milt Thompson climbed back into the M2-F1's cockpit and rolled out to Rogers lakebed. The changes in the controls worked wonders. On April 5th Thompson rose behind the Pontiac but this time found the aircraft "flew surprisingly well." In fact, to prove it to himself and the observers, he piloted the M2-F1 ten more times that day, all with satisfying results. Unfortunately, Thompson needed to attend to his Dyna-Soar duties and did not resume his lifting bodies work until the 19th of the month. When he did return he collaborated intensely with Reed and his team for three straight weeks, gaining valuable experience for himself and the program over the course of about 64 flights. Although at times the Pontiac raced as fast as 115 miles per hour across the lakebed, the M2-F1 rarely flew higher than 100 feet. During these car tow flights, technicians added instrumentation to the so-called "Flying Bathtub." Because no electronic systems existed on the M2-F1, radio signals from 15 sensors aboard the aircraft transmitted flight data to the control room in the FRC headquarters building relative to airspeed, altitude, and angle of attack; roll, pitch, and yaw; control position data from the elevator, rudders, and elevons; stability and control; and vertical, side, and longitudinal accelerations.[10]

---

Reed, *Wingless Flight*, 38-46; Syvertson, "Aircraft Without Wings," 48-49.
[10]Subsequent lifting body pilots also trained on the car tow system. Eventually, the Pontiac pulled some 400 flights. Lifting Body DEI Board to Paul Bikle, 4 April 1963, DFRC Historical Reference Collection; Reed, *Wingless Flight*, 45-46, 49-50; Thompson, handwritten essay on the early lifting bodies, 3, 5, Milt Thompson Collection, DFRC Historical Reference Collection; Thompson and Peebles, *Flying Without Wings*," 72-74; Flight Log (handwritten), 1963 lifting bodies flight program, n.d., File Number 002313, Headquarters NASA Historical Reference

After much preparation, the first air-tow flight occurred in mid-August 1963. These runs differed greatly from the car-powered ones. Tethered on a 1,000 foot towline to a C-47H aircraft, the lifting body and its pilot remained attached for 30 to 40 minutes as the "Gooney Bird" achieved altitudes as high as 13,000 feet. After release, Thompson had just three minutes to perform the programmed glide maneuvers and to return to the lakebed. The approaches left the uninitiated breathless. The lifting body, which flew little more than 100 miles per hour in normal free flight, required about 150 miles per hour for a safe landing. Consequently, Thompson and subsequent aviators picked up momentum by diving toward the desert floor at a 30 degree angle, ten times that of a conventional aircraft. Continuing on this seeming path to oblivion from the 1,000 foot level to a mere *200 feet* off the hard clay surface, only 15 seconds remained to perform a flare, level-off, and make any final adjustments before touching down. Early on the morning of August 16, 1963--the calmest part of the day at Rogers Dry Lake and the standard take-off time for all the air-tows--Milt Thompson began the process of learning these and other properties of the M2-F1. Many learned with him; the entire Flight Research Center staff stood on rooftops, on the aircraft parking ramp, even at the edge of the lakebed to watch the glide path of this bizarre machine. As expected, Thompson encountered wake flow disturbances from the C-47 as it accelerated on the ground and as the tow plane and then the M2-F1 lifted off. He quickly maneuvered well above the C-47 to avoid further turbulence and, keeping the bigger aircraft in sight through the nose window, remained in this position for most of the flight. Upon release he tried several turns without incident and then lined up for the runway at 2,600 feet. Just as he began to flare a pilot induced oscillation (PIO) of two to three cycles occurred. This phenomenon did not come as a complete surprise since the car tow flights also indicated oscillation problems in roll control. He eased back and tried the flare again at 2,250 feet. Once more, two to three cycles of longitudinal oscillation developed, but he continued his approach and landed safely. Thompson reported that the problem receded as his

Collection.

13

speed decreased. Overall, however, the plane handled with ease and agility. Many more M2-F1 flights remained until the last one in August 1966 and Milt Thompson would pilot the majority of them (45 of 77). But even as early as August 1963 one fact could not be disputed: an aircraft without wings not only flew, but flew well, and might hold the secret to safe and controlled re-entry from space.[11]

## BIGGER LIFTING BODIES

The M2-F1 glide flights yielded some invaluable data about the handling and the aerodynamics of wingless aircraft. But in order for lifting bodies to represent a true re-entry breakthrough, their characteristics in the turbulent air currents between the transonic and supersonic ranges needed to be ascertained. Dale Reed called this next giant step Configuration II. Due to of keen NASA interest in the subject, Bikle supported a proposal to build two heavyweight M2s, one for front-line use, the other as a backup in case of accidents. Headquarters NASA agreed to fund bigger and faster lifting bodies, but rejected Bikle's specific recommendations. Perhaps because the FRC director had kept Washington in the dark about the M2-F1, he now paid the price by being forced to accept some unpalatable terms. The headquarters allowed him to develop prototypes of operational lifting bodies, but these vehicles would serve only as *wind tunnel models*; Washington approved no funds for a flight research program nor money to develop a new powerplant. NASA also rejected the twin M2 procurement and instead insisted the Flight Research Center purchase one M2-F2, and a second vehicle

---

[11]Thompson, handwritten essay on the early lifting bodies, 5, Milt Thompson Collection, DFRC Historical Reference Collection; Syvertson, "Aircraft Without Wings," 49; Thompson and Horton, "Exploratory Flight Test," 233, 235; Milt Thompson, "Pilot Flight Notes, M-2 Flight, August 15, 1963," Milt Thompson Collection, DFRC Historical Reference Collection; Thompson and Peebles, *Flying Without Wings*, 74-88; Reed, *Wingless Flight*, xviii.

known as the HL (Horizontal Lander)-10. This machine sprang from the minds of Langley researchers. Beginning in 1957, the Hampton scientists conducted investigations parallel to those of Alfred Eggers, but they arrived at different conclusions. The Langley team decided a lifting body with negative camber and a flat bottom would be stable on all three axes and would provide greater hypersonic lifting capacity. It looked like a modified delta-wing aircraft. The initial Langley designs appeared in a paper presented by John Becker at the same 1958 High-Speed Aerodynamics Conference at which Eggers unleashed his half-cone theory. By 1962 the HL-10 had evolved into its familiar shape.[12]

During the last weeks of 1963 Dale Reed assembled a group to write a joint statement of work for the two vehicles. Through it, he asked prospective contractors to submit proposals (on a fixed price basis) for sequential fabrication of two subsonic gliders (M2 first, HL-10 afterwards) built to the exact external dimensions specified by NASA. "The vehicles," said the document, "will be of relatively low-cost construction, involving no system or hardware development. They are primarily aerodynamic research configurations designed to investigate the requirements and problems that may face future pilots and hardware designers." To adhere to a tight budget, potential contractors found it necessary to formulate designs using off-the-shelf items, as well as components already in NASA's inventory. Both machines were subject to the same weight limitations: 4,000 pounds empty, 7,000 pounds with full ballast tanks. Although NASA Headquarters denied the FRC a flight research program until a future time, the statement of work clearly anticipated one. While both models faced extensive wind tunnel tests, they also required the capacity to withstand air drops from a B-52 and to be fully equipped as flying vehicles, complete with sophisticated flight control mechanisms and adequate environmental control systems for "the pilot's safety and reasonable comfort...." The contract winner also needed to allow roughly six cubic feet for NASA instrumentation. Finally, like the M2-F1, the

---

[12]Thompson and Peebles, *Flying Without Wings*, 96-100; Reed, *Wingless Flight*, 69-71; Robert W. Kempel, Weneth D. Painter, and Milton O. Thompson, *Developing and Flight Testing the HL-10 Lifting Body: A Precursor to the Space Shuttle* (Washington, D.C.: NASA Reference Publication 1332, 1994), 7-9.

next generation of lifting bodies required landing assist rockets on the underbelly of the hulls in case of touchdown emergencies. The aircraft industry received request for proposals at the end of February 1964, to which five firms (Ryan, United Technology, Norair Division of Northrop, General Dynamics Astronautics Division, and North American Aviation) responded. The source evaluation board reduced the number to three when it eliminated United Technology and Ryan withdrew. The board rendered its verdict in mid-April when it chose Northrop/Norair of Hawthorne, California. Its bid of $1,200,000, while $200,000 more than North American, impressed the panel in several respects. Northrop offered to use tried and reliable T-38 components in assembling both vehicles, possessed a workforce experienced with experimental aircraft (recently, the X-21), and proposed a lifting bodies control system which satisfied the request for proposal. Clearly, despite the small return, Northrop really wanted the job because it assured the firm's entry into the manufacture of space products. The company invested about $1.8 million on the project and NASA paid over $4 million for the two completed vehicles with propulsion systems. On the scale of major space programs, this outlay constituted a bargain of the first order.[13]

During the remainder of 1964 the Flight Research Center and the Northrop teams sorted out their respective responsibilities and got down to work. During this initial phase Paul Bikle faced a hard choice. He reposed the highest confidence in Dale Reed's technical capacity; indeed, the center owed the very existence of the lifting bodies program to Reed. Without his personal commitment the concept may have remained one of the countless orphaned aeronautical ideas. At the same time, the FRC Director wanted an experienced program manager to run this technically complex, cost-conscious, concurrent project. Ultimately, Bikle made Reed the

_____

[13]The Statement of Work stressed the use of off-the-shelf systems wherever possible. Statement of Work (draft), FRC, "Design and Fabrication of Two (2) Research Lifting Body Vehicles," 21 January 1964, Milt Thompson Collection, DFRC Historical Reference Collection (both quoted passages, 1, 14); Request for Proposal, "Design and Fabrication of Two (2) Research Lifting Body Vehicles, Amendment Number 1," 27 February 1964 DFRC Historical Reference Collection; Hubert Drake to Headquarters NASA Director of Advanced Research and Technology, 21 April 1964, DFRC Historical Reference Collection.

project engineer and named John McTigue, an X-15 operations engineer, the project manager. McTigue's in-depth knowledge of the X-15's complicated propulsion system proved to be especially welcome; the powerplant chosen for the two heavyweight lifting bodies turned out to be the old XLR-11, withdrawn from museums and assorted locations and refurbished by Reaction Motors to deliver 2,000 pounds of thrust in each of its four barrels. In another example of frugal program management, Bikle demanded that virtually every component on either of the two aircraft pass the test of necessity. When Milt Thompson assumed he would have an altitude indicator during the glide flights, he learned otherwise. Bikle vetoed the purchase of the expensive instrument during this phase since all flights would occur in clear weather. Finally, McTigue and his Northrop counterpart Ralph Hakes decided from the start to avoid the combative, or at least adversarial relationship common among government and industry representatives during development programs. In their Joint Action Management Plan they chose instead to integrate the two sides into a single cohort.

> The keys to our success were mutual respect, trust, and cooperation. The Northrop engineers respected and trusted not only the expertise of the NASA engineers in aerodynamics and in stability and control analysis but also our operational experience with rocket-powered aircraft. Equally, the NASA engineers trusted and respected the outstanding ability of the Northrop engineers in fabricating airframes. Working one-on-one in small groups, we made on-the-spot decisions, avoiding the usual time-consuming process of written proposals and counterproposals in solving problems and making changes.[14]

Other signs of cooperation informed the lifting bodies project. The same Paul Bikle who attempted to outfit the two research vehicles for the lowest cost possible also sought economy in repair and maintenance. The USAF offered a way to reduce these outlays. Air Force Systems Command and Headquarters USAF committed the service to a lifting bodies program of its own, but it deemed it worthwhile to conduct some of the research in cooperation with NASA. In exchange for Air Force pilots sharing time in the cockpit with the NASA aviators, the Flight Test

---

[14]"Joint-Action Management Plan: A Report of the Management Plan used by NASA-NORTHROP to Design and Build the World's First Supersonic Lifting Body Vehicles--M2-F2 and HL-10," n.d., DFRC Historical Reference Collection; Thompson and Peebles, *Flying Without Wings*," 96-100; Reed, *Wingless Flight*, 77-78 (block quote).

Center Commander offered to assume responsibility for maintenance of the XLR-11 rocket engines, maintenance and operation of launch and chase aircraft, radar tracking, photographic support, crash and rescue, and medical services. Overtly patterned after the X-15 cooperative flight test program, a Memorandum of Understanding signed by Bikle and Major General Irving Branch of AFFTC on April 19, 1965, established a Joint FRC-AFFTC Lifting Body Flight Test Committee with Bikle as chair and Branch as vice-chair; one pilot, one engineer, and one project officer from each organization; one instrumentation expert from the FRC and a medical officer from AFFTC. Like the X-15 Joint Committee, it concerned itself with local flight test matters "which would normally fall under the jurisdiction of the NASA-FRC and the AFFTC." Another measure of fruitful collaboration involved the Langley Aeronautical Research Center. In February 1965 some lifting body researchers from Hampton visited the Flight Research Center and proposed a series of small but critical modifications of the HL-10. Their recommendation for *six* new control surfaces (elevator flaps on the elevons and outboard tip-fin flaps), announced more than halfway through the Northrop development program, did not generate much enthusiasm. But because the improvements grew out of new wind tunnel data suggesting a significant improvement in lift-to-drag ratio (rising from 3 to 3.4 out of a target 4) the FRC and Northrop team agreed to comply. Since program planners expected the M2-F2 to be ready before the HL-10, an eventual six month gap between the completion of the two constituted no surprise. But the late changes did yield some gratifying results; the lift-to-drag ratio rose as predicted, the added surfaces simplified the flight control system, and the modifications resulted in the installation of a simple switch whose activation allowed pilots to cross from subsonic to supersonic flight with minimal trim changes.[15]

At last, the M2-F2 arrived at Edwards on June 16, 1965. Since the wind tunnel tests

[15]Milt Thompson, telephone interview with Nancy Brun, 19 May 1976, File number 002313, Headquarters NASA Historical Reference Collection; Project Directive 65-53, "AFFTC/NASA - FRC Joint Lifting Body Flight Test Program," 19 February 1965, AFFTC/HO Historical Reference Collection; Memorandum of Understanding, AFFTC and NASA FRC on Joint NASA/FRC-AFFTC Lifting Body Flight Test Committee, 19 April 1965, AFFTC/HO Historical Reference Collection (quoted passage); Reed, *Wingless Flight*, 79-80.

proved so helpful to the M2-F1, late in July the new machine also went aboard a truck for conveyance to Ames and the 40 x 80 foot tunnel. If anything, these tests were more complex than the last. For example, responding to Langley worries that an upflow of air might push the released lifting body into a collision with the mothership, Reed and his cohorts used the full-scale tunnel to test the safety of the B-52 drop and pylon. In the end, the researchers returned to the Flight Research Center full of data on the machine's aerodynamic characteristics and its handling qualities. For months following, this information and a series of intensive ground checkouts occupied the lifting bodies staff. Each component and sub-component underwent individual inspections. For instance, the flight control machinery alone actually consisted of three distinct elements: the hydraulics system, which powered the actuators and moved the appropriate surfaces; the mechanical system, which animated the actuators; and the stability augmentation system. After each of the three passed muster, the flight control system was tested as an integrated unit for effectiveness and reliability. Finally, the engineers and technicians mated all of the main systems and checked the M2-F2 as a total aircraft. During one of the trials Milt Thompson noticed that depressing the radio microphone switch mysteriously disengaged the pitch stability augmentation system, a serious problem indeed. Only after extensive diagnostic work did the team find the failure in the electrical system. Even after the completion of the combined component tests the M2-F2 still faced evaluations while linked to the mothership, which itself had undergone step-by-step checkouts. Finally, intense scrutiny focused on the adapter which fastened the lifting bodies to the same hooks which carried the X-15 on the B-52 wing pylon.[16]

These essential reviews--which, like most center projects, lasted a solid year--occurred concurrently with flight plan preparations. The Lifting Body Flight Test Committee plotted out the basic approach, patterned after that of the X-15. The lifting bodies engineers designed each mission in keeping with a classic definition of their objectives: "the assembly of the various

---

[16]Reed, *Wingless Flight*, 80-82; Thompson and Peebles, *Flying Without Wings*, 102-104.

desired tests and maneuvers into a single and coherent flight from launch to landing which will

provide a maximum data return and simultaneously [e]nsure the highest possible level of flight

safety." Each individual flight plan arrayed the pilot's activities and the scientific experiments

in descending priority. Above all, the planners concentrated on measures to improve the safety

of flight in future operations. Next in line were the actual tests associated with the primary

objectives of the lifting bodies flights, then the evaluations of the aircraft and their subsystems,

and finally, the test bed experiments. In case of in-flight emergencies, detailed reviews

determined whether flight plans required modification to increase safety factors. Investigators

probing such occurrences centered their inquiries on subsystem malfunctions, errors in predicting

aerodynamic forces, atmospheric conditions, and combinations of the three. Those in charge of

M2-F2 and HL-10 data reduction provided immediate post-flight analyses in order to incorporate

knowledge gained from the preceding flight into the following one.[17]

Milt Thompson inaugurated the M2-F2 flights on March 23, 1966. He started with a

series of captive tests. The pylon and the adapter holding the little rocket plane underwent some

B-52 rudder kicks in order to assess their strength; both proved more than adequate. Also, the

FRC engineers got a good idea of the M2-F2's control surface loads during mated flight by

calibrating the readings on the lifting body's altimeter, speed, and rate of climb to the bomber's

instruments. These flights continued through July 6. On the morning of the first free flight (July

12) Paul Bikle seemed nervous. He had launched the M2-F1 program without the knowledge or

permission of his superiors and now the acid test of its successor had finally arrived. He told

Flight Research Center Public Affairs officer Ralph Jackson to be ready for failure, because "if

the M2-F2 crashed, the two of them would walk out the front door and keep on walking." The

flight planners set modest goals for the initial foray. Thompson half-jokingly stated his own

modest objective: to land safely on the designated runway. In fact, however, the landing

technique really did constitute the main objective. After that, Thompson would concentrate on

---

[17]Study Plan, "Flight Planning for the M2-F2/HL-10 Flight Test Program," 17 January 1966, AFFTC/HO Historical Reference Collection.

pilot induced oscillation during approach, verify whether the many subsystems operated as a cohesive unit, and evaluate the predicted launch behavior versus the real thing. The flight plan, reflecting Milt Thompson's influence, required a precise sequence of events: separation from the B-52 at 45,000 feet, an increase in airspeed to 220 knots, a 90 degree left turn during descent in altitude from 39,000 to 30,000 feet, and a pushover to minus three angle of attack in order to increase airspeed to 300 knots. Then, at 22,000 feet Thompson would start a flare producing level flight at around 18,000 feet. Decelerating below 200 knots, he would fire the landing rocket briefly to ascertain its proper functioning. Afterwards, Thompson would again push over, accelerate to 190 knots, and at 16,000 feet roll into a 45 degree bank to align the M2-F2 with runway 18 on the north part of Rogers Dry Lake. During this maneuver he would accelerate to 300 knots in preparation for the landing flare initiated 1,000 feet over the lakebed. Thus, in a total time of 3.5 minutes, the pilot would assess the lifting body's essential handling qualities, learning at the first turn the all-important lateral-directional control characteristics, and at the first flare the degree of pilot induced oscillation. During the remaining time, Thompson faced three main decisions: proceed as scheduled, modify the flight plan, or eject if he encountered uncontrollable handling. Unfortunately, the small aircraft's flight control system--especially the pilot-actuated linkage between aileron and rudder motions-- proved not to be his best friend in a crisis.[18]

The actual operation began at 6:30 a.m. on July 12, 1966, when Thompson, the M2-F2, and the B-52 took off. The separation proved to be mild with good pitch control. But just as the pilot pushed over to start the first flare (following the first turn), a disquieting lateral-directional oscillation began at low angle of attack. He responded by lowering the interconnect ratio (of ailerons to rudder) and increasing speed, which improved the situation. The flare then proceeded under good control, in a manner predicted by the simulator. But at the start of the push over into the final turn the lateral-directional oscillations started anew. Thompson again put his hand on

---

[18]Thompson and Peebles, *Flying Without Wings*, 104-107, 112-118 (quoted passage, 123).

the interconnect ratio wheel and turned it down--or so he thought. The oscillations increased to 10 degrees. On final approach, in a 30 degree dive, at 300 knots, and only 27 seconds from impact, Thompson found himself in terrible trouble. He again took the wheel and further reduced the ratio between rudder and aileron deflection. The lateral rotations increased to 45 degrees in both directions. One more crank and he lowered the ratio to zero. Yet, as if the control had been wired backwards, the aircraft now rocked *90 degrees* each way. Onlookers on the ground saw "the vehicle swinging madly from side to side." Thompson then commanded himself to do the hardest thing in a plane falling out of control: nothing. He simply let go of the stick, the accepted method of halting pilot induced oscillation. As he looked down, the source of the calamity dawned on him, and not an instant too soon. Because the ratio control on the simulator operated by a lever, and the actual aircraft's on a wheel, he had become confused, turning the control all the way up instead of down, causing the M2-F2 to become highly sensitive to control inputs which only aggravated his attempts to regain stability by maneuvering the stick. Turning down the ratio control ended the crisis. He flared at 1,000 feet and dropped down with a feeling of confidence in a machine that now obeyed all his commands. At 50 feet and 240 knots he lowered the landing gear and touched down with a little bouncing and a straight rollout. "Milt really pulled the flight out of the bag," wrote Joe Wilson, a young FRC engineer who witnessed the hair-raising approach.[19]

Milt Thompson flew his last lifting body mission on September 2, 1966, after which a number of men took his place. Bruce Peterson, a NASA research pilot and former Marine Corps aviator, became one of the lifting bodies stalwarts. His connection with the program began after Thompson's seventeenth consecutive M2-F1 flight, when Paul Bikle decided a second individual

---

[19] Flight Records, "NASA-FRC M2/HL10 Flight Initial Schedule," Flight Number M-1-8, 12 July 1966, DFRC Historical Reference Collection; Thompson and Peebles, *Flying Without Wings*, 118-123; Reed, *Wingless Flight*, 88 (first quoted passage); diary of Ronald "Joe" Wilson, Volume 1 (1965 to 1969, entry for 12 July 1966), DFRC Historical Reference Collection (second quoted passage). For an engineering assessment of the M2-F2 PIO see Robert W. Kempel, NASA Technical Note D-6496, "Analysis of a Coupled Roll-Spiral-Mode, Pilot Induced Oscillation Experienced with the M2-F2 Lifting Body," (Washington, D.C.: NASA, 1971).

needed to be checked out to fly the machine. Peterson got the nod because Thompson knew him from the Paresev flights the year before. During 1964 they alternated the M2-F1 chores, averaging one flight each per month. Therefore, when Thompson departed, Peterson's experience put him in line to be the project pilot and to take the controls of the third NASA lifting body, the HL-10. Northrop delivered the aircraft on January 18, 1966. As with the M2-F2, the wind tunnel, checkout, and simulator phases of the HL-10 occupied almost an entire year. The M2-F2 had virtually the identical dimensions to its predecessor, the M2-F1. On the other hand, the HL-10 gleaming in the FRC hangar, sprang directly from its drawings. Its delta planform swept back 74 degrees, featured three vertical fins, and measured just over 21 feet in length and 13.6 feet in span. Its weight far surpassed the original specifications: nearly 6,500 pounds in glide, just over 10,000 pounds at launch weight with propellants. By the time Northrop completed this machine, Dale Reed began to tire of the increasing bureaucracy associated with the large program now under his technical direction. He knew the demands would only become greater once the HL-10 joined the M2-F2 in the skies Moreover, Reed yearned to return to real engineering, particularly remotely-controlled vehicles. When his friend and fellow model aircraft enthusiast Gary Layton said he would like to take over as lifting bodies project engineer Reed agreed, and Paul Bikle ratified the transfer of responsibilities.[20]

The first indication of what Reed left behind occurred three days before Christmas, 1966, when the HL-10 flew for the first time. Bruce Peterson no sooner separated from the B-52 than he became aware of serious difficulties. He encountered instability in pitch as the aircraft oscillated from 15 to 12 degrees angle of attack. With more speed, the motions became worse. In coping with the control system he missed his turn at the designated altitude. He finally made the maneuver at 38,000 feet, saw the lakebed and aimed for it. Then to gain speed he pushed over to 6 degrees but had "an awful time holding it" and began to feel the first sensations of pilot induced oscillation as the aircraft accelerated. As he began to flare, the HL-10 rolled; he

[20]Reed, *Wingless Flight*, 55-57, 93-94; Kempel, Painter, Thompson, *Developing and Flight Testing the HL-10 Lifting Body*, 1994, 11-15.

counteracted it using the rudder. Peterson managed to land safely and actually found the machine possessed better steering than the M2-F2. But he could not deny the aircraft's predominant feeling of instability. "I was having a little trouble," he declared with understatement. "It [the pitch stick] was oversensitive and I really had to work at keeping it the way I wanted to, and I was trimming a little bit forward to keep a little back pressure to keep myself out of ...[pilot induced oscillation (PIO)]....I never did actually PIO the machine though." The problems proved to be more deep rooted than first thought. After much debate among the Flight Research staff, the principal parties met at Langley to discuss the main conclusion: the difficulties stemmed from massive airflow separation. To everyone's amazement, upon hearing this hypothesis Bob Taylor of the Langley aerodynamics team rose from his chair, threw his mechanical pencil to the ground, and uttered a string of obscenities. Taylor berated himself for not heeding his instincts; during earlier HL-10 tests in the 7 x 10 high speed tunnel he suspected that flow separation on the real aircraft would be worse than the model, but never took steps to design a remedy. But after considerable new Langley wind tunnel research, a solution finally emerged. By slightly extending and cambering the leading edges it seemed the instability experienced by Peterson could be minimized. The Flight Research Center contracted again with Northrop/Norair whose engineers designed fiberglass gloves of the desired shape to attach to the HL-10's leading edges. Unfortunately, *fifteen months* elapsed between Peterson's unpleasant encounter with the HL-10 and the delivery of the gloved version, leaving the flight research program still in its initial stages in March 1968.[21]

The M2-F2 suffered an even worse fate. Five months after Bruce Peterson piloted the HL-10's first flight--and with no lifting body experiences in between--he conducted a glide test of the M2-F2 with its rocket engine installed. On May 10, 1967, he and the aircraft dropped from the B-52 at 45,000 feet and began a standard descent, including two turns and three distinct

---

[21]Flight Records, First Flight of the HL-10, Flight number HL-1-3, 22 December 1966, DFRC Historical Reference Collection (quoted passage); Kempel, Painter, Thompson, *Developing and Flight Testing the HL-10*, 22-28.

phases. During the first and second parts he performed the planned research maneuvers; after the second turn he initiated his landing. But as he leveled out after the second turn, the old nemesis of lateral oscillation manifested itself and increased to violent proportions. Peterson got control of the roll in eleven seconds, but by then he had veered off of his heading on Runway 18 by 12 degrees. Too late to adjust, he began his flare immediately. He approached Rogers Lake without the runway markings or visual cues to guide him and because of his unorthodox glide path found himself closing with a rescue helicopter. Peterson became distracted, and with the chase plane at some distance to avoid any chance of a collision due to the earlier oscillations. he received none of the normal altitude callouts as he neared the desert floor. He fired his landing rockets to gain just a little time as he descended but this improbable chain of events ended when the vehicle struck the ground an instant before the landing gear locked into position. The aircraft skidded some distance, and then rolled over and over before coming to rest on its back. Peterson, lucky to be alive, nonetheless suffered severe head trauma--a fractured skull, serious facial injuries, and a damaged right eye (which he later lost to infection). Horrified by the crash and by the injuries to his friend, Milt Thompson put the blame squarely on the aircraft, calling it the lifting body with the worst performance and (due to PIO) the poorest flying qualities; in essence, an accident-prone machine. Whatever the cause, for the next ten months, the Flight Research Center had no lifting body to fly.[22]

## FROM NONE TO THREE

While the NASA lifting bodies underwent repairs at Northrop, the Air Force set about

[22]NASA News Release, "M2-F2 Lifting Body Accident Summary," Flight Research Center, 31 July 1967, AFFTC/HO Historical Reference Collection; Reed, *Wingless Flight*,106-109; Thompson and Peebles, *Flying Without Wings*, 144-150; diary of Ronald "Joe" Wilson, Volume 1 (1965 to 1969), entry for 10 May 1967, DFRC Historical Reference Collection.

producing its own candidate vehicle. During the early 1960s the USAF grew increasingly interested in high volume lifting bodies for re-entry from space. The FRC experience with the M2-F1 strengthened the service's resolve to develop such a vehicle. By December 1963 Martin Aircraft, under Air Force contract, selected a system known as the SV-5 for full-scale flight testing. The company believed the SV-5 fulfilled the USAF requirement for a lifting body capable of departing from its planned glide path during re-entry and then returning to its predetermined course before landing. Such an aircraft offered obvious advantages in surviving attack by potential aggressors. Martin received the go-ahead from Air Force Systems Command to design the vehicle in November 1964. About 18 months later the Baltimore, Maryland, company received a contract to fabricate one machine, designated X-24A by the USAF. After experiencing a beneficial relationship in the Joint FRC-AFFTC Lifting Body Flight Test Committee, the service representatives decided to seek NASA's advice and participation in the X-24A program. Paul Bikle and the incumbent Flight Test Center Commander Major General Hugh Manson agreed on October 11, 1966, to expand the existing lifting bodies Memorandum of Understanding to include the X-24A. Thus, the joint committee, chaired by Bikle, assumed jurisdiction over the fourth of the wingless aircraft. Six months later the two parties published a comprehensive Lifting Body Joint Operations Plan which codified the close collaboration developed between the parties during the M2-F2 and HL-10 programs and extended it to the X-24A.[23]

But the days when one engineer experimented by throwing paper gliders down a corridor had long since vanished. In just four years the lifting bodies research had evolved into a project of national consequence. Its importance impressed the leaders of the Defense Department and NASA to such a degree that they decided the joint agreements must not be confined solely to Edwards Air Force Base. Accordingly, in October 1967 NASA Deputy Administrator Robert

[23]Reed, *Wingless Flight*, 130-131; "Addendum to Memorandum of Understanding Between Air Force Flight Test Center and NASA Flight Research Center On Joint NASA-FRC - AFFTC Lifting Body Flight Test Committee," 11 October 1966, contained in "Lifting Body Joint Operations Plan," 1 May 1967, DFRC Historical Reference Collection.

Seamans signed an MOU with John Foster, Director of Defense Research and Engineering. It transformed the three lifting body flight research vehicles from a set of local cooperative projects into an organically unified, bilateral endeavor through the end of Fiscal Year 1970. It obligated the USAF to loan to NASA the X-24A and all necessary supporting equipment without charge and to combine the subsequent test program with that of the M2-F2 and the HL-10. "To realize the overall objectives" of the program, the MOU created a NASA-USAF Lifting Body Coordinating Committee co-chaired by the FRC Director and the Chief of the Research Projects Branch of the Aeronautical Systems Division, Dayton, Ohio, who together nominated its members. This national agreement did not alter the existing Joint Lifting Body Flight Test Committee at Edwards, which continued to implement "in detail" the program's research objectives under the direction of the Flight Research Center. Moreover, Seamans and Foster merely reiterated the division of resources by now common to the joint lifting bodies research. NASA provided full-scale wind tunnel facilities, instrumentation, vehicle maintenance, and ground support; the Air Force supplied base services, fuel oil, B-52 operations, chase and other support aircraft, and XLR-11 engine maintenance; and in cooperation, the USAF and NASA--the AFFTC and the FRC--shared test piloting, mission planning, data reduction, reporting, test operations, and range support. By adding its X-24A to the venture, the USAF also agreed to be responsible for the aircraft's logistics, spares, and contractor technical support. The execution of these terms happened sooner rather than later. Martin Marietta delivered the machine to Edwards Air Force Base on August 27, 1967, and an X-24A loan agreement between NASA and the USAF went into effect the following January. But, if anything, the lapse between the arrival of the aircraft and its first full flight test lasted even longer than that of the other lifting bodies. Between Ames full-scale wind tunnel tests, systems checkout, Flight Research Center modifications (strengthening the X-24's structure to cure persistent control system dynamic feedback), and extreme caution following Bruce Peterson's near fatality in the M2-F2, two years

27

passed before the X-24A took to the skies.[24]

While the X-24A underwent its preparations and the M2-F2 its repairs, the HL-10 returned to flight research after its long hiatus. Air Force Captain Jerauld Gentry flew it on March 15, 1968, and put it through a series of pitch and roll maneuvers so he could determine its basic stability. Although he found the longitudinal stick slightly sensitive, he considered it acceptable and encountered no roll tendency. Gentry actually said the aircraft performed as well as or better than the F-104 on approach and summarized its performance with the general comment, "the vehicle was solid." The FRC staff knew what this success meant for the lifting bodies and for the center. "People were standing on the roof, by the planes, lakebed, etc.," wrote an eyewitness. "I haven't seen so many observers for a first flight since I've been here. The day was almost absolutely clear and you could see the contrails of the B-52 and the chase." The beauty of the landing told them all they needed to know. "On the final turn to land the sun reflected off of the aircraft. It looked like a formation of fighters with the four chase and the HL-10. Gentry brought it in beautifully and made the comment, 'It flew like a champ.'"

On the 25th of May 1968, former Marine Corps aviator John Manke became the first NASA research pilot to fly the redesigned HL-10. He found it a pleasant experience, one for which the simulator prepared him fully. Longitudinal control at around 15 degrees angle-of-attack required constant vigilance; in that range the aircraft tended to drift slightly off trim. Also, the lateral sensitivity proved to be somewhat higher than he expected, but he encountered no rolling. Manke considered angle of attack easy to manage and on approach he banked, turned, and flared with no difficulty whatsoever. Only two features concerned him; he thought the

_____

[24]Memorandum of Understanding, "Provisions for the Use of the X-24A Vehicle in a Jointly Sponsored NASA-DOD (USAF) Lifting Body Flight Research Program," October 7, 1967, AFFTC/HO Historical Reference Collection; Reed, *Wingless Flight*, 132-133; "Working Agreement Between Research Projects Branch, Projects Division, Aeronautical Systems Division, Air Force Systems Command and NASA Flight Research Center on Joint USAF/NASA Lifting Body Flight Research Program Coordinating Committee," June 1968, AFFTC/HO Historical Reference Collection; "USAF/NASA Loan Agreement for X-24A Lifting Body Research Vehicle," January 1968, AFFTC/HO Historical Reference Collection; Thompson and Peebles, *Flying Without Wings*, 167-169.

cockpit glass distorted his depth perception and on landing he detected a very sharp oscillation as the nose gear touched down. These good reports on the re-designed HL-10 emboldened the program managers to resume a regular flight program designed to elicit detailed observations about the HL-10's handling qualities. In order to gear the stick for proper feel, the pilots were asked to describe longitudinal stability and control at subsonic and transonic speeds. Gentry, Manke and the others also reported on performance characteristics and on lateral-directional forces below and in the transition to Mach 1. As a result of the fiberglass glove modification proposed by the Langley aerodynamicists, the HL-10 emerged with "dynamics...significantly better than those of the M-2." Indeed, the HL-10 engineers proudly called their machine "the best flying of the lifting bodies." But more important than any parochial feeling, the return of the HL-10 not merely to the skies but to excellent reviews retrieved the reputation of the NASA lifting bodies just as the USAF fielded its own machine. Indeed, on the very day the X-24A rolled out for its first glide flight (April 17, 1969) John Manke pushed the HL-10, flying on three of its four rocket chambers, to Mach 0.99. Three weeks later, on only its seventeenth flight, the plane carried Manke to Mach 1.13, the first lifting body to cross the famous threshold.[25]

A few hours before Manke flew to 0.99, pilot Jerry Gentry acquitted himself well in the X-24A. He received the go-ahead from the X-24A Ad Hoc Committee on April Fool's Day, 1969. This panel, chaired by Milt Thompson, reviewed any possible weakness in the machine, the flight plan, the flight preparations, and any dubious indications from the captive tests. The fogging of the canopy was perhaps the foremost concern but apparently did not impair the pilot's vision; just to be sure, the technicians increased the flow of air in the cockpit forward of the pilot's head. Good as all the signs may have been, nothing instructed the aviators, the engineers,

---

[25]Reed, *Wingless Flight*, 116-117; diary of Ronald "Joe" Wilson, Volume 1 (1965 to 1969), entry dated March 15, 1968, DFRC Historical Reference Collection (first, second, and third quoted passages) ; Kempel, Painter, Thompson, *Developing and Flight Testing the HL-10*, 28-33, 39 (fourth and fifth quoted passages, 29, 33); Flight Record, Technical Debriefing: HL-10 Glide Flight H-7-11, 28 May 1968, AFFTC/HO Historical Reference Collection; Robert W. Kempel, "HL-10 Glide Flight Program," in Minutes of the Lifting Body Joint Operating Committee , 17 September 1968, AFFTC/HO Historical Reference Collection.

and the designers like the actual glide. Unfortunately, the small machine--a chunky looking vehicle 24.5 feet long, 11.5 feet at its widest point, and only 7.3 feet from top to bottom--held some secrets. Like Thompson and Peterson on the first flights of the M2-F2 and the HL-10, Gentry found the experience quite a trial. Unlike the earlier aircraft, the X-24's interconnect ratio between the rudder and aileron required no pilot management; it set itself automatically according to the angle of attack. When Gentry and his vehicle dropped from the B-52 everything appeared to be satisfactory as he went about maneuvers designed to measure lift-to-drag characteristics and longitudinal trim. But after a minute, the automatic interconnect system stuck in one position (too high at 35 percent) and resulted in lateral-direction instability during Gentry's landing approach. It reminded him of the behavior of the M2-F2. Plagued by roll oscillation as he reached 1,800 feet, he increased angle of attack to about five degrees, cut speed to 270 knots, fired the landing rockets, and avoided catastrophe. In eight more glide flights the engineers overcame this problem and another related to the control system on final approach. Then, on March 19, 1970, Major Gentry attempted the initial powered launch. Firing engine chambers two, three, and four just after separation from the mother ship, Gentry exclaimed with some relief, "it handles just like the simulator." He flew at Mach 0.87 but when he shut down the engines in preparation for a glide landing, he felt a sharp roll to the right. Otherwise, Gentry liked the handling, even when he turned off the roll and yaw dampers for a brief period. "Everything," said a close observer, "went like clockwork."[26]

The honor of bringing the X-24A into the supersonic realm fell to the FRC's John Manke not long after he completed the equivalent mission on the HL-10. On October 13, 1970, at 10 in the morning the X-24A dropped away from the B-52 in a gentle separation. The lighting of all four of the engine chambers could not have been smoother, the most fluid Make had yet

[26]Reed, *Wingless Flight*, 133-140; Milt Thompson to Alton D. Slay, Paul Bikle, and G.W. Bollinger, 10 April 1969, AFFTC/HO Historical Reference Collection; Flight Record, Flight Number X-10-15, 19 March 1970, DFRC Historical Reference Collection (second quoted passage, 8); Robert G. Hoey, *Testing Lifting Bodies at Edwards* (Lancaster, California: PAT Projects, 1994), 95; diary of Ronald "Joe" Wilson, Volume 2 (1969 to 1975), entry for 19 March 1970, DFRC Historical Reference Collection (second quoted passage).

experienced. When the fourth chamber came on, he felt "a pretty good roll trim" which seemed to subside when he manipulated the yaw trim. He climbed to 52,000 feet without problems and then pushed over, which seemed to initiate a little change in roll trim. In the transonic region he felt the same sensations rendered by the simulator: a rumbling feeling in the aircraft, the feel of "a drag of sorts." Just before the Mach jump he pulled up without difficulty. But during the subsequent push over (at around Mach .9) "it seemed like I PIO'd the airplane a little bit in roll as I got my angle of attack down. It is this old roll sensitivity problem that we have had before. It was there, and it surprised me just a little bit, because it was more sensitive than I had expected it to be." Manke found himself fighting to keep the plane level as he descended, but with higher angle of attack the danger seemed to pass. Passing through the Mach jump, at about Mach 1.05 he pulsed the rudder, then the ailerons, and performed a roll control maneuver. "This was almost exactly like the simulator," he reported with delight. "It was really beautiful. I had just the right amount of roll control." Manke shut off the engine at Mach 1.15 on his gauge. The glide down to the lakebed occurred without incident: "it was just as stable as a rock." He approached the runway just like the practices in the F-104. With a little spike of turbulence he touched down at 240 knots. Thus, in re-entry the X-24A exhibited superb handling characteristics; but as John Manke and every other pilot learned, dangerous longitudinal instability awaited them in the transonic range at angles of attack below four and above 12 degrees, a fact no less true as its flight envelope eventually expanded to Mach 1.6 and altitude of 71,400 feet..[27]

The story of the NASA's flight research on the first three heavy weight lifting bodies ends fittingly with the resurrection of the old warrior, the M2. After Bruce Peterson's crash, the M2-F2's team of engineers, pilots, and technicians endured a long and tortured trail before returning it to the runway. Some thought it should have been abandoned at the point of impact with the desert. Paul Bikle favored saving it, partly for something to use in case of another

[27]Flight Record, Flight X-18-23, 13 October 1970, DFRC Historical Reference Collection (quoted passages, post-flight, 2); Reed, *Wingless Flight*, 138; "X-24A Flight Test Program, April 1969 to June 1971," in John McTigue to Distribution List, 9 August 1971, AFFTC/HO Historical Reference Collection.

accident, but also because it possessed the *worst* flying qualities of the family.  He even told the plane's godfather, Alfred Eggers, that it should be retained because "If we can fly the M2, we can fly any of the other lifting bodies."   Bikle gave Milt Thompson, now Director of Research Projects, the formidable task of renewing the broken heap, so fundamentally damaged as to require a total reconstruction from inside out.  Project manager John McTigue contributed an essential ingredient to Thompson's task.  He persuaded Bikle to conceal the rebuilding from NASA Headquarters behind a cloak of  deception; by telling Washington that Northrop had disassembled the aircraft not necessarily to reconstitute it, but only to determine the extent of damage.  In actuality, as the contractor identified each of the ruined parts, the FRC ordered replacements and prepared to reassemble the plane at Edwards.  This sleight-of-hand saved a good deal of time.  Although Headquarters did supply some interim funding, it took 20 months for Thompson to finally persuade Washington to approve the full restoration and to release $700,000 for the project.  But the work did not merely restore the M2-F2.  The resulting lifting body, designated the M2-F3, differed significantly from its predecessor.  The control apparatus promised to better approximate that of a true re-entry vehicle, equipped with mixed reaction and aerodynamic controls, as well as a command augmentation system that at least a partially foreshadowed digital-fly-by-wire (see chapter 7).  Outwardly, the main change involved the addition of a third vertical fin between the existing outboard ones, designed to provide better roll control.  This important retrofit emerged from conversations among John McTigue, Northop's Ralph Hakes, and Ames' Clarence Syvertson.  Finally, subtle modifications were made to improve flying qualities and to strengthen the plane's structure in case of future ground accidents.[28]

Having prepared the M2-F3 for its return to the flightline, John McTigue and Milt Thompson needed someone to fly it and selected NASA pilot William Dana for the task.  Dana

---

[28]Written comments on a draft of this chapter by John McTigue, FRC lifting bodies program manager, DFRC Historical Reference Collection; Thompson and Peebles, *Flying Without Wings*, 149-153 (quoted passage, 150); Reed, *Wingless Flight*, 115,144-147, 150-153.

declined the honor; indeed, he denounced the M2 as an aircraft whose demonstrated hazards in pilot induced oscillation warranted its withdrawal from FRC service. Thompson tried to argue the M2-F3's weakness had been eliminated with the middle vertical fin, a claim substantiated by wind tunnel results and simulator flights. Moreover, Thompson felt the PIO was not a mystery, but a known phenomenon with well-established conditions of occurrence. He persisted with Dana because he knew him to be a fine pilot and a man of undisputed integrity. Dana also possessed an unusual combination of professional experiences: Naval Academy graduate, Air Force officer, recipient of a Master's degree in aeronautical engineering from USC, and after arriving at the Flight Research Center in 1958, a consultant on the Dyna-Soar project and an X-15 pilot. Dana finally agreed to fly the M2-F3, but never really trusted its characteristics. Its first glide flight seemed to vindicate Milt Thompson. On June 2, 1970 at 9:15 a.m. Dana and the lifting body fell away from the mother ship and he pronounced the separation "the easiest launch I have had in any vehicle, bar none." It simply rolled one way, then the other, then stabilized. Further, he described the flare as "smooth as silk" and said the "ailerons were beautiful...a copy from my friend the HL-10. It is just real solid. I just could scarcely believe it, because I had planned on nursing that baby all the way down to final." Dale Reed attributed Dana's good ride to two of the chief modifications of the aircraft which, in tandem, tamed its rude flying manners: the central tail fin prevented roll reversal; and the stability augmentation system (SAS) automatically damped the control surfaces when the pilot's overzealous inputs threatened oscillations. With the SAS system on, longitudinal and lateral-directional control proved to be excellent.[29]

The M2-F3's venture into the transonic range raised some of the same old questions about its flying qualities. Longitudinal instability occurred at its worst at Mach 0.85 and also affected angle of attack. Several expedients--moving ballast forward to the nose and increasing

---

[29]Thompson, *At the Edge of Space*, 19, 24; Thompson and Peebles, *Flying Without Wings*, 180-181; Flight Record, Flight Number M-17-26, 2 June 1970, DFRC Historical Reference Collection; Reed, *Wingless Flight*, 145, 150.

pitch damper gain to its maximum value--helped steady the vehicle. But Bill Dana and his colleagues realized that only by using the stability augmentation system could they maintain even marginal pitch control in the transonic region. On the other hand, Dana encountered no difficulties with roll and yaw. Even with the SAS turned off he could control both axes, although the machine reacted strongly to his slightest adjustments. But at this point the hard-luck M2 reverted to form. During Flight Number 22 on February 26, 1971, Dana reached altitude and fired the rocket chambers, but just two of them responded so he held the speed to Mach 0.77. Dana then silenced the engines and jettisoned the remaining propellant, only to find a small fire burning. The dumping of all fuel halted the flames, yet when he pulled the landing gear release handle during descent he found it immovable. Water had collected on the release and the resulting ice became frozen solid. A very hard pull by Dana finally sprung the wheels loose. These events grounded the aircraft for several months, during which time the entire flight program nearly went up in smoke. While the M2-F3 was being fueled for a return to the skies on May 6, one of the technicians happened to notice alcohol draining from an overflow tube in the liquid oxygen (LOX) tank. Multiple failures in the servicing line contaminated the LOX with the alcohol and water. The contact between LOX and any foreign substance usually caused an explosion as powerful as nitroglycerin but, inexplicably, in this case none occurred. Ground crews opened the LOX tank vent valves and cleared the area while FRC officials appealed to the Air Force to cancel all supersonic flights, fearing that a sonic boom might excite a detonation. The crisis ended the following day when the last of the LOX boiled out.[30]

Such events did not inspire confidence for the first M2-F3 supersonic flight but Bill Dana undertook this mission on August 25, 1971. The drop from the B-52 happened cleanly with a 5 degree right roll and a 7 degree heading change and the successful lighting of all four rocket chambers. While Dana did attain Mach 1.095, thus expanding the flight envelope, his maneuvers

[30]Reed, *Wingless Flight*, 151-152; Flight Record, Flight Number M-25-36, 9 August 1971, DFRC Historical Reference Collection; diary of Ronald "Joe" Wilson, Volume 2 (1969-1975), entry dated 6/7 May 1971, DFRC Historical Reference Collection; Thompson and Peebles, *Flying Without Wings*, 182-183.

also provided data relating to aileron adequacy and stability and control in the high transonic range (Mach 0.9 and 0.95, respectively). During the powered part of the flight the ailerons failed to control rolling to the extent predicted by wind tunnel tests, but above Mach 1 lateral-directional behavior closely matched the tunnel findings. For longitudinal control and trim between Mach 0.9 and 1.1, the flight paralleled conditions forecasted in the tunnel experiments. Dana experienced satisfactory handling qualities throughout, discovering no difficulties maintaining angle of attack during engine boost and no unexpected lateral-directional forces during the 6.5 minutes aloft. Trim change deviated most sharply from the wind tunnel model at Mach 0.95 due to transonic effects, and over the entire spectrum from Mach 0.94 to 1.0 (during climbout with all four chambers) there occurred an abrupt nosedown trim change, as expected. Much to the relief of a pilot ill at-ease with his mount, Bill Dana rated the transonic and supersonic flying qualities satisfactory, and even pronounced himself "very pleased" with the flight results, a significant concession from a man who "felt a sympathy for [the M2-F2], as one would toward a crippled child."[31]

Eventually, skeptics like Dana admitted the M2-F3 flew quite differently from its predecessor. Indeed, after adapting the standard Cooper-Harper handling qualities scoring system, Bill Dana quizzed the other lifting bodies pilots about this aircraft, even going to the extent of asking Jerry Gentry to return to Edwards and fly the M2-F3 on his way to service in Vietnam. (Gentry's evaluation carried special weight; he had also flown the M2-F1 and the M2-F2). As a whole, the pilots assigned the aircraft a general rating of satisfactory. In this context, the term satisfactory meant the person in the cockpit needed to compensate minimally for whatever "mildly unpleasant" flying properties the M2-F3 possessed in order to achieve the desired performance. Specifically, longitudinal flying characteristics, while better at subsonic

---

[31]Written comments on a draft of this chapter by John McTigue, FRC lifting bodies program manager, DFRC Historical Reference Collection; "Results of M2-F3 Flight M-25-37 by Lifting Body Project Group," 25 August 1971, DFRC Historical Reference Collection; diary of Ronald "Joe" Wilson, Volume 2 (1969-1975), entry dated 25 August 1971, DFRC Historical Reference Collection (first quoted passage); Thompson and Peebles, *Flying Without Wings*, 183 (second quoted passage, 181).

than at transonic or at supersonic speeds, received overall assessments ranging from "some mildly unpleasant deficiencies" to "minor but annoying deficiencies" requiring minimal to moderate pilot reaction. The same results prevailed for lateral-directional handling and for approach and landing flare. (Ninety percent of the pilots evaluated these flight conditions using the stability augmentation system, but even the ten percent who rated longitudinal and lateral-directional characteristics with the SAS *off* judged them to be satisfactory). The worst handling qualities involved the longitudinal control experienced during constant high angle of attack during powered boost. Here the pilots voted with less confidence, deciding the aircraft in such conditions exhibited "moderately objectionable deficiencies" requiring "considerable pilot compensation." But even this failing seemed within the grasp of improvement with minor adjustments in the command augmentation system (especially modifications in the command augmentation side stick). The most favorable assessments involved lateral-directional handling qualities during final approach which revealed negligible drawbacks on the part of the aircraft and required no pilot intervention. Thus, perhaps not as well regarded by pilots as the HL-10 and the X-24A, the M2-F3 still enjoyed a healthy measure of respect.[32]

## A CROWNING ACHIEVEMENT

Yet, a final chapter of the lifting bodies saga remains to be told. During 1969 two concepts emerged promising to produce higher lift vehicles than any yet conceived. One became known as the Hyper III design, pioneered by aerodynamicists at Langley. Radically different from its blunt lifting body precursors, this vehicle featured a flat bottom and a long, slender nose

---

[32]Robert W. Kempel, William H. Dana, and Alex G. Sim, NASA Technical Note D-8027, "Flight Evaluation of the M2-F3 Lifting Body Handling Qualities at Mach Numbers From 0.30 to 1.61" (Washington, D.C.: NASA, 1975), 19, 41-42 (quoted passages, 19).

cone. Its configuration rendered it unfit for extensive cargo, but it offered a hypersonic lift-to-drag ratio of 2.5, nearly twice that of its best predecessors. This feature rendered the Hyper III capable of landing almost anywhere on earth because its high lift potential permitted a deviation of up to 1,500 miles from the orbital re-entry path. Although he had retired from the lifting body projects, Dale Reed, accompanied by his friend Dick Eldredge, collaborated on the Hyper III, conducting radio-controlled model tests merely to satisfy their curiosity. Much like Reed's early M2-F1 model flights, his initial Hyper III tests were conducted without official sanction from Paul Bikle or the other FRC leaders. After some positive results, Reed approached Milt Thompson for support. Although retired from the cockpit, Thompson agreed to participate in simulator tests and Paul Bikle allowed Reed to draft volunteers. In December 1969, Thompson mounted a ground cockpit and "flew" a full-sized vehicle (35 feet long, 20 feet wide at the tail, and built in the Flight Research Center shops) as it glided to earth following release from a helicopter. It proved to possess an adequate degree of stability and damping and realized a subsonic lift-to-drag ratio of 4, lower than expected but far higher than anything yet experienced in a lifting body. Despite these favorable indications, NASA Headquarters denied Paul Bikle's request to put a pilot in and fly the Hyper III.[33]

Meantime, in Dayton Ohio, the Air Force Flight Dynamics Laboratory (AFFDL) came forward in 1969 with a bold proposal. For many years USAF engineers at AFFDL experimented with wind tunnel shapes designed to deliver high lift-to-drag and fly at hypersonic speeds with minimal re-entry heating. They named these designs Flight Dynamics Laboratory (FDL)-5, 6, and 7. An opportunity presented itself when the Commander of Air Force Systems Command (AFSC) made available to AFFDL two airframes called SV-5Js, fabricated by the Martin Company as jet-powered models during the construction of the X-24A. Apparently, Martin recently decided to loan them to the USAF. Consequently, during the same year Dale Reed experimented with the Hyper III, engineers at AFFDL announced plans to transform an SV-5J

---

[33]Reed, *Wingless Flight*, 155-166.

into an FDL-shaped lifting body. On reflection, Alfred Draper and other engineers at the Flight Dynamics Lab realized it would be easier to make the next generation FDL-8 from the existing, rocket-powered X-24A, rather than to modify the jet-propelled SV-5J. Accordingly, the Air Force engaged Martin to encase the pudgy, rounded X-24A and its internal workings in a stiletto-like fuselage much like the Hyper III's. Designated the X-24B (despite its radically different shape from its predecessor), it featured a 78 degree double delta planform tapering to a sharply pointed nose, as well as a flat underside. It yielded a machine with more than twice the planform area of the X-24A (330 versus 162 square feet, respectively) and nearly twice the body span (19 versus 10 feet). It also possessed a clear advantage over the Hyper III: as it landed, it retained a high lift-to-drag ratio (of at least 4) without the pivoting wings required of the Langley design.[34]

At first, the USAF declared itself "firmly behind total Air Force testing of the FDL-8...and extremely anxious to build up a total Air Force research vehicle test capability using an austere approach." But the Air Force abandoned its proprietary inclinations as the cost of the full scale X-24A conversion became evident and as this extraordinarily high-lift-to-drag vehicle finally won converts at NASA headquarters. The essential planning occurred during summer 1970 when Paul Bikle threw his full weight behind the project and veteran lifting bodies program manager John McTigue importuned his colleagues in Washington, D.C., to free money from *the existing budget* for collaboration with the military service on the X-24B. At the same time, representatives of the Flight Research Center, the Flight Dynamics Laboratory, the Air Force Flight Test Center, and Martin turned their minds to the technical details, the fiscal necessities, and to a joint memorandum clarifying the roles of the participants. The project jelled in March 1971, when NASA transferred $550,000 to the Air Force for X-24B development. The USAF agreed to match this contribution and in early February 1972, Martin received the modification

---

[34]Johnny G. Armstrong, AFFTC-TR-76-11, *Flight Planning and Conduct of the X-24B Research Aircraft Flight Test Program* (Edwards, California: Air Force Flight Test Center, 1977), 12; John A. Manke and Michael Love, "X-24B Flight Test Program," *The Society of Experimental Test Pilots 1975 Report to the Aerospace Profession,* September 24-27, 1975, pp. 146-147; Reed, *Wingless Flight,* 167-168.

contract. During the same month NASA and USAF representatives signed a memorandum of understanding to conduct the X-24B program as a joint venture. Much of the actual collaboration occurred at periodic Lifting Body Joint Coordinating Committee meetings--co-chaired by Dryden's De Elroy Beeler--at which the NASA staff participated in all phases of preparation for the aircraft's flight research program and its instrumentation.[35]

While the X-24A underwent re-tooling at Martin's Denver plant, the X-24B program came into focus. The Air Force and NASA each named a program manager--respectively, Johnny Armstrong of the Flight Test Center and Jack Kolf, an FRC figure who earned his spurs as an X-15 project engineer. Similarly, each side appointed its own chief program pilot: for NASA, the seasoned lifting bodies flier John Manke, and for the USAF, test pilot Major Michael Love. However, many of the duties could not be apportioned one-for-one. Due to practical considerations, the Air Force assumed more prominence in some aspects, NASA in others. For example, although the Flight Research Center's Norman DeMar controlled X-24B operations for NASA, AFFTC engineer Robert Hoey and his staff assumed overall responsibility both for mission planning and for the envelope expansion program. On the other hand, while Manke and Love enjoyed equal status as program pilots, because of his previous flying experiences in the M2-F3, the HL-10, and the X-24A, Manke flew all of the benchmark flights. As these roles became clarified, the objectives of the test program were agreed upon. The principal goal involved demonstrating the aerodynamics of the aircraft in the modes of low subsonic, transonic, supersonic, and landing approach flight. Subsidiary to these regimes were five considerations: handling qualities; measurements of pressure, vibration, and acoustical factors; loading of control surfaces; testing of landing gear capacities; and correlations between the flight test and

---

[35]Memorandum for the Record, Richard J. Harer, AFFTC Lifting Body Project Officer, 20 December 1968, AFFTC/HO Historical Reference Collection (quoted passage); Paul Bikle to Air Force Flight Dynamics Laboratory, 23 August 1970, AFFTC/HO Historical Reference Collection; John McTigue to NASA Headquarters (Code RV), 13 August 1970, AFFTC/HO Historical Reference Collection; Minutes of the NASA Flight Research Center/Air Force Lifting Body Joint Coordinating Committee, 9 August 1971, AFFTC/HO Historical Reference Collection; Reed, *Wingless Flight*, p. 168.

the wind tunnel data.[36]

The X-24B arrived at Edwards in the belly of a C-5 transport in October 1972 and the FRC technicians completed retrofitting the machine in four months. Meanwhile, this flight research program took a different turn than the rest. This time, the process did not begin with full-scale wind tunnel tests. Prior lifting body research in actual flight conditions suggested that scale models produced results roughly equal in accuracy to those obtained with the full-sized aircraft. But to substitute for the Ames wind tunnel, DeMar and the FRC crew needed to outfit the X-24B with instrumentation designed to yield data like that obtained in wind tunnel experiments. This approach imposed on the Flight Research Center's instrumentation cohort a complex series of ground and captive experiments before the X-24B's first flight, conducted during the six months between February and August 1973. One of the principal concerns of the X-24B design involved safe landings. The vehicle's elongated nose resulted in a center of gravity uncommonly far forward in relation to the aircraft's main landing gear. This situation threatened to inflict excessive loads, resulting in the gear's collapse on touch downs. By re-building the main gear's locking mechanism, engineers and technicians satisfied themselves that the system performed satisfactorily. They also subjected the nose gear to a sequence of tests in which the end of the lifting body was elevated to increasing heights and dropped on its front tires. In addition, the FRC team feared a new steering system on the X-24B might induce nose gear shimmying and structural failure once the tires touched down. Their worries ended when they placed the vehicle on the lakebed and fired two of its XLR-11 rocket chambers, sending the aircraft roaring down Rogers Dry Lake at 150 knots. No shimmying occurred, but a tendency to pull left due to an on-board weight imbalance did manifest itself, corrected easily by moderate right braking. Finally, the crew attached the dagger-shaped vehicle to the B-52 for captive tests. Unlike the other lifting bodies, the X-24B's new pylon hook-up prevented John Manke and Mike Love from ejecting while the aircraft hung from the B-52, but they could escape upon separation

---

[36]Robert G. Hoey, *Testing Lifting Bodies at Edwards* (Edwards, California: Air Force Flight Test Center, 1994), 125; Manke and Love, "X-24B Flight Test Program," 148.

40

from the bomber. No such emergencies occurred; the mated aircraft passed the structural resonance tests without incident.[37]

Adhering to the flight research tradition of "a cautious expansion of the Mach envelope," in the early morning of August 1, 1973, Manke flew the first of five unpowered missions aboard the X-24B. At launch altitude (40,000 feet) the experimental aircraft fell away from the mother ship in what the seasoned Manke called, " probably the smoothest launch I've ever had on a lifting body." The descent itself proved to be a good one for a first flight, but not without lessons for the future. Just after he dropped from the B-52, Manke detected some buffeting as the machine exhibited a mild tendency to pitch up, in contrast to the solid, level sensation in the simulator. The buffeting ended as the Mach numbers declined from the top speed of 0.65. He also found the aircraft required some trim adjustment in yaw and some aileron trim to maintain level wings. During a practice flare at 27,000 feet the aircraft handled "very nicely" and Manke declared it excellent in the roll axis. In light of the flight history of the lifting bodies, this observation seemed especially noteworthy and reassuring. Indeed, he later told the ground crew that he "looked for the PIO and there was absolutely no trace of anything like that--the airplane was doggone beautiful in roll." The approach provided a brief moment of anxiety, then a sense of relief and delight. Realizing he had flown a quarter mile farther down the flight path than planned, Manke also faced an uncommon wind pattern of east to northeast, rather than the prevailing wind from the west or southwest. But once he "started downhill ..I realized...we were home free." With a brief S-turn to the west of the lakebed to dissipate a little more descent energy, Manke pulled in his flaps and enjoyed a somewhat flat but comfortable and satisfying final approach at 290 knots. Distinct from those of the rest of the lifting body family, the landing gear dropped down almost imperceptibly, and at 240 knots Manke marveled that "I had just beautiful control of the airplane above the runway, no PIO tendency either in pitch or roll. It was just one of the most pleasant flying [experiences] right above the runway that I've ever flown."

---

[37]Reed, *Wingless Flight*, pp. 170-173; Thompson and Peebles, *Flying Without Wings*, 188-190.

Manke finished the X-24B's maiden journey by easing it toward the ground in a very smooth touchdown and a safe, 7,000 foot rollout.[38]

Fifteen weeks later Manke fired the X-24B's rockets for the first time, flying for almost seven minutes and achieving a top speed of Mach 0.92. The success of this and two other subsonic missions prepared the X-24B engineers for the all-important flight through and over the speed of sound. On a typical supersonic mission, Manke, Love and the other pilots followed a prescribed pattern. For the first minute after launch they flew the lifting body at a high angle of attack, guiding it to higher altitudes (roughly 65,000 feet) where the engines operated more effectively. Over the transonic range they reduced angle of attack to avert lateral-directional instability. Some stability and control maneuvers and heading changes also accompanied the powered part of the program. Then, achieving Mach 1.1 to 1.2, the pilots pushed over to a low angle of attack and accelerated to the maximum speed of the flight, at which point exhaustion of the fuel supply or intentional engine cutoff occurred. All of these events consumed about two minutes and culminated at about 70,000 feet. Subsequently, there occurred a three minute period of descent and glide before the approach and landing phase. Most of the flight test data emerged from this middle portion of the flight program. Immediately after shutdown, the pilots guided the machine through stability and control, loads maneuvers, and performance evaluations, all the while being alert to the Mach numbers and angle of attack necessary to achieve the required glide path. At about 30,000 feet the pilots typically changed pitch trim for subsonic, rather than transonic speeds. A 180 degree circling approach began at 25,000 feet followed by a flare at 1,000 feet, leveling off at about 100 feet, and landing at 180 knots per hour.[39]

John Manke flew the first of 21 supersonic flights on March 5, 1974, pushing the through

---

[38]Pilots and engineers often speak of "maintaining level wings" during lifting body flights. Of course, this is an impossibility on a wingless vehicle, but the expression is helpful to describe a maneuver otherwise difficult to explain. Manke and Love, "X-24B Flight Test Program," 132 (first quoted passage); John Manke, postflight interview after X-24B flight B-1-3, 1 August 1973, DFRC Historical Reference Collection (the remaining quoted passages); Hallion, *On the Frontier*, 343.

[39]Manke and Love, "X-24B Flight Test Program," 133-134, 138.

the entire transonic range and just over the speed of sound. Upon reaching a top altitude of 60,000 feet he experimented with a new technique. Previous flights told him the airplane exhibited a tendency to bank left due to greater weight on that side of the machine. To compensate, he found himself in an awkward and distracting situation which demanded he constantly correct for roll while simultaneously controlling pitch and trim using the same control stick and the same hand. This time, he simply placed his left thumb against the stick in the position which held the wings level and used his right hand solely for pitch, which "held in real well" at 14.5 degrees. He pushed over around Mach 0.9 and picked up acceleration to 0.95. "In this transonic region," he recalled afterwards, "I can feel changes in the airplane. I can feel little bits of buffet here and there and...some things going on in the airplane...I can't explain.... but they are there." After sensing the famous "Mach jump," Manke cut off the engines and undertook three sets of maneuvers: took his hands off the controls during deceleration from Mach 0.75 to 0.7; observed the effects of rudder sweep; and flew several push overs and pull ups. The approach and landing pattern proved to be his best yet in the X-24B. Although he paid no attention to his exact location in the descent pattern until he completed his maneuvers, Manke found himself positioned perfectly once he began to concentrate on this last part of the flight, which unfolded according to normal plan. Manke exulted as he described the experience, calling it a "superduper flight." aboard an aircraft "which sure does handle nicely. It was good or better than I had hoped all the way thr[ough]."[40]

Eventually, Major Love flew the X-24B to its maximum speed of Mach 1.75 and Manke attained the highest altitude at 74,000 feet. But the impressions gleaned during the first supersonic flight never wavered. The aviators marveled at the lifting body's steady handling properties and lack of lateral motion at all speeds, even with the dampers of the stability augmentation system disengaged. During subsonic flight in general and in landing approaches in particular, it demonstrated such fine flying qualities that the pilots rated it an extraordinarily high

---

[40]John Manke, postflight interview after flight B-9-16, 5 March 1974, DFRC Historical Reference Collection (all quoted passages); Reed, *Wingless Flight*, 173, 199-200.

2.5 on the Cooper-Harper flying qualities scale, far superior to the earlier lifting bodies. Indeed, those who flew it said the X-24B handled as well as an F-104 fighter. Only one important test remained, an essential one if this type of airplane might one day glide home from space; could it land safely on surfaces other than the Edwards lakebed? After Manke's and Love's collective experiences with 26 X-24B touchdowns, they felt confident of similar success on one of the base's concrete runways. Accordingly, Manke suggested he and Mike Love first fly F-104s and T-38s to simulate lifting body approaches and landings. Manke aimed these aircraft at the lakebed's mile marker more than 100 times, demonstrating to visiting political figures, astronauts, and sometimes doubtful engineers the technical validity of accurate glide landings. Then, on August 5, 1975, the veteran NACA pilot turned the needle-nosed X-24B toward a point 5,000 feet down runway 04/22 and made an unpowered landing within 500 feet of the target. Two weeks later Major Love made contact with the Edwards runway within the same margin.

Although Johnson officials decided in 1974--*before* the safe X-24B touchdowns on concrete--to forego engine landings and instead employ the unpowered approach (a choice they based on lakebed glides of other lifting bodies and some X-planes), Manke and Love nonetheless confirmed the validity of this mode of Shuttle descent. Of course, the Orbiter's winged design more closely resembled the Air Force's defunct Dyna-Soar more than any other vehicle, and its lift-to-drag ratio approximated that of the X-15, not the lifting bodies. Still, these timely X-24B flights, occurring at a moment when Shuttle managers grasped every opportunity to reduce the burgeoning weight of the Orbiter, helped transform American thinking about re-entry from space.[41]

INDIGENOUS PROJECTS

[41]Ronald "Joe" Wilson, interview with Michael Gorn, 11 April 1997, 22-23, DFRC Historical Reference Collection; Reed, *Wingless Flight*, 173-175, 199-200; Hoey, *Testing Lifting Bodies at Edwards*, 138; Thompson and Peebles, *Flying Without Wings*, 196, 210-214.

The lifting bodies projects represent nothing less than a crossroads in the NACA's and NASA's flight research history. Rather than continuing in the tradition of the Muroc Test Unit and the High-Speed Flight Research Station, the desert outpost found itself on a different path after the initiation of these strange-looking little aircraft. In a sense, the road wound back to Langley and an earlier time. Conceived by a few people and intended as an investigation of limited scope, the lifting bodies projects recalled the days when the Langley engineers invented worthwhile projects and pursued them under cover of some distantly related Research Authorization. During the FRC's entire formative period, the Research Airplane Program constituted the heart and the soul of NACA flight research. Many other worthy projects vied for money, attention, and time but none could match an undertaking which bore the imprimatur of *national* urgency. Its refreshingly simple objective--to conduct flight research on aircraft capable of ever -increasing speed and ever-loftier altitudes--also imparted a special élan and inevitability both to the projects and to the center's role. Nearly every American, typified by subscribers to the *National Geographic Magazine*, could comprehend FRC pilot Joe Walker's compelling account of jockeying the world's fastest airplane to the margins of space. Yet, at the very height of the X-15's popularity, the space race pulled hypersonics into the orbit of rockets and satellites. A question then loomed over the California desert like a cloud: what would become of flight research without the Research Airplane Program? Unknowingly at first, Dale Reed and his collaborator Milt Thompson supplied the answer. Most Americans recognized and appreciated the achievements of the X-15, but few understood the value or the purpose of wingless flying machines. Nevertheless, the lifting bodies offered a viable and an alternative style of flight research. In contrast to the great national enterprise exemplified by the X-15, the aircraft conceived by Dale Reed evolved locally, both in concept and in fabrication. Moreover, rather than following the familiar strategy of enticing the nation's military and civilian aeronautics authorities to loosen their pursestrings, the lifting bodies team survived, at least initially, by budgetary legerdemain and by the tight-fisted use of resources. Yet, despite their more subdued character, the lifting body programs ultimately unleashed as much engineering and scientific imagination as ever existed in the high speed airplane projects.

# CHAPTER 7

## A Tighter Focus:
## The Pursuit of Practical Projects

### REASSESSING FLIGHT RESEARCH

The lifting bodies projects left a deep impression on NASA flight research. Instructed by the lessons of local initiative and scrupulous cost-control, the Flight Research Center continued to channel its energies towards practical programs directly applicable to civil and military aviation. But the new patterns of NASA flight research did not merely borrow from recent experiences with the wingless aircraft; powerful forces external to the space agency also governed the choices. When President Kennedy announced to Congress in 1961 the initiation of a lunar flight program, those involved in NASA aeronautics realized that their work faced inevitable curtailment. Indeed, aeronautics had been experiencing declining support for some time. Under the stewardship of NACA Director Hugh L. Dryden, hypervelocity and space-related activities absorbed a rising proportion of the NACA's time and attention, even as the agency's overall appropriations rose slowly during its final years.[1]

The bureaucratic standing of flight research did not improve with the realization of President Kennedy's goal of planting human footsteps on the moon. On the contrary,

---

[1] Arnold S. Levine, *Managing NASA in the Apollo Era* (Washington, D.C.: NASA SP-4102, 1982), 11, 255; Roland, *Model Research*, 2: 475; Jane Van Nimmen and Leonard C. Bruno with Robert L. Rosholt, eds., *NASA Historical Data Book Volume I: NASA Resources, 1958-1968* (Washington, D.C.: NASA SP-4012, 1988), 305.

after the first lunar walk by former Flight Research Center pilot Neil Armstrong in July 1969, the leaders of the space agency found themselves pressed by sharp budget reductions on one hand, and a search for a successor to Apollo on the other. Just at the moment when the lunar program and NASA basked in glory, the agency suffered its third consecutive year of fiscal shortfalls and personnel layoffs. Terminations befell such important programs as the Voyager Spacecraft, the NERVA II nuclear rocket, and the so-called Apollo Applications project (a euphemism used by Administrator James Webb to describe a small orbiting laboratory known as Skylab, a hoped-for precursor to a full space station). Just two months after the great triumph on the moon, President Richard Nixon's Space Task Group issued a report entitled The Post-Apollo Space Program: Directions for the Future. It offered three options: a piloted Mars mission complete with space stations orbiting both the earth and the moon; the Mars mission itself; and a space station served by space shuttles. In January 1972, the President selected the third and cheapest alternative, committing his administration to a reusable space shuttle and to Skylab. Still, the decision to sustain the space program on a big scale while continuing to reduce NASA's budgets suggested even leaner times for flight research. Of course, no one could deny the indispensable contributions of the Flight Research Center to the X-15 research airplane and to the lifting bodies, the two types of aircraft that taught the nation to fly in space, to re-enter the atmosphere by gliding, and to land reliably and safely on runways. Cognizant of these achievements, NASA headquarters tried repeatedly during the 1970s to find an institutional formula by which flight research might retain its vigor, yet conform to the general NASA pattern of retrenchment and austerity.[2]

Aeronautics expenditures fell under headquarters scrutiny after James Webb's successor, Thomas O. Paine, resigned in September 1970. James Fletcher followed Paine as NASA Administrator, and it became clear that unlike his predecessor who resisted the

---

[2]Levine, *Managing Apollo*, 25, 255-261, 328; Roger D. Launius, *NASA: A History of the U.S. Civil Space Program* (Malabar, Florida: Krieger Publishing, 1994), 97.

budget ax, Fletcher accepted cost-cutting as a necessary measure. Accordingly, prompted by Office of Management and Budget (OMB) suspicions that the agency owned more aircraft than it needed, he sanctioned investigations into the practices of flight research at all of the affected NASA centers. Actually, the concept of centralization of aeronautical assets was hastened primarily by the termination of the X-15 program. Because of its great popularity with the public and its absorption of so much of the Flight Research Center's resources, the X-15's cancellation inevitably raised questions about the survival of the FRC itself, a suggestion heard in such high places as the Senate Appropriations Committee. At the same time, other aeronautical programs faced reduction or elimination under the intense pressure to further the space program. Sensitive to these considerations, headquarters instructed Associate Administrator for Aeronautics and Space Technology (OAST) Roy Jackson to launch a comprehensive review of all experimental flying in NASA "with the objective of improving...aircraft operations management, modernizing [the] aircraft fleet, and minimizing...recurring costs." In August 1972, Jackson asked De Elroy Beeler, the Flight Research Center's Deputy Director and one Muroc's early arrivals, to chair the OAST Committee on Flight Operations. Not only did a prominent FRC figure head the probe; Jackson's instructions for conducting the inquiry were modeled on a memo written to headquarters by FRC Director Lee Scherer two months before Beeler began his task. Jackson directed Beeler to initiate a few critical reforms desired by Washington: establish the FRC as the lead center for OAST flight operations and as the sole facility for high risk flight programs; reduce the recurring costs of aircraft operations; minimize duplication of flight programs and aircraft by designating the Flight Research Center as the one location where NASA tested whole aircraft as integrated systems (as opposed to science platforms or testbeds for single components); and modernize the NASA fleet by retiring obsolete aircraft and

3

replacing them either through military channels or through the lease or purchase or more efficient ones.[3]

But Beeler soon found himself in a cross-fire. The headquarters made its objectives plain, yet at a meeting in June 1972, the center directors of four flight research complexes (Langley, Ames, Lewis, and the FRC) exhibited "strong parochial feelings....None can be expected to offer more than token changes," wrote one observer. "It is clear that any significant changes in responsibilities must be directed from OAST after careful consideration of all issues." Luckily, those involved understood the etiquette of the issues; they had been debated at least since 1960. In that year NASA's Associate Administrator Richard E. Horner directed all centers to concentrate flight testing at Edwards. But Horner served only a short time and the directors of Ames, Langley, and Lewis offered successful resistance, transferring only a token number of their vehicles and programs to the desert. Meanwhile, Ames and Langley continued to undertake research on helicopters and on vertical/short take-off-and-landing aircraft while Lewis retained the high performance planes required to test flight propulsion. The controversy flared again during the period between Paine and Fletcher. Acting Administrator George M. Low asked Major General John M. Stevenson, the Associate Administrator for Organization and Management, to review the operations of NASA's aircraft fleet and offer suggestions to improve efficiency. The subsequent report proposed "that all NASA...sponsored flight research be centralized at the NASA Flight Research Center." But once he took charge, Fletcher wanted to draw his own conclusions and to include the center directors in the process. He thus empowered De Beeler to lead the investigation.[4]

[3]De Elroy Beeler to J.D. Hunley, 7 May 1999, DFRC Historical Reference Collection; Launius, *NASA*, 95-96; Lee Scherer to NASA Deputy Associate Administrator for Management, 27 June 1972, Milt Thompson Collection, DFRC Historical Reference Collection; Roy F. Jackson to OAST Management Council, 10 October 1972, Milt Thompson Collection, DFRC Historical Reference Collection (quoted passage).
[4]Lee Scherer to NASA Deputy Associate Administrator for Management, 27 June 1972,

Conducted over five months, Beeler's inquiry proved to be thorough and far-reaching. Each of the four centers inventoried every one of its aircraft, including the remotely-piloted ones, and provided estimates of the total costs to keep them flying. Then, in an effort to establish the fundamental programmatic requirements of flight research, Beeler and his six committee members undertook a grueling circuit of briefings at Langley, Lewis, Ames, and Edwards from January 23 to February 1,1973. At each place they heard detailed descriptions of virtually every flight research project, plans for future programs, the methods of acquiring and de-commissioning aircraft, the procedures of flight research, and the training of pilots. The Beeler committee also saw most of the equipment involved in NASA flight testing. By the end of the ten day ordeal they declared themselves "saturated from the extensive material presented..." and decided to digest the data independently. The recommendations, issued on April 20, 1973, envisioned a hybrid institutional arrangement making the FRC the focus of NASA flight research. It provided for an OAST Aircraft Operations Office staffed by Flight Research Center personnel and responsible to headquarters through the FRC Director. The Aircraft Operations Office assumed a pivotal role. It supervised all budgetary matters relative to NASA's aircraft inventory, evaluated all research proposals, determined which center and which aircraft should undertake specific projects, recommended the "acquisition, allocation, and disposition" of all aircraft, and advised the other centers about flight operations and safety. The Beeler Report also assigned aircraft to the centers by type. Henceforth, the Flight Research Center would fly all experimental, general aviation, proficiency, and supersonic planes. The FRC would also operate remotely piloted vehicles, aircraft whose test configurations differed markedly from their original designs,

---

Milt Thompson Collection, DFRC Historical Reference Collection (first quoted passage); Briefing, "Flight Research Within NASA," n.d., Milt Thompson Collection, DFRC Historical Reference Collection; George M. Low to Associate Administrator for Organization and Management, 27 April 1971, DFRC Historical Reference Collection; Stevenson Report, Fall 1971, Milt Thompson, DFRC Historical Reference Collection (second quoted passage).

and high risk projects. Langley won the franchise for all rotary wing aircraft. All other vehicles would be assigned to the centers according to such factors as programmatic requirements, unique facilities, safety factors, and available manpower.[5]

During 1976, the Flight Research Center won an unmistakable distinction; in March it became known officially as the Hugh L. Dryden Flight Research Center (DFRC), in memory of the NACA's last director and NASA's first Deputy Administrator. Despite this honor and the accompanying recognition of the center's fine work, a sense of disquiet affected the flight research mission. By this time it became clear the Beeler initiative had collapsed, symbolized by the disappearance of the Aircraft Operations Office from the Dryden organization charts. Beeler himself realized "that our effort would die on the vine," the victim of frequent turnover in OAST leadership and insufficient authority to enforce its will on the center directors. Still, during 1976 OAST's Acting Associate Administrator Robert E. Smylie tried to revive the 1973 reforms but he, too, failed. The Langley, Ames, and Lewis directors simply refused to relinquish their flight research prerogatives to Dryden, and Smylie left headquarters to become Deputy Director of the Goddard Space Flight Center. As a result, few aircraft designated for DFRC under the 1973 arrangement actually arrived at Edwards. In light of the failure of consolidation, morale at Dryden faltered for a time. Some at DFRC wondered if the other centers successfully resisted centralization because OAST lacked confidence in Dryden's effectiveness and disapproved of its existing schedule of projects.

---

[5]Donald Bellman to Those Concerned, 28 November 1972, Milt Thompson Collection, DFRC Historical Reference Collection; Minutes of the OAST Flight Research Operations Review Committee, 23-24, 25, 26, 29-30 January, and 1 February 1973, Ken Szalai Collection, DFRC Historical Reference Collection (first quoted passage from 1 February Minutes); De E. Beeler to Distribution, 15 December 1972, Milt Thompson Collection, DFRC Historical Reference Collection; Proposed Policy and Implementation Plan for OAST Aircraft Operations Office, 20 April 1973, Milt Thompson Collection, DFRC Historical Reference Collection (second quoted passage); Briefing, "Flight Research Activity Within NASA," n.d. (probably 1977), Milt Thompson Collection, DFRC Historical Reference Collection.

Others questioned whether the center's heavy workload merely reflected internal preferences, rather than a real understanding of the needs of its many clients in industry and in the armed forces. When NASA headquarters requested data about manpower and facilities, rumors flew about reductions-in-force and about the center reverting to test station status. Dryden's Chief Counsel summed up the feeling of institutional slippage "If we don't come out of our shell," he warned, "there is...very little chance that DFRC will remain a NASA Center for more than another three years. We simply are not a viable and vital part of NASA at this time; and if we don't become so, we leave NASA little choice but to abolish DFRC as the least...valuable NASA Center."[6]

Gradually, the Dryden workforce shook-off the self-doubts. Former D-558 and X-15 research pilot Scott Crossfield, now Technical Consultant to the House Committee on Science and Technology, urged a senior member of the DFRC staff to think of means to re-capture the prestige enjoyed by the center during the X-15 program. Crossfield proposed that Dryden reconstitute a Research Airplane Committee and plan for a flight vehicle to fill "the void between the X-15 envelope and space...." Chief Engineer Milt Thompson, one of Crossfield's fellow pilots during the X-15 days, led the forces of

---

[6]Not only did Dryden fail to become the focus of aircraft consolidation; Langley never assumed preeminence in rotorcraft flight research, despite the Beeler Committee's recommendation. Hugh L. Dryden Flight Research Center Organization Chart, May 1976, DFRC Historical Reference Collection; Briefing, "NASA Roles and Missions-- DFRC's View," n.d. (probably 1977), Milt Thompson Collection, DFRC Historical Reference Collection; Hallion, *On the Frontier*, 235; Personal Notes, author unknown, "DFRC' (sic) Most Significant Problem," n.d. (probably 1976 or 1977), Milt Thompson Collection, DFRC Historical Reference Collection; DFRC Acting Director to Philip Culbertson (draft), n.d. (probably 1976 or 1977), Milt Thompson Collection, DFRC Historical Reference Collection; Personal Notes, "Potential Manpower Reductions," author unknown, n.d. (probably 1977), Milt Thompson Collection, DFRC Historical Reference Collection; Personal Notes, "Dryden Management Philosophy," author unknown, n.d. (probably 1976-1978), Milt Thompson Collection, DFRC Historical Reference Collection; De Elroy Beeler to J.D. Hunley, 7 May 1999, DFRC Historical Reference Collection (first quoted passage); Isaac T. Gillam to Leonard Jaffe, 10 January 1978, Milt Thompson Collection, DFRC Historical Reference Collection; John C. Matthews to DFRC Chief Engineer, 7 July 1977, Milt Thompson Collection, DFRC Historical Reference Collection (second quoted passage).

renewal inside the center. He wanted to restore a highly successful program begun during the early 1950s by Air Force General Laurence C. Craigie when he served in the Pentagon as Deputy Chief of Staff for Development. Craigie thought all new high performance USAF aircraft should receive an independent evaluation by the NACA flight research specialists, and consequently loaned early production planes--usually the Number 6 "A" model--to the NACA with no expectations other than an overall flight assessment and the willingness to help solve critical inadequacies should they materialize. For instance, on the F-100A, the very first aircraft procured under this plan, NACA engineers discovered a tendency toward inertial coupling and devised a way to protect USAF pilots from its harmful effects. The flight researchers, in return, acquired a state-of-the-art vehicle to instrument and experiment with as they pleased. This informal, unwritten agreement continued through the 1950s and yielded one each of the Century Series fighters for the NACA runways. The practice ceased when the volume of advanced Air Force fighters declined during the 1960s. While NASA did receive other military planes in succeeding years, the arrangements were made on an ad hoc, rather than a systematic basis.

Thompson pressed DFRC Director David Scott in 1976 to resuscitate the loan agreements with the military services in order to acquire two high performance aircraft: the Air Force's F-16 and the Navy's F/A-18. Scott made an effort to accommodate the request. While he did not raise Thompson's idea of a formal loan agreement between the NASA and the services, he did inform NASA headquarters of the research opportunities possible if NASA could have on loan two Air Force YF-16s about to be retired from USAF testing. Scott persuaded Ames Director Hans Mark to back this arrangement in principal, provided the loaned vehicles were used for specific research purposes (like high spin tests and advanced controls work). Scott also succeeded in prompting NASA headquarters to establish an intercenter study group to consider the validity of military production vehicles for NASA flight research. Meantime, Milt Thompson waged the battle on two fronts. First, he contacted former X-15 pilot and friend Forrest S. "Pete" Petersen, who had achieved the rank of Vice Admiral and commanded Naval Air Systems Command. Thompson was plain with his old comrade; under the pressure of budget cuts, he feared NASA research on high performance military aircraft might be curtailed severely. He made a similar appeal to another X-15 alumnus, Robert Rushworth, now a major general and the vice commander of the USAF's Aeronautical Systems Division. Unfortunately, this campaign failed to resurrect the 1950s relationship Thompson wished to restore. But it did elevate the debate about the decline of high performance testing, opened the way for future loans of military aircraft on an ad hoc basis, and offered a necessary diversion from the frustrations of unsuccessful flight research consolidation.[7]

In fact, Thompson's efforts may have emboldened the DFRC leadership to try again for centralization. Using the existing climate of tight NASA budgets and manpower ceilings, David Scott urged headquarters to think again about flight research consolidation under Dryden as a means of improving overall effectiveness, reducing duplication of effort, making the most efficient use of the flying fleet, and saving money. Once again, DFRC compiled a catalog of its advantages as the lead flight research

---

[7]A. Scott Crossfield to Gene Matranga, 30 November 1977, Milt Thompson Collection, DFRC Historical Reference Collection (quoted passage); Briefing, "Joint NASA/DOD New Production Aircraft Program," Milt Thompson Collection, DFRC Historical Reference Collection; Milt Thompson to David Scott, 2 January 1976, Milt Thompson Collection, DFRC Historical Reference Collection; David Scott to NASA headquarters, 5 January 1977, Milt Thompson Collection, DFRC Historical Reference Collection; David Scott to Hans Mark, 5 May 1977, Milt Thompson Collection, DFRC Historical Reference Collection; Hans Mark to David Scott, 3 June 1977, Milt Thompson Collection, DFRC Historical Reference Collection; handwritten note, David Scott to Hans Mark, 27 September 1977, Milt Thompson Collection, DFRC Historical Reference Collection; A.M. Lovelace to David Scott, 28 September 1977, Milt Thompson Collection, DFRC Historical Reference Collection; Milt Thompson to Forrest Petersen, 29 March 1977, Milt Thompson Collection, DFRC Historical Reference Collection; Milt Thompson to Robert Rushworth, 21 September 1977, Milt Thompson Collection, DFRC Historical Reference Collection; Robert Rushworth to Milt Thompson, 6 October 1977, Milt Thompson Collection, DFRC Historical Reference Collection.

facility: the safety offered by a 15,000 foot runway situated next to a 44 square mile lakebed under clear, uncrowded skies; a range of test facilities--for rocket engines, heat and loads, weight and balance, and data tracking and acquisition--unavailable anywhere else; and 300 square miles of government property around Edwards, effectively eliminating complaints about noise and pollution. Dryden staffers also reminded those who would listen that some major programs, such as the Tilt-Rotor project being pursued at Ames, should have been assigned to DFRC under the terms of the 1973 consolidation. But these efforts proved ineffective. James Kramer, the Associate Administrator of OAST, visited DFRC in November 1978 and made his position plain: he rejected the past efforts to have Dryden "take over the traditional role of other centers. This approach made the other centers uncomfortable." Rather, "the proper role for DFRC," said Kramer, "is to provide flight support to the rest of the agency."

Failing to enlarge its role, DFRC turned inward to review its own practices. Milt Thompson launched a Dryden Image Committee to improve the center's profile outside the agency, and also within it. The panel reported its frank recommendations in April 1979, most of which admitted some recent lapses. Implicitly, it blamed management for failing to inform the workforce of the center's fundamental external commitments and to allocate resources accordingly. It suggested DFRC had fallen into the habit of trying to interest outsiders in projects which had no direct constituencies beyond the confines of Edwards. It insisted that flight research programs should end when completed, not soldier on long past their usefulness. It observed that the center tended to pivot its relationships with headquarters and with industry on airplanes, rather than on the technology gleaned from flying them. Thompson's committee also emphasized the need for long range planning to lift the center's sights above the present problems. Perhaps most important of all, the panel recommended inviting the private sector into DFRC's deliberations in order to better understand and to more fully accommodate their needs. One leading Dryden engineer with more than twenty years seniority elaborated on the all-

10

important relations with the manufacturers. Compared to past years, he observed, by the late 1970s DFRC confined itself too much to "specific configurations" while neglecting the aircraft design trade-offs so crucial to the engineering staffs at firms like Northrop, Douglas, and Lockheed.[8]

## AN EXTRAORDINARY TESTBED, A PROMISING HYBRID

During this interlude in which Dryden's ambitions for a wider role in flight research rose and fell, the center concentrated on important, but perhaps narrower work than it had in the past. Meantime, encouraged by James Kramer's guidance, the other aeronautical centers pursued their programs with renewed confidence. At the Langley and the Ames Research Centers two projects of consequence to aeronautics emerged during this period. During mid-1971 engineers at Hampton, Virginia, received instructions from officials at NASA Headquarters to begin research on the nation's air transport operations. With the advent of cheap air travel during the 1960s, U.S. airports experienced unprecedented congestion. To relieve it, a NASA-Department of Transportation report called the Civil Air Research and Development (CARD) policy study examined measures to cope with the heavy traffic in the air and ancillary traffic on

---

[8]David Scott to James Kramer, 20 September 1977, Milt Thompson Collection, DFRC Historical Reference Collection; Briefing, "NASA Roles and Missions--DFRC's View," n.d. (probably 1977), Milt Thompson Collection, DFRC Historical Reference Collection; Briefing, "Flight Research Activity Within NASA," n.d. (probably 1977 or 1978), Milt Thompson Collection, DFRC Historical Reference Collection; Briefing, "Consolidation of Flight Activity Within NASA," n.d. (probably 1977 or 1978), Milt Thompson Collection, DFRC Historical Reference Collection; Briefing, "Seventeen Years of Flight Research Consolidation," n.d. (probably 1978), Milt Thompson Collection, DFRC Historical Reference Collection; Berwin Kock to James Kramer, 30 December 1978, Milt Thompson Collection, DFRC Historical Reference Collection (first quoted passages); Berwin Kock to DFRC Director, 6 April 1979, Milt Thompson Collection, DFRC Historical Reference Collection; Handwritten notes to interview with Gene Matranga by Richard Hallion, n.d. (probably 1976), Hallion Collection, DFRC Historical Reference Collection (second quoted passage).

the ground. Meantime, Langley's research program director Jack Reeder--a former Langley research pilot--developed his own set of technical and procedural proposals that, combined with the CARD suggestions, set in motion an ambitious flight research project called the Terminal Configured Vehicle (TCV)/Advanced Transport Operating Systems (ATOPS) program. To undertake TCV/ATOPS, the NASA researchers needed a jetliner for their testbed and purchased from Boeing Aircraft (at low cost) the original 737 prototype aircraft.

It arrived on the Hampton runway--fully refurbished by the manufacturer--in May 1974, at which time it became known as the Transport Systems Research Vehicle (TSRV). As such, it played a pivotal role in American flight research well into the 1990s. Pressed into service for many roles and tasks, the TSRV served for 20 years as the iron horse of the TCV/ATOPS initiative, contributing to civil airliners and military transports such marquee innovations as electronic flight displays and the so-called "glass cockpit"; microwave-based landing systems capable of navigating complex airport approaches; and adapting the Global Positioning System (GPS) for approaches and automated touch-downs.

During the latter part of these crucial investigations, the Langley 737 flew in another, and perhaps even more memorable incarnation. In August 1985, a Delta Airlines L-1011 approaching Dallas-Fort Worth Airport crashed and killed 137 passengers and crew. Subsequent investigation determined that the episode resulted from atmospheric conditions associated with microburst windshear. While this phenomenon burst on the public with the Delta Airlines tragedy, researchers at such institutions as the FAA and the NACA had been studying it for decades, in between which a number of other jetliners were lost due to windshear. Some progress in understanding its characteristics had actually been made. For instance, after a careful study of the prevailing knowledge of the physics of microburst and windshear, during the early 1980s Langley engineers programmed super computers to simulate the impact of these

12

meteorological forces on flight. But the Delta Airlines disaster really catalyzed windshear research and elevated its priority, in part because of intense news coverage of the event, and in part because the system in place at Dallas-Fort Worth--the Low Level Windshear Alerting System--failed to warn flight controllers until *after* the fatal impact. With fresh funding from Congress, the FAA and NASA officials agreed to a joint venture during summer 1986, with research to be undertaken by Langley's Flight Systems Directorate. The Hampton team won a leading role not only because of the center's previous simulation work, but because of access to the 737 transport research aircraft.

Over the next seven years of flight research, the program's engineers and pilots pursued three objectives: to devise a means of expressing the danger that each windshear incident posed to aircraft; to perfect forward-looking airborne detection systems; and to develop and test measures by which the data could be converted to usable flight management for pilots. The overall strategy was simple: create a system that detected and warned of the intense downdrafts characteristic of microburst windshear, especially on approaches to airports. In the end, the flight research aboard the 737 demonstrated-- perhaps more persuasively than any ground-based investigation might have--that Doppler Radar reliably predicted windshear at least 40 seconds before aircraft entered it, giving the FAA, the aircraft manufacturers, and the airline industry high confidence in the effectiveness of its application.[9]

The Ames Research Center sponsored another flight research project of long duration and wide impact. Here, more than two decades of wind tunnel testing, system development, and full-scale flight research were devoted to an aircraft of uncommon promise. Since the founding days of icing investigations (see chapters 4 and 8), Ames flight research assumed its own particular emphases and style. Its aircraft inventory grew impressively in kind and in numbers. During the late 1940s to mid 1950s four P-51s

---

[9]Lane E. Walllace, *Airborne Trailblazer: Two Decades with NASA Langley's 737 Flying Laboratory* (Washington, D.C.: NASA SP-4216, 1994), 9-14, 55-73, 119-120.

underwent wing flow flight tests. Pilots also flew these aircraft to satisfy the curiosity of aerodynamicists interested in comparing drag measurements gathered in the wind tunnels to drag data recorded during instrumented flights. Forty-one different types of vehicles participated in Ames' extensive stability and control research conducted concurrently with the P-51 tests. The Ames engineers also undertook intensive flying qualities research for the military services, including an especially elaborate project in which three pilots flew ten different aircraft in 41 configurations to determine the minimum speed required for aircraft carrier landings. As a consequence of these tests, Ames pilot George Cooper derived a standard system for rating flying qualities which assessed the difficulty of the maneuvers, the aircraft's behavior, and the pilot's accuracy The resulting Cooper Pilot Opinion Rating Scale published in 1957 (modified in 1969 to incorporate new research by Cornell University's Robert Harper and subsequently known as the Cooper-Harper Handling Qualities Rating Scale) represented a permanent and an internationally recognized contribution to the technique of flight research.

Perhaps the only rivals to flying qualities investigations at Ames were the research programs devoted to rotorcraft, Short Take-Off-and Landing (STOL), and Vertical STOL (V/STOL) flight vehicles. The helicopter experiments, initiated in 1959 with the appearance of the H-23C, included such advances as fully automatic flight (tested on the UH-1H), two bladed helicopter aerodynamics (AH-1 Cobra), and a range of improvements to military helicopters to allow low level flight at night and during adverse weather (UH-60 Blackhawk). The first of the STOL/VTOL machines (the FJ-3) arrived at Ames in 1954 and the last left the center at the end of 1995. Of the 22 models tested at Ames during these 40 years, none was so much a creature of the center as the Experimental-Vertical Take-off and Landing (XV-15) Tilt-Rotor aircraft. The XV-15, however, did not appear in the 1970s without precedent. Its predecessor, known as the XV-3, originated with Bell Aircraft engineer Robert Lichten. Bell hoped to interest the Army and the Air Force in a vehicle capable of vertical flight like a helicopter and horizontal flight like a fixed-wing airplane. Lichten's model looked more like a rotorcraft than an aircraft. It featured a complement of two 20-foot-long rotor blades mounted at the end of each wing tip. Positioned parallel to the ground, the rotors raised and lowered the small craft; with the blades tilted forward like conventional propellers, the lifting power was transferred from the rotors to the wings and the aircraft flew horizontally. Its future looked bright during the ground development phase, which ended when it rolled out of the Bell Helicopter plant in 1955. But the flight test at the Bell factory in Texas revealed persistent weaknesses. Underpowered and restricted in its payload capacity, the XV-3 also exhibited some dangerous aerodynamics, including rotor-nacelle-wing whirl instability at higher speeds. Even when the machine merely hovered the Bell test pilot experienced intense cockpit vibration. The company confined the vehicle to ground testing for a year but failed to solve the instability dilemma. Bell engineers then tried lengthening the rotor masts and other minor adjustments and decided to resume flights of the XV-3. But the test model crashed when the aircraft's pylons-- mounted at the end of the wings for the purpose of swinging the rotors into position-- rotated forward fifteen degrees from vertical, resulting in a cockpit vibration so strong it caused the pilot to lose consciousness. At this point the Ames Research Center entered the world of the Tilt-Rotor. The center's Full Scale (40 by 80 foot) Wind Tunnel probed its aerodynamic flaws during 1957 and 1958 and a flight research program began the following year. Fred Drinkwater flew it often in the aerodynamically unstable high speed region and his impressions and the data derived from the instrumentation helped clarify some of its mysteries. Ames pilots like Don Heinle succeeded in taming the vehicle to the extent he could safely tilt the rotors in-flight, converting the vehicle from the helicopter to the cruise mode and back again with little trouble. But Ames' participation in the XV-3 ended when the vehicle's pylons tore loose during additional Full Scale Wind Tunnel tests. By extraordinary luck, no damage to persons or property occurred. The Ames flight research group returned the machine to Bell in mid-1965. By then, the

NASA researchers concluded that high speed stability, flight performance and dynamics, and controls needed to be improved before Tilt-Rotor realized its promise.[10]

They wasted no time attempting to perfect this tantalizing flight concept, attractive not only because of its military value, but for commercial purposes as well. The services recognized its utility as an airborne assault and as a direct-delivery logistics vehicle; civil and military authorities both liked the concept of being able to airlift isolated or injured parties from rough terrain and to evacuate them at high speeds; and airport officials and airline executives realized a successful Tilt-Rotor might alleviate the intense ground and air congestion anticipated at large airports in the decades to come by augmenting feeder, interurban, and regional transport. But the technical hurdles remained formidable, probably higher than most realized in the mid 1960s. None realized this better than the Ames aerodynamicists who continued to experiment with

---

[10]The most comprehensive attempt at a history of the XV-15 Tilt Rotor program to date is found in a manuscript by Martin D. Maisel, Demo J. Giulianetti, and Daniel C. Dugan, "The XV-15 Tilt Rotor Research Aircraft: From Concept to Flight," 1999, Ames Research Center, DFRC Historical Reference Collection. As of this writing, it is in draft form. Although written by three veterans of the Tilt-Rotor program who provide a capable synthesis, it is lacking in personal observations and instead draws from published technical reports, proceedings, and journal articles. For a broader overview of Ames flight research, see Paul F. Borchers, James A. Franklin, and Jay W. Fletcher, "Flight Research at Ames: Fifty-Seven Years of Development and Validation of Aeronautical Technology," Ames Research Center, 1998, 2-7, 10, 17-24, DFRC Historical Reference Collection; anon., "Fifty Years of Excellence: Ames Research Center, 1939-1989," Ames Research Center, 1989, 4-12, DFRC Historical Reference Collection; NASA Ames Research Center News Release, "New X Series Research Aircraft in Final Assembly," 5 November 1975, Ames Research Center; Robert R. Lynn, "The Rebirth of the Tiltrotor--The 1992 Alexander A. Nikolsky Lecture," *Journal of the American Helicopter Society* (January 1993): 3-11; anon., "Review of NASA-Ames Research Program on VTOL/STOL Aircraft Concepts," 2-3, contained in carton labeled Minutes of and Reports to Automatic Stabilization and Control Subcommittee and Research Advisory Committee on Control, Guidance, and Navigation, 1954-1961, folder for control, guidance, and navigation committee, 9-10 February 1960, San Bruno, California, Federal Records Center Accession Number 255-69-0140, copy in DFRC Historical Reference Collection. (The Dryden Flight Research Center Historical Reference Collection contains copies of all of the San Bruno Federal Records Center documents mentioned in this chapter).

various Tilt-Rotor configurations in the years after the XV-3 returned to Bell. By the end of the decade the Ames Flight Research Systems Division endorsed a new wing-pylon design to improve stability and declared the research "far enough along in technology development to justify projects to fabricate...aircraft to these configurations and conduct flight research investigations...." Others also saw Tilt-Rotor's possibilities. In 1968 NASA and France's Office National d'Études et de Recherches Aerospatiales (ONERA) signed a joint wind tunnel agreement to share data on Tilt-Rotor. Moreover, these two research institutions were joined by the XV-3's chief sponsor, the U.S. Army, which also contributed research facilities and staff. The resulting theoretical analyses, completed in spring 1971, persuaded NASA and the Army to pursue the construction of a prototype vehicle designed in accordance with the new data. Consequently, NASA included in its Fiscal Year 1973 budget submission the funds to launch a joint Tilt-Rotor aircraft program with the full cooperation and the partial budgetary support of the Army. The partnership became effective in November 1971 when both sides, familiar with other each after years of XV-3 collaboration, issued a vague statement to potential contractors obliging NASA and the Army Air Mobility Research and Development Laboratory to "endeavor to provide the funding levels required to develop and operate [Tilt-Rotor aircraft] in accordance with plans to be mutually developed...." Despite this less than complete budgetary profile, Bell Helicopter and Boeing-Vertol raced to participate in the prototype work; Bell offered to construct a simulator and Boeing issued an unsolicited proposal to fabricate a testbed aircraft.[11]

---

[11]Lynn, "The Rebirth of the Tiltrotor," 15; anon., "History of the [Ames] Flight Research Division, 1966-1978," 1, n.d., Ames Research Center Library (first quoted passage); Woodrow Cook to Director, Ames Research Center, 14 April 1970, San Bruno Federal Records Center Accession Number 255-93-25, Box # RMO-22, DFRC Historical Reference Collection; Wallace Deckert to Director, Ames Research Center, 20 July 1970, San Bruno Federal Records Center Accessions Number 255-93-25, Box # RMO-22, DFRC Historical Reference Collection; William Aiken, Jr. to Leonard Roberts, 16 July 1970, San Bruno Federal Records Center Accession Number 255-93-25, Box# RMO-22, DFRC Historical Reference Collection; John Boyd and Leonard Roberts to NASA

17

The Army and NASA agreed to situate the Tilt-Rotor program office at Ames. Here a team of research engineers structured the program to culminate in a series of detailed proof-of-concept flights to evaluate "the technical feasibility and operational suitability of the Tilt-Rotor approach to high-speed [minimum 300 knots per hour] VTOL." In the pursuit of these objectives, Ames attempted its first procurement of a flight vehicle for the express purpose of proving a particular aeronautical technology. In sharp contrast to most flight research endeavors, the Ames engineers decided they had such a wealth of experience going into the project that they would not only furnish in-house theoretical and experimental research, but actually manage contractor fabrication according to their own specifications. Accordingly, after request for proposals were published the two firms which expressed immediate interest--Bell and Boeing--each received $500,000 design contracts in September 1972. Perhaps because of its proven experience with the XV-3 (not to mention the personal relationships established between Ames and the contractor during the earlier program) Bell won the competition in April 1973. Four months of negotiations between the contractor, NASA, and the Army yielded formal terms: a four year $26.4 million project culminating in a final design, in the construction and delivery of two aircraft, and a in flight test program pursued in conjunction with Ames and the Army Air Mobility Laboratory. Although the contract was structured as a cost-plus-incentive-fee agreement, Bell accepted clauses by which it

Headquarters Acting Director, Aerodynamics and Vehicle Systems, 5 August 1971, San Bruno Federal Records Center Accessions Number 255-93-25, Box# RMO-22, DFRC Historical Reference Collection; Roy Jackson to Dr. Hans Mark, 10 September 1971, San Bruno Federal Records Center Accessions Number 255-93-25, Box# RMO-22, DFRC Historical Reference Collection; "An Agreement Between [NASA] and the Department of the Army for Joint Development and Operation of Rotor System Test Vehicles...," 1 November 1971, San Bruno Federal Records Center Accessions Number 255-93-25, Box# RMO-22, DFRC Historical Reference Collection (second quoted passage); Bell Helicopter, Fort Worth, Texas to NASA Ames, 5 November 1971, San Bruno Federal Records Center Accessions Number 255-93-25, Box# RMO-22, DFRC Historical Reference Collection; D.A. Richardson to W. Cook, 17 November 1971, San Bruno Federal Records Center Accessions Number 255-93-25, Box# RMO-22, DFRC Historical Reference Collection.

received bonuses for completing the project below the agreed upon price and a penalty of 50 percent of any overruns. In the interim, Ames engineers proceeded with further theoretical studies of the Tilt-Rotor's dynamics, necessary before flight research occurred. One aerodynamicist, for example, developed equations for a rotor mounted on a cantilevered wing and examined the problems of whirl flutter caused by a rigid propeller spinning on a pylon. Meantime, Bell activated its assembly line, as well as that of some subcontractors. It formed partnerships with Rockwell International for the tail assemblies and fuselages and with AVCO-Lycoming for modified T-53 engines. For its part, NASA signed a contract with Sperry Rand for the design and installation of the XV-15's avionics; that is, its electronic navigation, guidance, and control systems operated and integrated by an on-board digital computer. At last, on October 22, 1976, aircraft number 1 wheeled out of the Bell plant in Fort Worth, Texas, to the accompaniment of congratulations from NASA and Army officials. Bell's pilots then initiated a long series of test flights at Fort Worth, beginning with ground and hover maneuvers and followed by envelope-expansion flights of aircraft number 2. Indeed, XV-15 number 1 did not arrive at Ames until March 23, 1978, when the press corps watched it disgorge from an Air Force C-5 transport aircraft.[12]

The year in which the Tilt-Rotor first appeared at Ames proved to be one of great

[12]Kenneth G. Wernicke, "Mission Potential Derivatives of the XV-15 Tilt-Rotor Aircraft," *AGARD Conference Proceedings No. 313: The Impact of Military Applications on Rotorcraft and V/STOL Aircraft Design*, 1981, 19A-1 (quoted passage); Borchers, et al., "Flight Research at Ames," 19; NASA Ames New Release, "NASA-Army Design Contracts, 5 September 1972, ARC; Ames News Release, "Tilt-Rotor Research Aircraft," 26 December 1972, ARC; Ames New Release, "Tilt-Rotor Contractor Selected," April 13, 1973, ARC; Ames News Release, "Contract Signed for Research Aircraft," 1 August 1973, ARC; Wayne Johnson, NASA TN D-7677, *Dynamics of Tilting Proprotor Aircraft in Cruise Flight* (Washington, D.C.: NASA TN-D 7677, 1974); Ames News Release, "Avionics System for Tilt-Rotor," 13 February 1975, ARC; Ames News Release, "New X Series Research Aircraft in Final Assembly," 5 November 1975, ARC; Ames News Release, "Rollout Scheduled for Advanced NASA-Army Research Aircraft," 13 October 1976, ARC; Robert Burns, compiler, "Ames Research Center Aircraft Inventory," 13, 3 March 1992, ARC; Ames News Release, "Note to Editors," 20 March 1978, ARC.

promise for aeronautical research at the center. The XV-15 represented one of the most important flight research projects ever undertaken there. But still more good news materialized. Recognizing that the Beeler initiative, which awarded Langley all rotary wing aircraft, was indeed moribund, Hampton transferred five vehicles and their research programs to Moffett Field, strengthening Ames' claim to supremacy in NASA rotorcraft. Ames received the Rotor Systems Research Aircraft, the small UH-1 and AG-1G for rotor experiments, and the SH-3 and CH-47 for operational technique studies. The Ames Director of Aeronautics Leonard Roberts decided that the addition of these machines-- combined with such existing advantages as the wind tunnels, the simulation facilities, the flight research infrastructure, and the close proximity to the Army Air Mobility Research and Development Headquarters--warranted the creation of a new Helicopter Technology Division. Indeed, in the overall scheme of NASA flight research, Ames' increased prestige could not be denied. One highly placed visitor from NASA headquarters instructed a group of Dryden engineers to place a "high priority [on] getting a flight support capability established" in order to assist Ames in its rotorcraft duties.[13]

Encouraged by these developments, Roberts' staff prepared to probe the soundness of the XV-15 in flight. But first, the aerodynamicists needed to arrive at some preliminary judgments. During May and June 1978 the aircraft, modified for remote-control operation, underwent intensive testing in the 40 by 80 foot tunnel. Within the admitted limitations of these operations, the researchers attempted to gauge the XV-15's flight envelope and evaluate its airworthiness prior to the flights of aircraft number 2 by Bell in early 1979. Before a pilot stepped foot in it, they wanted to determine the Tilt-Rotor's overall aerodynamic and structural profile: its performance, its stability and

---

[13]Leonard Roberts to A. Scott Crossfield, 6 July 1977, File Folder number 000403, Headquarters NASA Historical Reference Collection; Berwin Kock to James Kramer, 30 December 1978, with attachment "Highlights of Discussion with Dr. Kramer-November 16, 1978," Milt Thompson Collection, DFRC Historical Reference Collection (quoted passage).

control qualities, its stall and loads factors, as well as its noise and vibration characteristics. After subjecting the XV-15 to 54 hours in these artificial gales the engineers "unearthed no insurmountable problems...no fundamental reason why the tilt-rotor concept should not fulfill its promise..." and only suggested that in the cruise mode the aircraft might benefit from some drag clean-up work. The Ames staff considered several modifications to satisfy this recommendation, as well as remedies to surmount two other deficiencies: a low maximum lift coefficient, and high tail loads as the vehicle converted from vertical flight to cruise mode during non-level flight. Bell began its developmental flight tests on April 23, 1979, in order to demonstrate the XV-15's capacity to fly safely in both helicopter and airplane configurations, a necessary step before transferring the Tilt-Rotor to Ames and to the Army for more in-depth analysis. Just four months later, NASA, Army, and Bell engineers reported the Forth Worth flights appeared favorable enough to plan for the release of the vehicle to Ames.[14]

These results by the Texas contractor represented a great deal of in-flight experience. The Bell pilots and their Army and Marine counterparts flew the aircraft for 60 hours in 140 separate missions. The instrumentation records gained from their maneuvers began to yield a picture of the XV-15's aerodynamic and structural characteristics. The aviators satisfied the speed requirement by achieving 301 knots and flew as high as 14,000 feet. Within these performance limitations they collected data on

[14]Borchers et. al., "Flight Research at Ames," 19; anon., "Outstanding ARC Achievements for 1978: Aeronautics Directorate," *The Astrogram* (28 December 1978): 1, ARC; Ames News Release, "Research Aircraft Due at Ames," 17 March 1978, ARC; John P. Magee and Kenneth Wernicke, "XV-15 Tilt-Rotor Research Aircraft: Program Report," presented at the Atlantic Aeronautical Conference, Williamsburg, Virginia, March 1979, 6 (first quoted passage); Robert L. Marr, Sheppard Blackman, James Weiberg, and Laurel Schroers, "Wind Tunnel and Flight Test of the XV-15 Tilt-Rotor Research Aircraft," presented at the 35th annual national forum of the American Helicopter Society, Washington, D.C., 79-54-2 to 79-54-3; Kenneth Wernicke and John P. Magee, "XV-15 Flight Test Results Compared with Design Goals," presented at the American Institute of Aeronautics and Astronautics Aircraft Systems and Technology meeting, New York, New York, 20-22 August, 1979, 9-10.

rotor loading, which appeared well within safe limits; on the aerodynamic and aeroelastic relationships among the wings, the fuselage, and the engine nacelle pylons as the rotors turned; on the effect of engine motion on aircraft performance; and on the loads borne by the tail structures. Because of its uniqueness, the experience of flying the XV-15 required some familiarization. The transition from vertical to horizontal flight depended on a combined pylon-engine nacelle at the end of each wing, which pivoted upwards until the plane of the rotor blades paralleled the ground, and reversed course until the rotors assumed a position like a conventional aircraft. Thus, beginning at the top of the arc of motion, the pylon swung from 95 to 75 degrees for helicopter flight, 75 to 0 degrees during conversion, and locked in the front position as an airplane. Despite its transfiguration in the air (and a tendency to behave strangely in wind gusts), pilots found the XV-15 possessed excellent handling qualities in its helicopter and its standard aircraft configurations, as well as during the conversion from one mode to the other. Yet, its advanced design belied the fact that the Ames and Bell engineers conceived the vehicle with many off-the-shelf components. Capable of hovering for an hour, yet also capable of horizontal flight using standard control surfaces, it combined in one airframe the two main kinds of powered flight but with less noise and vibration than a helicopter and better fuel efficiency than a standard turboprop aircraft. It also enjoyed some distinct advantages in size and bulk. Light at 9,076 pounds empty weight, it measured only slightly more than 46 feet in length, nearly 13 feet from the ground to the top of the tail, and about 32 feet across its forward swept wings. Even with its pair of three-bladed rotors each measuring 25 feet in diameter, the entire vehicle fit comfortably into the Ames 40 by 80 foot wind tunnel, allowing detailed and reliable analysis before the pilots attempted their maneuvers.[15]

---

[15]J.M. Bilger, R.L. Marr, and Ahmad Zahedi, "In-Flight Structural Dynamic Characteristics of the XV-15 Rotor Research Aircraft," *Journal of Aircraft* (November 1982): 1005-1011; Anon., "Tilt-Rotor Set For Government Test," *The Astrogram* (26 December 1980): 3, ARC; Anon., "XV-15 Tilt-Rotor Research Test," *The Astrogram*, 21

In spite of the Tilt-Rotor's many uncommon characteristics, the Ames flight research program concentrated on just two overriding objectives: "to demonstrate an aircraft free of structural aeroelastic instabilities and also to demonstrate one ...able to achieve a 300-knot airspeed with enough maneuvering envelope for the military to evaluate the aircraft for both potential and existing mission suitability." The center's first outward sign of the crucial new phase manifested itself with the construction of a Tilt-Rotor tie down facility in late spring 1980. Here engineers could run the vehicle's rotors in all flying configurations (including aircraft mode by elevating it on a hydraulic lift) either for preflight or post-flight operations. But when the Bell flight program ended on July 23 after more than a year of testing, the number 2 aircraft appeared at Dryden for government acceptance tests. Technicians uncrated it on August 13, 1980, and reassembled and checked out the peculiar-looking machine over the next seven weeks. The pilots made sixteen flights from October 3 to 30, in which they opened the maneuvering envelope beyond that of the contractor and assessed its performance and operational suitability in light of its ultimate military and civilian roles. Subsequently, both aviators and engineers reported the XV-15 satisfied the joint proof-of-concept evaluation guidelines and pronounced it fit for government flight. Dryden Director Ike Gillam officially transferred this Tilt-Rotor to Ames and Army representatives on October 30. With the number 2 aircraft on its books, Ames shipped its wind tunnel model

August 1980, 1, ARC; David D. Few, *A Perspective on 15 Years of Proof-of-Concept Aircraft Development and Flight Research at Ames-Moffett by the Rotorcraft and Powered-Lift Projects Division, 1970-1985* (Washington, D.C.: NASA Reference Publication 1187, 1987), 9; Daniel C. Dugan, Ronald G. Erhart, and Laurel G. Schroers, "The XV-15 Tilt-Rotor Research Aircraft," *Symposium Proceedings of the Society of Experimental Test Pilots 1980 Report to the Aerospace Profession*, Beverly Hills, California, 24-27 September 1980, 176; anon., "Tilt-Rotor Tie Down Facility," *The Astrogram*, 26 June 1980, 2, ARC; Ames News Release, "Tilt-Rotor Tunnel Tests Underway," 2 May 1987, ARC; Anon., "Second XV-15 Arrives at NASA Dryden," Dryden *X-Press*, 13 March 1981, 13, DFRC Historical Reference Collection; Martin Maisel, *NASA/Army X-15 Tilt-Rotor Research Aircraft Familiarization Document* (Washington, D.C.: NASA TM X-62, 1975), 2-11; anon., "The XV-15: Bell's Tilting Testbed," *Aerophile* (October 1979): 20.

(number 1) to Dryden for similar tests. After 35 flights in March, April, and May, 1981 (with a break to perform at the Paris Airshow) this vehicle also received the go-ahead from DFRC.[16]

After a almost a decade of development, thirteen more years of combined flight research and wind tunnel experiment still awaited the XV-15. Starting in 1981, investigators probed its handling qualities, stability and control, side stick control, performance in all flight configurations, acoustics, aerodynamic flow, loading limitations, structural dynamics, and aeroelastic stability. John Magee became project manager during the first full year of XV-15 flight research at Ames and collaborated with two pilots, Daniel Dugan and Ronald Gerdes, as well as a number of military aviators. The flight research program got a good beginning when an Army general with long helicopter experience flew the Tilt-Rotor and called it, "smoother than any helicopter and even faster than many light airplanes...." Naturally, flight research involved much more than quick impressions, however well-informed. The first serious tests involved the vehicle's hovering characteristics. An XV-15 instrumented to record rotor torque, fuel consumption, aircraft attitude, and control positions was raised over the large Ames VTOL pad at five different wheel heights: 50, 25, 12, 6, and 2 feet. At each level the researchers sought to discover such important factors as the influence of ground effect on hover performance, downwash phenomena, handling qualities at each altitude, and acoustics. After reviewing the data and interviewing the pilots, the project engineers reported that outside the range of ground disturbance the vehicle offered no control

---

[16]Few, *A Perspective on 15 Years*, NASA Publication 1187, 9 (quoted passage); anon., "Tilt-Rotor Tie-Down Facilities," *The Astrogram*, 26 June 1980, 2, ARC; anon., "XV-15 Tilt-Rotor Research Test," *The Astrogram*, 21 August 1980, 1, ARC; Anon., *Annual Report on Research and Technology, FY 1981* (Washington, D.C.: NASA TM 81356, 1981), 15; anon., "Tilt-Rotor Set for Government Test," *The Astrogram*, 26 December 1980, 3, ARC; Burns, "Compilation of Ames Aircraft Inventory," 14; anon., "Second XV-15 Arrives at NASA Dryden," Dryden *X-Press*, 13 March 1981, 2, DFRC Historical Reference Collection.

problems. Lower down, the pilot found his workload increased significantly in order to maintain position, but the handling qualities remained adequate. Downwash appeared moderate at the aircraft's sides but high in the front and back. Moderate noise levels prevailed, and acoustics experts described the sound quality as acceptable.

By 1982 the Navy, Marine Corps, and even the Air Force recognized the warfighting potential of the Tilt-Rotor and joined the Army to form a multi-service program office. As a consequence, not long after the hover tests Army Lieutenant Colonel Ronald Carpenter and Navy Lieutenant Commander John Ball took turns flying the XV-15. Despite the experimental nature of the aircraft and the project, NASA agreed to deviate from tradition and allow the service pilots to fly the XV-15 in mock combat environments. The Army operated one of the vehicles at barren Fort Huachuca, Arizona, simulating a special electronics mission in which the XV-15 maneuvered against air defense threat systems. Colonel Carpenter found the aircraft comfortable to fly under required conditions (up to 180 knots). It followed the earth's contour easily in low level flight, responded nimbly, "well, and with seemingly little effort." No control coupling manifested itself at any air speed.. Maneuvers close to the ground allowed for "low pilot workload, good field-of-view, good control response, and good terrain masking ability." Moreover, the Tilt-Rotor appeared free of pilot induced oscillation and responded effectively to lateral and longitudinal control. Carpenter did note that he did needed to pay close attention at maximum bank angles in order to avoid scraping the long rotor blades on the ground. But otherwise, he felt the XV-15 increased the chances of flying safe and effective combat missions due to the aircraft's unusual flight characteristics, its relatively low demands on pilot attention, and its fine cockpit visibility. The Navy's chose to make a shipboard evaluation of the Tilt-Rotor and conducted maneuvers on the deck of the USS Tripoli. Navy pilot Ball found the XV-15 behaved like a helicopter but promised "to open up new missions in flight operations far beyond those of

25

helicopters."[17]

With the accumulation of more flight research data and the increased experience of the Ames pilots, a full appreciation of the subtleties of Tilt-Rotor flying became more evident. After a period of initiation, Ames project pilot Daniel Dugan wrote that its short take-off and landing performance "can only be described as remarkable." At a weight of 15,000 pounds fully loaded the aircraft rose from the ground in just 200 feet, and could surmount a 50 foot barrier after only 400 feet of flight. More remarkable to Dugan, it climbed as quickly on one engine as on two. Dugan also praised the lateral and longitudinal stability of the Tilt-Rotor. After countless experiences at the controls, he found "the magnitude of the stick input is not critical and the pilot soon finds the proper-size inputs to keep pitch or roll attitudes and the resulting translations comfortable." Overall, the aircraft won Dugan's admiration.

> From the hovering efficiency of the highly twisted Tilt-Rotor through the maneuverability, fuel efficiency, and quiet operation of the high performance turbo-prop, the XV-15 has "proven the concept." Short take-offs in the tilt mode have demonstrated remarkable performance for the maximum gross weight condition. Aeroelastic stability has been investigated through the critical-flight envelope and tests will continue with the installation of advanced composite rotor blades. A three-axis side-stick controller has been developed and evaluated by a broad cross-section of pilots, and has been found to be suitable for a tilt-rotor aircraft. New techniques for deriving the open-loop dynamic response of an aircraft have been developed and applied to the XV-15. These are only some examples of the many flight tests that have been and will be conducted with the XV-15 to further develop Tilt-Rotor technology....[18]

Research pilot G. Warren Hall likewise praised the Tilt-Rotor's qualities, but also

---

[17]Borchers, et.al., "Flight Research at Ames," 19; Few, *A Perspective on 15 Years*, 9; Anon., "Army General Flies Tilt-Rotor," *The Astrogram*, 13 November 1981, 2, ARC (quoted passage); M. Maisel and D. Harris, "Hover Tests of the XV-15 Tilt-Rotor Research Aircraft," *Proceedings of the American Institute of Aeronautics and Astronautics Conference*, Las Vegas, Nevada, 11-13 November 1981, 1, 4; Ronald Carpenter, John Ball, and Chris Becker, "XV-15 Experience: Joint Service Operational Testing of an Experimental Aircraft," *Symposium Proceedings of the Society of Experimental Test Pilots*, 1982, 5-11 (quoted passages, 7, 11).
[18]Daniel C. Dugan, *The XV-15 Tilt-Rotor Flight-Test Program* (Washington, D.C.: NASA TM 86846, 1986), 24-30 (quoted passages, 26, 28, block quote, 30).

noticed some fine points about which aviators needed to be alert. "The first thing a pilot notices," wrote Hall, "is that on the ground the airplane is sensitive to lateral control inputs, and during ground taxi there is a tendency for the airplane to lean into turns, thereby requiring a small amount of lateral control to keep the wings level." Tilting the nacelles two to three degrees, on the other hand, eliminated the need for any longitudinal corrections and, at 10 knots, resulted in a smooth ground speed. Flying with the Tilt-Rotor configured as an airplane, Hall described the stall characteristics as "very docile and conventional," preceded by mild buffeting or a shudder five knots above the danger zone of 95 to 110 knots and easily recovered by standard means. "From a piloting viewpoint," Hall thought the in-flight conversion from helicopter to aircraft "is the most interesting feature." The relationship of appropriate speed to appropriate tilt angle (for example, accelerating during take-off to between 60 and 80 knots, positioning the nacelles at 70 to 80 degrees, and retracting the landing gear), yielded smooth flight with low pilot effort and only slight longitudinal trim change. But even if air speed exceeded that recommended for a particular tilt angle the aircraft flew with no problem. For these reasons, Hall joined in the chorus of satisfaction, declaring the XV-15 to be an "outstanding vehicle resulting in major improvements in the field of vertical and short take off aircraft." It also marked a noteworthy success for the Ames flight research team, the guiding light in the vehicle's development from the earliest engineering concept until the center returned vehicle number 2 to Bell Helicopter in April 1994. Although a long time germinating, the XV-15 not only proved a complicated flight concept which failed in an earlier incarnation, but led to the military's V-22 Osprey and to much enthusiastic discussion about Tilt-Rotor as one solution to the world's inundated airports and commuter highways.[19]

---

[19]G. Warren Hall, *Flight Test Research at NASA Ames Research Center: A Test Pilot's Perspective* (Washington, D.C.: NASA TM 100025, 1987), 3-5 (quoted passages); Burns, "Compilation of Ames Aircraft Inventory," 14; John Magee, NASA Contractor Report 166440, "The Tilt-Rotor Research Aircraft (XV-15) Program," in *Proceedings of the*

# MARRYING COMPUTERS TO AIRCRAFT

As the Ames flight researchers drafted their initial specifications for the XV-15, a few hundred miles south at Edwards Air Force Base a small group of engineers undertook a project at least as far-reaching in its implications for global aeronautics. Like Tilt-Rotor, it developed over a long period of time; but unlike the Ames project, its origins were obscure. Known first as fly-by-wire, it stemmed from the urgent necessity to give pilots the means of controlling high performance aircraft whose speed and agility threatened to outstrip the capacities of the human beings in the cockpit. Paradoxically, the development of fly-by-wire increased aircraft capabilities all the more. But to succeed, this new technology depended not on aeronautical breakthroughs so much as advances in computerization. Because these discoveries happened at their own pace, their application to flight hinged on events outside of aviation circles. But once digital breakthroughs did occur, cross-pollination required a person or persons with sufficient knowledge and insight to marry the new computer technology to the practice of aeronautics. Defined as "the complete replacement of the mechanical linkages between the pilot's stick and the control surface actuators by electrical signal wires," fly-by-wire did not easily replace the cables and moving parts relied upon since the Wright Brothers lifted-off at Kitty Hawk.

The fledgling attempts started when engineers installed in inherently unstable machines like the B-49 flying wing a stability augmentor, designed to make this

*Monterey Conference on Rotorcraft and Commuter Air Transportation* (Washington, D.C.: NASA, 1983), 24-33; anon., "Inclined Planes," *Flight International* (26 September 1987) 34-38; W.H. Deckert and J.A. Franklin, *Powered-Lift Flight Technology* (Washington, D.C.: NASA SP-501, 1989), 14-16; William Decker and Rickey Simmons, "Civil Tiltrotor One Engine Inoperative Terminal Area Operations," in *Research and Technology 1995: NASA Ames Research Center* (Washington, D.C.: NASA TM 110419, 1995), 7-8.

particular airplane handle as if it possessed a tail surface. Then, during a long

interregnum in which all of the mechanical systems remained in place, manufacturers of

high-speed, highly maneuverable military aircraft began to install a parallel control

system operated by "black boxes"; that is, on-board analog computers which corrected or

modified pilot inputs to the control surfaces in vehicles prone to instability in flight.

Because these early contrivances supplemented, but did not replace the existing

mechanical controls, some called this evolving art "pseudo digital fly-by-wire," although

in fact, they all remained analog systems. The Flight Research Center's most famous

celebrity, the X-15 research airplane, furthered the relationship between computing and

aeronautics. Because of its immense flight envelope, the machine's designers assumed

from the start that it would require some kind of analog automatic stability augmentation.

Under contract to prime contractor North American Aviation, Honeywell developed for

X-15 number 3 a so-called adaptive control system for the re-entry phase of flight,

consisting of rate gyros; pitch, roll, and yaw servocylinders; an electronic case assembly;

and gain selector and function switch assemblies. These complements to the mechanical

systems shared authority with the pilot on decisions affecting pitch and yaw, but

possessed twice the human input on roll maneuvers. Unfortunately, the X-15's adaptive

control system did not lack problems. By late 1962 the aircraft outfitted with this system

malfunctioned on fully 25 percent of its free flights. Pilot skepticism proved to be an

even more stubborn obstacle to its acceptance. At the end of the X-15's flight research

program, Milt Thompson reported to a meeting of the NATO Advisory Group for

Aerospace Research and Development (AGARD) some disturbing patterns in adaptive

control practice:

> Our flight research indicates that...the gain-changing logic can be fooled and a
> number of environmental factors such as turbulence, structural modes, pilot
> control activity, and electrical interference can compromise the performance of
> the     system and can restrict the usable range of variable gain. The loss of the X-15
> aircraft [number 3] with the adaptive flight control system cannot be attributed

solely to the adaptive flight control concept, since there were [sic] a number of other factors involved. Elimination of the mechanical backup system and the use of an electric stick could have prevented saturation of the servo-actuators. Higher servo-actuator rates or separate control surfaces for pitch and roll would also have precluded this particular problem. Yet the adaptive feature cannot be absolved, since it is theoretically supposed to operate regardless of these practical or real-

life     compromises. The system functioned as designed, but the design did not consider this particular and unique combination of conditions and pilot response [author's italics].[20]

It appeared that the path to successful fly-by-wire still held some mysteries.

Spaceflight technology supplied some of the answers. Until this point, the NACA's and NASA's aeronautics programs acted as the nurturing mother of the U. S. space program  But in the case of digital-fly-by-wire, the roles reversed themselves. In their initial proposals to establish a fly-by-wire project, two Flight Research Center engineers suggested adapting the simple on-board analog computer used to curb oscillations on the lifting bodies. They also witnessed the extensive testing of the Lunar Landing Research Vehicle (LLRV) over Edwards. Not only did the flight research pilots maneuver these ungainly looking machines using fly-by-wire electronics, but the Gemini 2 spacecraft operated on a similar control system. Yet if these two workhorses of the space program offered inspiration, a third one provided the needed software. Shortly after his epochal walk on the moon in July 1969, former NASA research pilot Neil Armstrong accepted a position at NASA headquarters which afforded him the opportunity to return to his stick and rudder days, if only vicariously. Until he became Deputy Associate Administrator for Aeronautics, fly-by-wire for aircraft attracted little

---

[20]For a description of the crash of X-15 number 3 with pilot Michael Adams, see chapter 5. J.P. Sutherland, "Fly-By-Wire Flight Control Systems," 10 August 1967, 1-5, SAE-18 Aerospace Vehicle Flight Control Committee, Boston, Massachusetts, Hallion Collection, DFRC Historical Reference Collection; Duane McRuer, interview with Lane Wallace, Hawthorne, California, 31 August 1995, 2; William Elliott, *The Development of Fly-By-Wire Flight Control*, (Dayton, Ohio: Air Force Materiel Command History Office, 1996), 13; Robert A. Tremant, *Operational Experiences and Characteristics of the X-15 Flight Control System* (Washington, D.C.: NASA TN D-1402, 1962), 1-2, 12; Milton O. Thompson and James R. Welsh, "Flight Tests Experience With Adaptive Control Systems," Summary of Papers Presented at AGARD Guidance and Control and Flight Mechanics Panels, Oslo, Norway, 3-5 September 1968, 5 (block quote).

interest in a Washington obsessed by space exploration. Armstrong's appointment

opened possibilities. By this time the FRC engineers--Melvin Burke and Calvin Jarvis--

succeeded in convincing their bosses to back a modest fly-by-wire flight investigation. In

search of funding, they traveled to NASA headquarters to brief the new deputy for

aeronautics. Burke and Jarvis told Armstrong they wanted to connect an analog fly-by-

wire system to a highly maneuverable aircraft and undertake a full flight research

program to demonstrate the influence of complete electronic control over the behavior

and the design of an agile vehicle. To their surprise, Armstrong objected. Why analog

technology, he asked? Rather than a system that sent impulses, he proposed they employ

the more advanced digital system, one based on counting--digital fly-by-wire (DFBW).

Burke and Jarvis knew of no flight-qualified digital computer. "I just went to the moon

and back on one," said Armstrong. "Have you looked at the Apollo system?" They

admitted with embarrassment they had not even thought of it. He told them to contact the

Charles Stark Draper Laboratory in Cambridge, Massachusetts, whose designers

conceived of the Apollo system, an electronic network which mediated between

Armstrong's commands and the spacecraft's reaction controls. His approval constituted

the birth certificate of the project. But rather than rely on a specially designed

experimental vehicle to fly DFBW, the F-104 Starfighter offered an economical

alternative since the FRC had a number of them in its inventory. However, after some

discussion at the Flight Research Center among engineers, pilots, and maintenance crews,

a consensus emerged about using a different aircraft, a standard Navy F-8C Crusader.

This proposal seemed sensible since the center also possessed an F-8, a supersonic

vehicle well-known for stability throughout its flight envelope.[21]

---

[21]Gary Krier, interview with Michael Gorn (by telephone), 11 March 1999, DFRC
Historical Reference Collection; Kenneth J. Szalai and Calvin R. Jarvis, interview with
Lane Wallace, 2-7, 30 August 1995, DFRC Historical Reference Collection (quoted
passage, 5); Hallion, *On the Frontier*, 217-218; Elliott, *Development of Fly-By-Wire*, 13;
Ronald "Joe" Wilson, interview with Michael Gorn, 11 April 1997, 31-32, DFRC
Historical Reference Collection.

31

On a subsequent journey to confer with the Draper scientists, a young FRC electrical engineer accompanied Cal Jarvis, now the project manager. Still under 30 years of age, Kenneth J. Szalai, a graduate of the University of Wisconsin, had become the principal investigator--in effect, the technical director--of the DFBW project. A fellow researcher familiar with Szalai remarked, "I've never see anybody that could work so hard, so strong, and with so much imagination." His appointment could not have been better timed; in March 1969 Szalai seemed ready to re-focus his career. "I have very little responsibility for anything really important," he wrote candidly. "I am losing my initiative and enthusiasm for my work...." Indeed, just before DFBW materialized he actively considered leaving Edwards for other opportunities.

Led by Cal Jarvis, the DFBW team began their investigation with a historic departure from the first 70 years of powered flight: the project engineers instructed the mechanics to disconnect the aircraft's mechanical cables, linkages, and push rods running from the cockpit to the control surfaces. Using actuators and electrical wire, the researchers then interposed the Apollo computer-- fabricated by Raytheon, designed by Draper Lab, and re-programmed for the F-8 by the Draper staff--between the pilot and the flight controls. Of course, the events leading to the early flights of DFBW did not occur without adversity and even intense frustration. Although the programming of the computer occurred at the Draper Lab, Jarvis and his cohorts actually designed the control laws and converted them to software specifications before transferring them to Cambridge. Sometimes, the software engineers in Massachusetts found logical errors or gaps in the Dryden specifications. On other occasions, the NASA investigators contacted the Draper engineers about unexplained difficulties. So each side could "see" the problems and better diagnose them, the Draper contingent constructed a mockup of the F-8. Jarvis, Szalai, and their lieutenants spent countless hours on the telephone trying to puzzle out the software. Eight hour conversations were not uncommon; one stretched to ten hours. During one marathon, an operator broke in to make sure the parties realized

32

the charges being run up. On another such occasion, the Flight Research Center team reached an impasse in which they found it impossible to prevent the Apollo computer from quitting and re-starting when it detected programming errors. Naturally, such behavior could not be tolerated when the control of the aircraft (and the survival of the pilot) depended on the electronic circuitry. The answer, after much angst, turned out to be a redundant system to handle the flight chores while the main machine recycled itself. Subsequent simulations suggested the problem had been solved. Consequently, research pilot Gary Krier (an engineer and aviator who later practiced law briefly) completed the first flight of the F-8 with DFBW on May 25, 1972. He soon discovered, however, that in the interval between shutdown of the Apollo computer and engagement of the backup, a one second pause occurred in which the control surfaces ceased to respond. Krier experienced this phenomenon as a pitch up, so he responded by pushing the stick forward almost to its limit, restoring control of the F-8 which continued to fly satisfactorily. But the lesson had been learned; in a dynamic system such as DFBW, the dialogue between Draper Lab and the FRC required the input of the human being in the cockpit.[22]

This first phase of DFBW ended with a satisfactory integration of the Apollo computer into the F-8 machine. Just as important as the actual flight success, no one could doubt the feasibility of the technology nor question the value of the software verification or the data amassed on control law design and mechanization. Before its completion, however, some important refinements were required. For example, because

[22]Gary Krier, interview with Michael Gorn, 11 March 1999, DFRC Historical Reference Collection; NASA Biographical Data, Dryden Flight Research Center "Kenneth J. Szalai," March 1994, DFRC Historical Reference Collection; Jim Skeen, "Dryden Director Rose Through the Ranks," *Antelope Valley Daily News*, 7 January 1991, DFRC Historical Reference Collection; Ronald "Joe" Wilson, interview with Michael Gorn, 11 April 1997, 32-33, 36-38, DFRC Historical Reference Collection (first quoted passage, 32-33); handwritten note by Kenneth Szalai, 7 March 1967, Kenneth Szalai Collection, DFRC Historical Reference Collection, (second quoted passage); Kenneth Szalai and Calvin Jarvis, interview with Lane Wallace, 30 August 1995, 3-4, 13,19, 22, DFRC Historical Reference Collection; Hallion, *On the Frontier*, 218.

the computer processed the flight control data in digital segments, when the pilot first moved the stick he felt the subsequent movements in the flight surfaces as short, repetitive bumps, rather than as a smooth response. Szalai and his staff imparted a sensation of even flow over the entire range of stick motion by changing the position of the stick sensor on the digital path. Ultimately, however, the bigger improvements lay in the installation of a computer system better suited for aeronautical flight. By late 1973 the market began to witness the arrival of the first actual flight control computers, stimulated in part by the immense digital requirements of the Space Shuttle.

This fortunate development coincided with a stream of obligations crossing the desks of Jarvis, Szalai and their group as word of the successes with digital control became known and as the demands of the project increased. As the engineering team completed the final updates of the software specifications for Phase I, invitations invited those involved in DFBW research to present papers on their preliminary results at AGARD and at other conferences. They spent hour upon hour re-programming the simulator with the latest flight data and devoted still more time on the machine to coaching such able FRC research pilots as Einar Enevoldson in the ways of the new system. In addition, FRC personnel served as consultants to the Space Shuttle program office at the Johnson Space Center, advising about the spacecraft's elaborate DFBW needs, which comprised not one, but four on-board computers for Orbiter flight control. Meanwhile, Jarvis and Szalai made three important decisions about Phase II. First, in consultation with Johnson engineers they selected a computer, the fully programmable IBM AP-101. Then, after several months of negotiations they reached agreement with the Langley Research Center on joint participation. Finally, after "strong request[s]" from the Johnson Space Center they fashioned a more formal relationship in which Szalai and his team offered advice and shared the Flight Research Center's data, in exchange for about one million dollars to support DFBW Phase II.[23]

The next stage of DFBW involved the integration and programming of three AP-101s to be used in tandem, followed by redundancy and reliability flight research designed to prove the practicality of the system. Working in cooperation with Langley, the FRC contingent devised a triplex computer system based on three separate machines which operated like three wholly independent units to the programmer, but to the pilot functioned like one seamless entity. The achievement of these complex, interdependent requirements caused untold hours of grinding work. Ground simulations failed frequently and ultimately the project went through nine of the IBM machines, which often aborted due to faulty mechanical construction. This situation surprised few (except the manufacturer, who predicted better performance). After all, the AP-101 represented the vanguard of computers designed for aircraft; indeed, the Flight Research Center began Phase II using the first three ever made. The system operated on the basis of two levels of electronic redundancy, in addition to three layers of gyroscopes and accelerometers. As a consequence, after flight research began in August 1976, when one computer failed in flight (no more than one ever did), the pilot noticed no consequences in the vehicle's operation or handling; the two good computers simply ruled out the inoperable one and carried the load themselves. But even had the digital malfunctions been catastrophic, a back-up analog system existed as a fail-safe. For safety, however, the research pilots landed at the first sign of any computer malfunction. Once the early Phase II flying ended, the F-8 saw double duty as a testbed for the Shuttle Orbiter flight

---

[23]Kenneth Szalai, (draft) "Digital Fly-By-Wire Benefits and Applications," FRC, 12 July 1973, 20, DFRC Historical Reference Collection; Kenneth Szalai, handwritten notebook entitled F8 DFBW with entries from 30 July 1973 to 2 November 1975, Kenneth Szalai Collection, DFRC Historical Reference Collection (quoted passage, entry dated 14 March 1973); Ronald "Joe" Wilson, interview with Michael Gorn, 11 April 1997, 39-40, DFRC Historical Reference Collection; Duane McRuer, interview with Lane Wallace, 31 August, 1995, 4, DFRC Historical Reference Collection; Kenneth Szalai and Calvin Jarvis, interview with Lane Wallace, 30 August 1995, 17, 18, 20, DFRC Historical Reference Collection.

control system.

The researchers found themselves even more busy than the F-8 machine. Now the daily telephone calls came less from Draper than from aerospace firms wanting to know about the process, requesting technical reports, and asking for assistance. The Dryden engineers presented many briefings, often to skeptical listeners. One crew chief who pressed Ken Szalai for answers reflected this "show me" attitude. "I want to see what's in this [black] box," he insisted. "I know how much you want to see what's in this box," said Szalai, "but it's not going to help....What's running this airplane is software....And it's invisible. It has no weight, takes up no volume, and takes no power." This psychological, or perhaps conceptual hurdle inhibited many in his audiences--even the technically sophisticated--from embracing the new process. Moreover, those in charge of corporate finances, who would be asked to approve DFBW on production aircraft, also expressed reservations. But painstaking flight research, which proved the practicality of flying with no mechanical back up systems, could not be ignored. During summer 1977 NASA headquarters officials learned the essential results of the DFBW investigation: the triplex computer approach operated like a single channel system; longitudinal flying qualities under DFBW proved to be generally excellent; and "no inherent obstacles to [the] practical use" of digital flight control materialized during the Dryden research. Leaders of the DFBW project later characterized it as "one of the model programs...[b]ecause we had a tremendous amount of freedom to seek out the real problems. We weren't held to:...'your objective is to have 40 flights by September.'" Rather, they received simple marching orders "go find the problems of digital fly-by-wire."

During the succeeding years, the resistance to DFBW gradually diminished. Indeed, in 1978 McDonnell Douglas became the first manufacturer to integrate it in a production aircraft, the U.S. Navy's famous F-18 Hornet. Shortly thereafter, the later models of the Air Force's F-16 front line fighter rolled off the General Dynamics assembly line with DFBW, succeeded by such diverse military vehicles as the F-117 Stealth Fighter, the B-2 Bomber, and the YF-22 Advanced Tactical Fighter. In commercial aviation, Airbus Industries acted first to make digital-fly-by-wire the standard for airliner equipment, followed in the 1990s by the Boeing 777. Experimental aircraft have also profited from DFBW. One aerodynamicist calculated that in the event of the highly maneuverable X-29 forward swept-wing demonstrator losing DFBW, it would fall out of control in less than two seconds. Finally, Dryden's borrowing of computerization from the space program came full circle; the Shuttle Orbiter flew safely and reliably in part because of the installation of DFBW, tested first at DFRC. (See chapter 8 for DFRC's contributions to the Shuttle program).[24]


## SOMETHING NEW IN AERODYNAMICS


During the very period in which the Dryden team attempted to perfect DFBW, a lone figure at the Langley Research Center used his extraordinary imagination to add inherent efficiency to airframe design. Richard T. Whitcomb arrived at Hampton in 1943, a 22 year old with a Bachelor of Science degree in mechanical engineering from the Worcester Polytechnic Institute. Whitcomb showed backbone from the very start of

---

[24]Kenneth Szalai and Calvin Jarvis, interview with Lane Wallace, 30 August 1995, 20-23, 25-26, 32, 35, 39, 48, DFRC Historical Reference Collection (first and third quoted passages, 39, 48); briefing text, prepared by S.R. Brown, R.R. Larson, K.J. Szalai, and R.J. Wilson, "F-8 Digital Fly-By-Wire Active Control Law Development and Flight Test Results," 26 August 1977, sections 1.0 and 10.0, Kenneth Szalai Collection, DFRC Historical Reference Collection (second quoted passage, section 10.0); Wallace, *Flights of Discovery*, 116-118; Edwin J. Saltzman and Theodore G. Ayers, *Selected Examples of NACA/NASA Supersonic Flight Research* (Edwards, California: NASA SP-513, 1995), 48-49.

his career at Langley. He loved airplanes and had been a model enthusiast since boyhood; thus, when his superiors attempted to place him in the Instrumentation Division, he insisted on working in aerodynamics. His bosses relented and in future years no one would regret the decision. It soon became apparent that the laboratory had found a young man of exceptional abilities, but one better left to his own reckoning than to close supervision. Able in mathematics, Whitcomb possessed keen intuition and a mind bent on speculation, as well as an unusual gift for coaxing answers from the materials he shaped on his workbench. He also threw himself headlong into whatever he conjured in his private thoughts. Whitcomb specialized in transonic flow and the eight-foot high-speed wind tunnel became his companion and collaborator. Here he experimented with aerodynamic solutions to the problems of drag encountered at speeds just below Mach 1. He gradually concluded that the answer to reducing drag lay not merely in more efficient airfoil shapes, but in some new mating of the wing and the fuselage. Whitcomb's subsequent tunnel tests during the late 1940s confirmed his suspicion, revealing hitherto unknown shockwaves formed at the fuselage and the wings. During a moment of contemplation in his office, a random thought crossed his mind with such force that it actually catapulted him from his desk: low transonic drag depended on reducing the combined lateral span of the body, the wings, and the tail. Compressing this cross-section by narrowing the fuselage where the wings joined it would reduce the profile and hence the amount of drag. Whitcomb's new design challenged the classic bullet-shaped body as the aerodynamic ideal. John Stack instructed Whitcomb to prove his case, which he succeeded in doing through more tests in the high speed tunnel, culminating in the publication of a 1952 NACA Research Memorandum entitled "A Study of the Zero Lift Drag Characteristics of Wing-Body Combinations Near the Speed of Sound." American aircraft manufacturers received copies immediately and almost as quickly began to apply its principles to aircraft design. "The basic idea," he remarked with characteristic understatement, "was as simple as giving the air someplace to go so it

38

wouldn't push back on the wing. It was as simple as putting wings on a coke bottle. I had a coke bottle shape that day. Then, the next day, I arrived at a rule of thumb...:transonic drag is a function of the longitudinal development of the cross-sectional area of the entire [air]plane." Combining "rule of thumb" with "cross-sectional area" resulted in the now famous area rule. Whitcomb's discovery received its initial exposure to formal flight research during trials conducted by the NACA. These tests originated with the development of the Convair F-102 interceptor, christened the Delta Dagger. In the midst of designing this machine, their first supersonic aircraft, Convair's engineers discovered that the vehicle produced so much transonic drag that it might be unable to penetrate Mach 1. However, a glimmer of hope appeared in 1952 when Whitcomb himself introduced a delegation from Convair to the area rule, even before his findings appeared in print. After extensive wind tunnel experiments and consultations with Whitcomb, the company supplied an F-102A and a YF-102 to the NACA: the first with the area rule planform, the second without it. Verification flights conducted by Walt Williams' staff during 1956 and 1957 pitted the two models against one another and proved conclusively the superiority of the area rule design in the transonic range, a demonstration that won the concept a place on almost every future supersonic aircraft.[25]

Once his work on area rule subsided during the late 1950s, Whitcomb devoted the following four years to researching designs for an American supersonic transport (SST) able to compete for passengers with subsonic airliners. Disillusioned because he found no airframe light enough and no engine efficient enough to offset the high costs of fuel,

[25]For a general discussion of the area rule discovery, see Lane E. Wallace, "The Whitcomb area rule: NACA Aerodynamics Research and Innovation," in Pamela Mack, ed., *From Engineering Science to Big Science: The NACA and NASA Collier Trophy Research Project Winners* (Washington, D.C.: NASA SP-4219, 1998), 135-148 (see especially pages 144 to 147 for a discussion of the F-102); Hansen, *Engineer in Charge*, 331-341; Brian Welch, "Whitcomb: Aeronautical Research and the Better Shape," *Langley Researcher*, 21 March 1980, 1-2, Hallion Collection, DFRC Historical Reference Collection (quoted passage); Wallace, *Flights of Discovery*. 182.

he decided "to quit the field. I'm going back where I know I can make things pay off," back to transonic research. During 1964 Whitcomb began to consider a method to further reduce the drag induced as aircraft--especially transport aircraft--flew toward Mach 1. This time he concentrated only on the airfoil shape. For the next two years Whitcomb acted on a hunch: that a smoother flow of air would be achieved above wings configured not in the bird-like shape traditional since the drawings of Sir George Cayley appeared 150 years earlier. Rather, in the flight regime approaching the speed of sound Whitcomb decided a wing virtually flat on top would produce less drag than the customary upper surface which curved downwards from the mid-point to the leading and trailing edges. Indeed, he essentially turned the time-honored airfoil upside down. He evolved the design slowly, making his own calculations and systematically modifying the wind tunnel model by shaping and filing its subtle contours with his own hands. When the Langley technicians finally mounted his new airfoil on an F-8 model, he spent countless hours in the tunnel, preferring to sleep on a cot throughout the machine's 24 hour-a-day operation rather than drive home and back. In the thick of research Whitcomb often worked around the clock, explaining, "when I've got an idea, I'm up in the tunnel." He called his concept the supercritical wing (SCW)--an airfoil which delayed significantly the onset of high transonic drag. He derived the term from "critical" Mach number (the speed at which supersonic flow manifested itself above the wing), and "super," meaning "beyond" (illustrated in the word supersonic). Whitcomb unveiled his preliminary findings in May 1966 during a conference on aircraft aerodynamics held at Langley. Without flourish, he described an unprecedented wing shape "incorporat[ing] a slot between the lower and upper surfaces near the trailing edge with negative camber of the airfoil ahead of the slot and substantial positive camber rearward of the slot." Swept back at 35 degrees and tested to Mach 0.90, the supercritical wing generated five percent less drag than conventional designs in the tunnel tests. Whitcomb distinguished between the conventional airfoil, over whose upper surface a powerful shock wave and a separated

40

boundary layer developed in transonic flight; and the supercritical wing, above whose upper surface a weak shock wave and less pronounced boundary layer separation occurred. By lessening the intensity of the shock wave and, even more important, reducing the disruption in the boundary layer, the Whitcomb wing not only promised to fly towards Mach 1 with less drag. but with greater stability and reduced buffeting.[26]

By this point, Whitcomb had amassed considerable experimental evidence to support his claims. But his boss at Langley, Director for Aeronautics Laurence K. Loftin, wanted the theory of perhaps his most original thinker to be tested outside the confines of the center. As a consequence, Loftin turned the problem over to researchers at the Courant Institute at New York University to develop an analytical method by which the pressure distribution and drag characteristics of SCW could be predicted and verified. But this alone did not satisfy Loftin. "This thing is so different from anything that we've ever done before," Whitcomb quoted him as saying, "that nobody's going to touch it with a ten foot pole without somebody going out and flying it." In other words, no aircraft manufacturer would consider it for the production line just on the strength of wind tunnel results, however convincing. Accordingly, he called a meeting of the principals in March 1967, at which the parties discussed the essential objectives of a supercritical wing flight research program. They agreed that the airfoil appeared promising, but also felt it raised design questions answerable only by building the wing, mounting it, and flying it. Could its complex contours be machined to meet the necessary tolerances? How would the

---

[26]Richard Whitcomb, interview with unknown interviewer, 27 March 1973, 7, DFRC Historical Reference Collection (first quoted passage); Welch, "Whitcomb: Aeronautical Research," 2, Hallion Collection, DFRC Historical Reference Collection (second quoted passage); Richard T. Whitcomb, "The State of Technology Before the F-8 Supercritical Wing," in volume 1, *Proceedings of the F-8 Digital Fly-By-Wire and Supercritical Wing First Flight's 20th Anniversary Celebration at NASA Dryden Flight Research Center, Edwards, California, 27 May 1992* (Washington, D.C.: NASA Conference Publication 3256, 1996), 81-84; Richard T. Whitcomb and James A. Blackwell, Jr., "Status of Research on a Supercritical Wing." Presented at a Conference on Aircraft Aerodynamics, Langley Research Center, Hampton, Virginia, 23-25 May 1966, 367-368, DFRC Historical Reference Collection (quoted passage).

wing behave in rough air, during sideslip, and during maneuvering? Would the wing experience deflections or deformations under flight loads? Did the wing offer a margin of safety under conditions in which pilots found it necessary to exceed the recommended cruising speed? Finally, the attendees, who included Loftin and Whitcomb, selected the Vought F-8A Navy fighter as the project's testbed. The choice entailed a compromise since Whitcomb intended his innovation for transports and for a new generation of commercial airliners. But the F-8 got the go-ahead for practical reasons; not only did NASA have access to one through a Naval Air Systems Command contact, but the machine featured such practical advantages as a wing removable in one piece, landing gear which retracted into the fuselage rather than the wings, enough thrust to cover the necessary speed range, and far cheaper operating costs than, say, a Boeing 707. Loftin subsequently contacted Vought in order to initiate a design study on the supercritical wing size appropriate to the F-8 and to launch plans for contractor airworthiness flight tests. Vought swung into action during the summer of 1967 and a period of collaboration with Whitcomb followed on such matters as testing a model of the F-8/supercritical wing combination in the Langley tunnels.[27]

The fortunes of the supercritical wing project took an unexpected turn in January 1968. Headquarters NASA imposed a freeze on all planned contracts, eliminating Vought's offer to perform the F-8 modifications. In the breach, Loftin and the others determined the project should enter a collaborative phase with the Flight Research

---

[27]Richard T. Whitcomb, "Research on Methods for Reducing the Aerodynamic Drag at Transonic Speeds," The Inaugural Eastman Jacobs Lecture, NASA Langley Research Center, Hampton, Virginia, 14 November 1994, 5-6, DFRC Historical Reference Collection; Whitcomb, "The State of Technology Before the Supercritical Wing," 85 (quoted passage); Meeting Notes, Laurence Loftin, "Discussions of Supercritical Wing Research Airplane," 21 March 1967, Hallion Collection, DFRC Historical Reference Collection; Warren C. Wetmore, "New Design for Transonic Wing to be Tested on Modified F-8," Aviation Week and Space Technology (17 February 1969): 23; Tom Kelly, interview with Richard Hallion (handwritten notes), 24 April 1978, Hallion Collection, DFRC Historical Reference Collection.

Center. The FRC received its first formal notification in April and a joint meeting was held the following month. Consultations ensued from spring to fall, during which time the Langley team members informed their Fight Research Center counterparts about the main features of the project and expanded the wind tunnel experiments to accompany the upcoming flight tests. As delegated responsibilities became clearer, the FRC participants undertook the required preliminaries such as designing instrumentation, contracting for the modification of the testbed aircraft, and enlisting project engineers, pilots, and a flight crew for work. Laurence Loftin retained the role of Langley manager and Whitcomb served as chief consultant. But rather than trust to spoken understandings, the two centers agreed to their respective roles in a written memorandum. Just before Thanksgiving 1968 they agreed upon Langley's responsibilities: to define the essential flight research objectives (subject to FRC augmentations); to determine the contours of the modified wing and fuselage, as well as its construction tolerances; and to conduct collaborative wind tunnel tests during all phases of the flight program. Meanwhile, the Flight Research Center concentrated on determining the wing size, weight, and balance (in cooperation with the contractor and with Loftin's staff); acquiring the F-8A and supervising its in-house and contractor modifications; undertaking flight research, ground tests, data acquisition and reduction; analyzing the flight data in collaboration with Langley; and preparing, with Hampton's approval, a flight research plan and an instrumentation suite. Clearly, at least in its early stages, there could be no doubt about who owned the keys to the supercritical wing project and concept.[28]

Nonetheless, the Flight Research Center assumed its role with enthusiasm, conceiving of it as an opportunity to participate in a pivotal joint program, one which center director Paul Bickle felt might presage many more. It also opened the possibility

[28]Tom Kelly, interview with Richard Hallion (handwritten notes), 24 April 1973, Hallion Collection, DFRC Historical Reference Collection; Associate Chief, Full-Scale Research Division, Langley Research Center to Assistant Director, Group 3, Langley Research Center, 20 November 1968, Hallion Collection, DFRC Historical Reference Collection.

of demonstrating the historic imperatives of Langley and of the Flight Research Center: respectively, the predictive power of the wind tunnel, and the verification and correction of ground testing through flight research. The spirit of cooperation in search of these objectives proved to be better than many anticipated. The appointment of John McTigue as SCW project manager, a function he also served for the lifting bodies, reflected the project's importance to the FRC. Tom McMurtry, the lead project pilot and a former naval aviator, was backed up by DFBW's Gary Krier. The Flight Research Center issued request for proposals to industry in February 1969. North American Rockwell's Los Angeles Division won the competition and, for a cost of $1.8 million, delivered the finished wing to Edwards in November 1970. Over the next four months the Flight Research Center technicians fitted it to a Navy surplus TF-8A trainer, acquired for the program in lieu of the originally desired F-8A. During this period, McMurtry and many others at the Flight Research Center found themselves in frequent contact with the illustrious Richard Whitcomb. One encounter involved suggestions by the FRC engineers to mount flaps and ailerons on his wing. The originator of SCW resisted their request, asking instead for a rolling tail to control lateral motions. This concession, said Whitcomb, would leave his airfoil shape uncompromised and free to be tested "like a wind tunnel in the sky...." Unfortunately, at slower speeds the F-8's tail structure failed to produce the necessary roll power and Whitcomb conceded the ailerons. "Dick Whitcomb," observed Tom McMurtry, "is...a supersmart guy as far as design is concerned, but he's also a very practical guy, too. He's adamant about some things, but he'll back down if you point out it's just not feasible, it's just not practical." Indeed, over the course of the project the FRC staff "really got along well with him." Once the flight research got underway the FRC group often asked him for data requiring new wind tunnel tests and even though these imposed additional burdens on his solitary research, he did his best to oblige.[29]

Whitcomb was present at Edwards on March 9, 1971, when McMurtry flew the supercritical wing for the first time, testing its basic airworthiness, its low-speed handling, and its performance as high as 10,000 feet and as fast as 300 knots. Other pilots shared the duties during the later phases, but during its first six months in the air McMurtry piloted the F-8 in all but one flight. These tests proved the merit of the new airfoil under many conditions and yielded valuable data which began to substantiate Whitcomb's predictions. But flight research accomplished more than simply proving Whitcomb a prophet. For instance, McMurtry discovered an abrupt pitch up at 11 degree angle of attack, compared to 13 degrees in the tunnel tests. Because the stability augmentation system (SAS) controlled the phenomenon, the subsequent flight program included maneuvers with and without the SAS to assess its effectiveness and its necessity. By the time McMurtry completed his 13th flight on September 15, 1971, he succeeded in opening the performance envelope to Mach 0.99 and an altitude of 46,000 feet. By February 1972 a clearer picture of the SCW in flight started to emerge. McMurtry and FRC engineers Neil Matheny and Donald Gatlin reported the F-8 displayed good stability and control at cruise velocities and satisfactory handling over Mach 1, including banking maneuvers. Throughout the operating envelope it demonstrated conventional flying qualities, including the landing approach. Even without the stability augmentation system the aircraft exhibited acceptable pilot control. However, some divergences from the wind tunnel data also materialized. Flight data revealed greater longitudinal forces in the transonic range than predicted, although not significantly different and not noticeable in the cockpit. Similarly, while the tendency

---

[29]Hallion, *On the Frontier*, 203-205; Wetmore, "New Design for Transonic Wing," 22; anon., "SCW Technology Chronology," 1969-1974, Hallion Collection, DFRC Historical Reference Collection; flight log, F8 Supercritical Wing, Hallion Collection, DFRC Historical Reference Collection; Thomas McMurtry, interview with Richard Hallion, 3 March 1977, 1-8 (quoted passages, 5, 8); Einar Enevoldson, interview with Richard Hallion, 4 March 1977, 8, Hallion Collection, DFRC Historical Reference Collection.

towards aileron-induced yaw appeared more pronounced in flight than in the tunnel

predictions, it remained too small to adversely affect the aircraft's response and

McMurtry described excellent roll control at transonic speeds. Midway through its flight

program the SCW received high marks.

> than
>
> the
>
> The F-8 supercritical wing program has indicated that the piloting tasks and procedures at cruise speeds in the vicinity of Mach 1 should be no less routine in present day transport operations. Some differences do exist between flight and wind-tunnel measurements of the stability and control characteristics; however, handling qualities were predicted well. No unexpected or violent control characteristics have been encountered. This brief assessment...can perhaps be summarized in one overall observation: The introduction of the supercritical wing is not expected to create any serious problems in day-to-day transport operations.[30]

During the balance of the SCW flight investigation researchers turned from

questions about the Whitcomb wing's safety and stability to ones of its inherent value to

commercial aviation. Once the last flight occurred on May 23, 1973, some remarkable

generalizations could be gleaned from the 86 SCW flights and the 87 hours aloft. Above

all, the data confirmed the wisdom of Laurence Loftin's insistence on flight research, just

at it broadly confirmed Whitcomb's wind tunnel forecasts. Afterwards, aircraft

manufacturers and the airlines expressed a keen interest in the new airfoil which could

increase by 15 percent the efficiency of commercial jets. Indeed, in a business where a

fraction of a percentage added or subtracted million of dollars on the balance sheet, here

was a proven discovery which offered a 2.5 percent increase in profits over aircraft with

conventional wings. One estimate translated the savings to $78 million per year in a fleet

---

[30]Anon., "SCW Technology Chronology," Hallion Collection, DFRC Historical Reference Collection; flight log, F-8 Supercritical Wing, Hallion Collection, DFRC Historical Reference Collection; McMurtry, interview with Hallion, 3 March 1977, 6, Hallion Collection, DFRC Historical Reference Collection; Weneth Painter, interview with Richard Hallion, 8 August 1977, 8, Hallion Collection, DFRC Historical Reference Collection; Thomas C. McMurtry, Neil H. Matheny, and Donald H. Gatlin, "Piloting and Operational Aspects of the F-8 Supercritical Wing Airplane," in *Supercritical Wing Technology: A Progress Report on Flight Evaluations, A Compilation of Reports Presented at a Symposium Held at NASA Flight Research Center, Edwards, California, February 29, 1972,* 97-102, DFRC Historical Reference Collection (block quote, 102).

of 280 jets carrying 200 passengers each. Some calculated the net gain to the world's airlines at almost half a billion dollars annually. The predicted windfalls resulted from fuel economies. General aviation pioneer William Lear reckoned that the supercritical wing--which required no increase in engine thrust and no additional airframe weight-- would raise the cruising speed of his machines by 10 percent and their range by 20 percent.

Unlike many in history, Richard Whitcomb did not suffer the fate of unrecognized genius. His area rule concept won the prestigious Collier Trophy in 1954. During 1974 Whitcomb received a reward for the supercritical wing, as well as its aftermath, when Dr. James Fletcher presented him with the largest cash prize ($25,000) ever received by a NASA employee. Later that year, the National Aeronautic Association recognized the achievements of both the area rule concept and the supercritical wing by bestowing on him the Wright Brothers Memorial Trophy. Perhaps the greatest honor, however, is paid to Whitcomb in airports the world over, where his design insights grace commercial aircraft of every description.[31]

## A NEW ANALYTICAL TOOL

While Richard Whitcomb evolved the concept of the supercritical wing, which eventuated in a three year flight verification of his wind tunnel analyses, a young

---

[31]Another bearer of the supercritical wing is the AV-8B Harrier, as well as many other subsonic and transonic aircraft. Flight Log, F8 Supercritical Wing, Hallion Collection, DFRC Historical Reference Collection; NASA Facts, "F-8 Supercritical Wing," Dryden Flight Research Center, n.d., DFRC Historical Reference Collection; NASA Press Release, "Supercritical Wing," Dryden Flight Research Center, n.d., Hallion Collection, DFRC Historical Reference Collection; Anon., "SCW Technology Chronology," Hallion Collection, DFRC Historical Reference Collection; Wallace, *Flights of Discovery*, 13-14; Hallion, *On the Frontier*, 209; Wallace, "The Whitcomb Area Rule: NACA Aerodynamics Research and Innovation," in *From Engineering Science to Big Science*, 135.

researcher at the Flight Research Center named Kenneth Iliff trained his mind on a different facet of the relationship between wind tunnel data and the information amassed during flight research. But this junior engineer, abetted by Lawrence Taylor--an extraordinarily able senior FRC engineer--sought to invert the accepted practices. Rather than use flight research to verify wind tunnel findings, Iliff and Taylor devised a means of estimating the essential flight derivatives from the flight data itself. Iliff arrived at the Flight Research Center in the early 1960s at the age of 21, after taking a double major in mathematics and aeronautical engineering from at Iowa State University. Having spent his first two years in physics, he received a well-rounded undergraduate education and entertained several job offers. The NASA opportunity paid the least of all, but he accepted it, in part because of its fame as the home of the X-15 program. As it turned out, Iliff started his career on the X-15 staff, assigned the task of analyzing a variety of flight data. Two or three weeks after arriving at the center, however, he paused on his way back from the library one day, looked in on Dale Reed and his cohorts, and asked about the lifting bodies project. When he learned the details of their exciting work, he volunteered to join its ranks. He soon assumed an important role. Despite Iliff's youth and inexperience, Reed assigned him some complicated tasks. His principal duties at the FRC still involved the X-15, but true to center tradition at the time, so long as he kept up with his "day job," his supervisors raised no objections to "moonlighting" for Reed.

Iliff's first task for Reed involved sifting the small archive of wind tunnel experiments on the M2-F1, adding some library research, and arriving at an analysis of the aerodynamics of the upcoming towed flights. (See chapter 6). Then came the far bigger role of making a mathematical model of the M2-F1. He and Taylor rigged up an ad hoc dynamic wind tunnel in the older man's garage in nearby Lancaster and derived the essential data. Luckily, Taylor's neighbors were slow to realize that the routine dimming of lights in the evening could be traced to the machine in his home. In any event, with the tunnel results in hand, Iliff applied a technique not yet known in aeronautics; using FORTRAN in conjunction with FRC's IBM 704 computer (a cast-off from Ames) he analyzed the tunnel data and derived aerodynamic moments and a lift and drag profile for the M2-F1. (The aerodynamic derivatives were determined from the dynamic wind tunnel data with the same techniques that would later be used for flight data). Programming of this kind did not seem extraordinary to Iliff. His alma mater specialized in computer studies; indeed, it trained many an Ames aerodynamicist in the earliest form of computational fluid dynamics. Moreover, Iliff had experienced personally the sheer tedium of obtaining results by the existing methods, having participated during the X-15 program in the painstaking process of coaxing stability and control derivatives from analog computers. Other factors also contributed to the climate of change. As flight control surfaces fell under computer control and started to move continuously rather than sporadically, it would become all but impossible to render a portrait of their motions using the standard techniques. Just as Ken Szalai's impending fly-by-wire advances depended on digital computing, Iliff realized the same electronics offered a means of speeding up the laborious process of analyzing massive amounts of flight data and rendering the whole into a coherent, consistent set of flight data analysis rules.[32]

Beginning in 1964, Iliff and Taylor began to develop techniques to integrate the new computing into flight research. They analyzed the flight data from such experimental aircraft as the X-15, the XB-70, and the lifting bodies and during the fall of 1966 extracted stability and control derivatives using the maximum likelihood parameter

---

[32]Kenneth Iliff later earned a Ph.D. at the University of California at Los Angeles and assumed faculty status there. Kenneth Iliff, interview with Michael Gorn (by telephone), 1 December 1998, DFRC Historical Reference Collection; written comments on a draft of this chapter by Kenneth Iliff, DFRC Chief Scientist, DFRC Historical Reference Collection; list of DFRC accomplishments, "World Standard Aircraft Parameter Estimation Technology," n.d., Ken Szalai Collection, DFRC Historical Reference Collection; Reed, *Wingless Flight*, 24-26, 31, 58, 59; Wallace, *Flights of Discovery* 56-57; Saltzman and Ayers, *NACA/NASA Supersonic Flight Research*, 38-39.

estimation technique. This technique employed an algorithm essentially the same as that found in modern parameter estimation computer programs. In essence, the programs themselves have also remained much the same, although many features have been added to simplify the application of the techniques.

But the full evolution of the concept, not to mention its acceptance among flight research practitioners, took some time to develop. Building on their mathematical modeling of the M2-F1, they took a giant step forward. Rather than converting wind tunnel data into a computer format, they devised a computer program by which the flight data itself could be analyzed and, finally, yield the stability and control derivatives. The power of the computer, hitherto unavailable in its more flexible digital form, now rendered this a possibility. Iliff added to the M2-F1 flight data additional information gleaned from some applicable X-15 maneuvers. Fortunately, stability and control phenomena could be expressed accurately in linear differential equations, which also went into the program. Finally, collaborating with Professor A. V. Balakrishnan at the University of California at Los Angeles, Iliff and Taylor modified the maximum likelihood estimation theory to fit their needs. Known as parameter estimation or parameter identification, the resulting process devised by Iliff and Taylor not only proved to be highly accurate in its own right, but because the derivatives were obtained independently of wind tunnel analyses, the two methods could be used as mutual cross-checks. In effect, one technique bolstered the other. While data gotten from wind tunnels could be erroneous due to scale effects and other causes, nothing could replace its invaluable capacity to predict the behavior of a vehicle in flight prior to placing a human being in its cockpit. But, the predictive quality also divided the two methods since parameter estimation relied not on prediction, but on the actual motions of a vehicle in flight. On the other hand, parameter estimation suffered from inaccuracies in the sensing devices which measured the flight motions and the control surface positions. These errors often foiled the efforts of the most careful researcher from obtaining entirely

50

correct readings. Poor samplings inevitably reduced the value of parameter estimation, but other factors might also intervene to subvert its effectiveness. Flight vehicles always operated at the mercy of unpredictable disturbances in the atmosphere. Moreover, errors in the model itself presented an inherent weakness for which no satisfactory answer existed other than evolutionary improvements. But the tightly integrated world of flight research offered some assurance of quality control.

> In the flight test environment, results are subject to detailed critical review. If our results disagree with predictions, someone will ask where we erred; we need to convincingly defend our results before an often skeptical audience. If we suggest that the simulator be revised based on our results, we must demonstrate why the update is worth the work (and hope the pilot notices that the revised simulator flies more like the airplane). If we suggest that instrumentation errors have occurred, someone will test it and contradict us. If we request more test data, the schedulers will complain about milestones and cost. In some flight regimes the controls and handling qualities group wants assurance that our results are very accurate because they have little margin for error; in other flight regimes they may insist that we must be wrong, because if our results are correct, the control system needs to be redesigned. Throughout this process, few people care if we have an elegant, sophisticated, and innovative method; they simply want good results and they want them immediately.[33]

Iliff advised colleagues interested in applying parameter estimation methodology in institutions such as NASA to be wary of non-technical complications which, if not heeded, might vitiate the whole process. Many of his suggestions took a page directly out of the NACA approach to flight research. He cautioned them to avoid the common error of pursuing a narrow flight program with the intention of achieving success;

---

[33]Kenneth Iliff, interview with Michael Gorn, 1 December 1998, DFRC Historical Reference Collection; written comments on a draft of this chapter by Kenneth Iliff, DFRC Chief Scientist, DFRC Historical Reference Collection; list of DFRC accomplishments, "Parameter Estimation," Ken Szalai Collection, DFRC Historical Reference Collection; Kenneth W. Iliff and Richard E. Maine, "NASA Dryden's Experience in Parameter Estimation and Its Uses in Flight Test," Preprint of AIAA Paper 82-1373, to be Presented at the AIAA Atmospheric Flight Mechanics Conference, San Diego, California, August 9-11, 1982, 1-2; Richard E. Maine and Kenneth W. Iliff, *Application of Parameter Estimation to Aircraft Stability and Control: the Output-Error Approach* (Washington, D.C.: NASA Reference Publication 1168, 1986), 1 (block quotation); Wallace, *Flights of Discovery*, 56-57.

51

planning a few flights with a single maneuver at the portions of the envelope in question doomed the whole effort to failure. Instead, "the flight test plans should reflect a problem-oriented [author's italics] philosophy. Assume," wrote Iliff, "that there will be problems and design the tests to maximize the chances of finding and fixing the problems." As it pertained specifically to the successful pursuit of parameter estimation, the flight test programs required a number of specific ingredients. The statement of objectives needed to be transparent; for example, it had to state whether one model sufficed for the entire envelope or whether several were needed to offer a composite of the full flight regime. The requirements for the predicted derivatives and their sources of data also required a clear statement. Flight maneuvers required careful consideration, including the instrumentation and data systems necessary to measure and record the phenomena desired for analysis. Finally, effective parameter estimation demanded a definition of the analytical methods and the differential equations before the flights occurred. But Iliff tempered the need for planning with an admonition about flexibility. "In many cases," he said, "you will need to revise earlier decisions based on later results. For instance, unexpected trends in the estimates might justify extra instrumentation, additional maneuvers, or alternative analysis methods. Inflexibility and refusal to reevaluate previous decisions invite poor results."

Iliff's cautionary suggestions enjoyed an international audience; the techniques developed in collaboration with Taylor and later with engineer Richard Maine assumed such stature that they found acceptance in flight research organizations across the globe. Moreover, beginning solely as a method to determine stability and control derivatives, each new generation of computer programs broadened in scope to account for the whole range of aerodynamic effects (such as flying qualities, maneuverability, and safety), as well as structural dynamics and performance. Aircraft improved by parameter estimation exist the world over. Among the 70 or so flight vehicles flown at Dryden that profited from this advanced analytical technique, the XB-70 Valkyrie bomber, the SR-71

52

Blackbird, the Space Shuttle Orbiter, NASA's High Angle-of-Attack Research Vehicle (HARV), the X-29 forward swept-wing technology demonstrator, and all of the lifting bodies exemplify just a few of the beneficiaries.[34]

## A PAGE TURNS

Flight research entered a difficult period during the early 1970s. As NASA struggled to cope with steep budgetary reductions and attempted to re-fashion itself after the end of the Apollo program, the practice of flight research underwent intense scrutiny without a fundamental transformation. As a consequence, the discipline witnessed no new programs of national scope and urgency (like the X-15) and no revolutionary flight vehicles birthed from the loins of NASA itself (like the lifting bodies). Instead, smaller projects with large return on slight investment--but also with low profile outside the aeronautics community--crowded the flight logs during the 1970s. Thus, while the value of the undertakings remained high and their influence remained great, a decline occurred in the visibility of NASA flight research inside and outside the institution. During the tense days at Dryden in the mid 1970s when the center found itself searching for an appropriate role and unsure of its essential constituencies, some predicted the

[34]Interview, Kenneth Iliff with Michael Gorn, 1 December 1998, DFRC Historical Reference Collection; list of DFRC accomplishments, "Parameter Estimation," Ken Szalai Collection, DFRC Historical Reference Collection; Maine and Iliff, "Applications of Parameter Estimation," 5-6 (quoted passages); Wallace, *Flights of Discovery*, 56-57. See also Chester Wolowicz, Kenneth Iliff, and Glenn Gilyard, "Flight Test Experience in Aircraft Parameter Identification," Presented at AGARD Flight Mechanics Panel Symposium on Stability and Control, Braunschweig, Germany, 10-13 April 1972, 1-11; Richard E. Maine and James E. Murray, "Application of Parameter Estimation to Highly Unstable Aircraft," Collection of Technical Papers, AIAA Atmospheric Flight Mechanics Conference, 18-20 August 1986, Williamsburg, Virginia, 25-30; Kenneth W. Iliff, "AIAA Dryden Lecture in Research For 1987: Aircraft Parameter Estimation," AIAA 25th Aerospace Sciences Meeting, 12-15 January 1987, Reno, Nevada, 1-17; Kenneth Iliff, "Parameter Estimation for Flight Vehicles," *Journal of Guidance, Control, and Dynamics* (September-October 1989): 609-622.

disappearance of the Flight Research Center from the NASA organization charts. Events in the 1980s and 1990s proved these people to be prophets for a time, but false prophets in the fullness of time. Paradoxically, during the very period in which flight research suffered an eclipse in status, it planted the seeds of its own regeneration by pursuing important new missions and responsibilities.

# CHAPTER 8

## New Directions

## A PROPHECY FULFILLED

Despite declining agency budgets and lower profile projects during the 1970s, NASA flight researchers nonetheless undertook many impressive investigations that yielded highly beneficial results. Ames Research Center demonstrated the practicality and efficiency of the tilt-rotor concept. Langley fostered the invention of the supercritical wing and tested its value in cooperation with Dryden. Finally, NASA's most remote center pioneered parameter identification and disseminated its techniques internationally, developed the world's first completely non-mechanical flight control system in digital-fly-by-wire, and began a series of crucial contributions to the fledgling Space Transportation System (more about which in this chapter). Yet, entering the 1980s, there remained an uneasy feeling that the best days of flight research might not lie ahead. The number of civil service employees at DFRC did not lend encouragement. During the five years from 1977 to 1981 the rolls fell steadily from 520 to 450, a reduction of 70 positions and an almost 14 percent decline. By themselves, these losses did not constitute a catastrophe. In the midst of the reductions, however, new tensions developed. Early in 1979, after 15 months of study and amid rising anxiety among the staff at Dryden, a reorganization occurred that consolidated the six top-line directorates--Research, Data Systems, Flight Operations, Aeronautical Projects, Shuttle Operations, and Administration--into three. Henceforth, Data Systems, Shuttle Operations, Aeronautical Projects (as well as the Office of the Chief Engineer) disappeared from the organizational chart. Only Administration survived, the new Directorates of Engineering and of Flight Operations absorbing what remained.  More than mere "musical chairs," this realignment, intended "to better carry out [the center's] goals," caused turmoil and affected many adversely.  Of the almost 300 civil service employees affected, mostly in the mid-upper grade technical fields, nearly ten percent experienced actual reductions in grade.  Although Civil Service regulations protected the incumbents from loss of pay, subsequent holders of the positions received the lower grades and the lower salary, provided these jobs were not among those eliminated before, during, or after the reorganization of 1979.

No doubt stimulated by these events, enthusiasm for DFRC's fundamental work diminished, but not merely among the rank and file.  Some senior engineers and pilots also expressed concern.  Milt Thompson posed the question in a handwritten memorandum entitled, "Why Dryden is not doing more innovative research."  He sensed a pattern in which NASA headquarters and Langley "chastized" [sic] Dryden when it initiated imaginative work.  Acceding to this pressure, DFRC launched many worthwhile projects with definable, near-term applications.  But as a consequence, the center all but abandoned broader, more expansive research (like the lifting bodies) whose pursuit offered no definite payoff at the time, but later resulted in unanticipated and substantial benefits.  Thompson associated these trends with a gradual erosion in the autonomy of the center directors, especially in their capacity to tap discretionary funds to undertake riskier projects.  Paul Bikle planted the seeds of the lifting body program in just that way, prompting Thompson to argue for "meaningful levels" in the discretionary pool: a million dollars and authority to use multi-year funding if necessary.[1]

But rather than winning greater decisionmaking autonomy, the center saw the very opposite occur; it lost control of its own affairs.  Like a thunderbolt on a cloudless day, a stunned Dryden staff learned on April 27, 1981, that effective October 1 DFRC would assume facility

[1]Handwritten DFRC Manning Chart, September 1977 to September 1986, n.d., Milt Thompson Collection, DFRC Historical Reference Collection; Hallion, *On the Frontier*, 273; DFRC Organization Charts, May 1976, August 1981, DFRC Historical Reference Collection; anon., "Dryden Research Center Undergoing Reorganization," *Antelope Valley Press*, 4 February 1979; Handwritten Paper, "Why Dryden is not doing more innovative research," n.d., Milt Thompson Collection, DFRC Historical Reference Collection.

status and be consolidated as an entity of the Ames Research Center. The secret had been kept so well that even the Dryden Director of Public Affairs, whose business depended on inside knowledge, told reporters, "It hit us by surprise. We have no idea what it's going to mean." The passage of a little time instructed everyone about the circumstances of the shotgun marriage. Evidently, NASA headquarters found itself in budgetary straits due to Congressional demands for austerity and the agency decided to demonstrate its commitment to efficiency by closing some of its centers. But instead of actually shuttering any of its field operations, NASA authorities chose to preserve the smallest organizations through consolidation with larger ones. Accordingly, headquarters officials incorporated Dryden into Ames *and*, at the same time, merged the Wallops Island, Virginia, Flight Center into the Goddard Space Flight Center in Greenbelt, Maryland. The stated reasons for the DFRC-Ames marriage seemed to some at Dryden to lack conviction, promising predictably to "focus the resources of each of the installations on what it can do best. The close relationship between Ames' and Dryden's efforts in aeronautical programs as well as the unique facility capabilities and the physical proximity of the installations provides an opportunity to improve overall program effectiveness...." Although institutional survival animated the amalgamation, and although care was taken to preserve Dryden's dignity and separate identity, one bald fact remained. A new sign hung above the entrance to the former DFRC main office building now read "Ames Research Center" on its top line, and "Dryden Flight Research Facility" (DFRF) below.

To accommodate the realities of the situation, the 1979 reorganization was itself transformed. After several months of deliberations, a proposed re-structuring emerged from the counsels of a task team composed of members from headquarters OAST, the affected centers, as well as DFRC Director Isaac Gillam and Ames Director C.A. Syvertson. To begin with, the DFRC Flight Operations Directorate expanded to include Ames rotorcraft and science platform aircraft, as well as traditional Dryden research engineering. Secondly, the entire DFRC administrative machinery received instructions to re-locate to Moffett Field, leaving only a Site Manager's office as a local caretaker. Dryden's Project Management Office, including all of its project managers, made plans to move north to join the Ames Aeronautics Directorate.  No forced lay-offs or relocations resulted from these events.  However, in the cold light of experience, two of these changes were reversed: the transfer of the Project Management Directorate proved to be a short-lived experiment, and the tilt-rotor and the Quiet Short-Haul Research Aircraft remained at Ames.  But, perhaps most significantly, the amputation of the total Dryden administrative apparatus remained unaltered.  Syvertson put the best face on a situation not of his making when he informed the two staffs about the accomplished deed.  "I am firmly convinced," he stated, " that Ames and Dryden can be merged into a single effective and efficient organization for the conduct of advanced aeronautical research."  With that, Isaac Gillam received a transfer to NASA headquarters to serve as Special Assistant to the Administrator and John Manke, who had achieved fame as a lifting bodies pilot and later became Director of Dryden Flight Operations, assumed the position of director of a combined Ames/Dryden Office of Flight Operations.  In this capacity he not only managed flying activities at both centers but acted as on-site manager of DFRF on behalf of Ames.[2]

## THE SALVE OF IMPORTANT WORK

[2]Essay by Milt Thompson on the Ames/Dryden Consolidation, n.d., 1-2, 6, Milt Thompson Collection, DFRC Historical Reference Collection; anon., "Takeover Surprises Dryden Officials," Public Affairs clipping from unidentified newspaper, April 29, 1981, DFRC Historical Reference Collection (first quoted passage); Dryden New Release, "NASA Consolidates Center Operations," n.d., DFRC Historical Reference Collection (second quoted passage); Donald L. Mallick, unpublished memoir, 1995, 302, File number 001421, Headquarters NASA Historical Reference Collection; anon., "Ames/Dryden Consolidation to Begin October 1, 1981," Dryden *X-Press*, 7 August 1981, 2, DFRC Historical Reference Collection; anon., "Gillam Named to NASA Position," *The Bakersfield Californian*, 23 September 1981, B6, DFRC Historical Reference Collection; NASA Ames News Release, "Manke Named to Head Dryden and Ames Flight Operations," 6 October 1981, DFRC Historical Reference Collection.

Despite the often awkward accommodations required of Ames and Dryden personnel to give life to this bureaucratic hybrid, the Dryden staff found little time for complaints. For a number of years the DFRC workforce and budget became entwined in a program of high national visibility, the U.S. Space Shuttle. Indeed, on the very day in April 1981 when news of consolidation washed over the center, *Columbia*, the first Orbiter to go into space, took off from Dryden aboard its special 747, bound for the Kennedy Space Center (KSC) to be readied for another flight. Almost two weeks earlier the spaceship accomplished its first mission, carrying pilots John Young and Robert Crippen around the earth. It returned them to a perfect landing at DFRC on April 14, 1980. Yet, because of the unfortunate, and undoubtedly unintentional timing of the consolidation announcement, Dryden employees experienced conflicting feelings; jubilation at the technical achievement they contributed to, but disappointment at the way NASA Headquarters seemed to repay their hard work. Nonetheless, these sentiments faded with the need to carry on the mission, not least on the Shuttle itself. Indeed, in a briefing in December 1981 in which Dryden leaders presented 20 of their biggest projects, the Space Shuttle assumed the top position.

Dryden's role in Shuttle Orbiter development originated well before this period. Even prior to the 1969 recommendation by President Nixon's Space Task Group for a new space transportation system to follow the Apollo-Saturn rocket combination, programs such as the lifting bodies (see chapter 6, especially the coverage of X-24B) and the X-15 (chapter 5) suggested conceptual frameworks for the next generation of spaceflight. Indeed, the concept of power-off, runway landings of the Orbiter originated with Flight Research Center engineers who worked actively to persuade the Johnson Space Center (JSC) planners of its validity. The Houston team initially wanted to return the ship to earth using special landing engines. But the glide concept could not be dismissed; it had been proven in X-15 and in lifting body flight research and documented fully in instrumented data, pilot commentaries, and in technical

5

reports. This approach also offered the advantages of precious savings in weight and fuel and the ultimate benefit of higher payload capacity. To win the point, the Flight Research Center held a symposium in June 1970 to relate the lifting bodies' potential to the Shuttle program. Perhaps the most convincing remarks were uttered not by any NASA personnel, but by the Air Force's Jerauld Gentry, one of the most experienced of the lifting bodies pilots. He did much to reassure the Johnson visitors about the feasibility of glide flight.

> The criticality of our lifting body approach, flare, and landing is really much less than you might realize. The USAF Aerospace Research Pilot School at Edwards graduates approximately 30 students every year. Each of those pilots must demonstrate proficiency in accomplishing unpowered approaches and landings in the F-104 airplane that are much more critical than the lifting body task. Assuming that the shuttle vehicle will have reasonable stability and handling characteristics, I cannot foresee any significant problems with an unpowered approach and landing. In addition, although the shuttle vehicle is intended to operate somewhat like a commercial airliner, I seriously doubt that the first shuttle pilots are going to be ex-airline captains. Rather, I imagine they will be experienced test pilot/astronauts.[3]

Because the USAF insisted on a large 60 by 15 foot cargo bay for the Orbiter, the smaller capacity lifting bodies were eliminated from consideration. Ultimately, Houston chose a more traditional fuselage with the delta wing, but its developers did accept the unpowered landing concept. Six months after President Nixon authorized NASA to proceed with the Shuttle in January 1972, Rockwell International of Downey, California, won the Orbiter prime contract. During the five years in which the company fabricated these hypervelocity aircraft and integrated

---

[3]Wallace, *Flights of Discovery*, 138; Ronald "Joe" Wilson Diary, vol. 5, entries for 14, 15, and 27 April 1981, DFRC Historical Reference Collection; essay by Milt Thompson on the Ames/Dryden Consolidation, n.d., 6-7, Milt Thompson Collection, DFRC Historical Reference Collection; briefing (to Ames representatives by Milt Thompson?), 23 December 1981, Milt Thompson Collection, DFRC Historical Reference Collection; Dryden Fact Sheet, "Space Shuttles and the Dryden Flight Research Center," November 1994, 1-3, DFRC Historical Reference Collection; anon., "DFRF Technical Support of STS Program," Ken Szalai Collection, DFRC Historical Reference Collection; Dryden Fact Sheet, "The Space Shuttles," August 1995, 2, DFRC Historical Reference Collection; Wen Painter, interview with Michael Gorn, 15 May 1997, 60, DFRC Historical Reference Collection; William H. Dana and J. R. Gentry, NASA TM X-2101, "Pilot Impressions of Lifting Body Vehicles," *in Flight Test Results Pertaining to the Space Shuttlecraft, A Symposium held at Flight Research Center, Edwards, California, June 30, 1970* (Edwards, California: NASA, 1970), 77, DFRC Historical Reference Collection.

the components of its subcontractors, the Flight Research Center continued to offer important counsel to the project. As the prototypes emerged in the Downey plant Ken Szalai and the DFBW group advised the Rockwell design team about the installation and operation of the four on-board IBM AP-101 computers, identical to the electronic brains purchased a year earlier for the F-8's flight control system. A Honeywell four-channel fly-by-wire subsystem connected the computers and linked them, in turn, to the flight control surfaces. Unlike the Crusader's predictable and mannerly handling qualities, however, the Orbiters possessed "terrible flying qualities" and exhibited dangerous instability if not guided by an absolutely dependable flight control system. This fact lent an urgency to the computerization of the spacecraft. Well paid (one million dollars) by Johnson to infuse the Orbiter development with the F-8's experience (see Chapter 7), Szalai and his associates did just that, forwarding the results and data from the F-8 flights, testing the Orbiter system, and otherwise supporting the program as requested.[4]

Meanwhile, the Orbiter awaited flight research in 1977. During the previous year Edwards witnessed the construction of two 100-foot vertical structures linked at the 80 foot level by a horizontal arm. Known as the Space Shuttle Mate-Demate Device, its tripartite design was conceived to hoist the Orbiter into position to receive post-flight maintenance and repair, and to lift the spacecraft atop a transport aircraft for shipment to its next destination. To avoid the charges of extravagance almost certain to result from procuring specially built aircraft, two existing jumbos--the Lockheed C-5 cargo plane and Boeing's 747 airliner--underwent scrutiny as

---

[4]The decision by the Dryden DFBW team to buy the AP-101 computers was influenced strongly by engineers at Johnson who realized that their use in flight would provide valuable lessons for the developing Space Shuttle. Also, the Shuttle posed a far greater challenge for DFBW than the F-8 simply because the spaceplane, unlike the Crusader, was an all new airframe which needed to be capable of flying home from space. See comments on a draft of this chapter by Mary Shafer, DFRC engineer, 13 July 1999, DFRC Historical Reference Collection; Dryden Fact Sheet, "The Space Shuttles," August 1995, 2, DFRC Historical Reference Collection; William H. Dana, interview by Peter Merlin, 14 November 1997, Dryden Flight Research Center, 17, DFRC Historical Reference Collection; Dryden Fact Sheet, "Space Shuttles and the Dryden Flight Research Center," 4, September 1995; Duane McRuer, interview with Lane Wallace, 31 August 1995, 12, DFRC Historical Reference Collection; Kenneth Szalai and Calvin Jarvis, interview with Lane Wallace, 30 August 1995, 20, DFRC Historical Reference Collection; Elliott, *Development of Fly-By-Wire*, 28.

the Orbiter carrier. After studies to determine ease of separation from the mother ship, the 747 won the assignment and the program bought a used, obsolete model for $15.6 million. But in the days following this decision, a debate flared about the upcoming flight research. Wrangling over the flight path prompted FRC Director and former astronaut David Scott to write to NASA headquarters and express concern about Johnson Space Center's plans to drop the Orbiter from its carrier ship at 18,000 feet. Scott, supported by the observations of John Manke and Chuck Yeager, warned that a launch from this altitude left only the barest margin of safety and allowed no time for the pilots to familiarize themselves with, or to evaluate, the handling qualities of the Orbiter during descent. Scott argued persuasively that below 300,000 feet the Orbiter flew not as a spacecraft, but as an aircraft, "a new vehicle being subjected to the non-linear environment of the atmosphere....This I believe requires knowledge and techniques not available in the vast storehouse of space experience...."[5]

Scott won the argument for higher drop altitudes, but Johnson maintained its lead role. During final preparations for the flight research program Donald "Deke" Slayton, JSC's manager of the upcoming approach and landing tests, apparently walked into a meeting of Dryden engineers and said, "We're gonna fly in three days, I don't care what you guys say." Slayton's inflexible remark placed him at odds with the Dryden custom of flying not merely to satisfy a deadline, but to pursue a clearly defined research objective, consistent with the safety of the pilot and with the completeness of the preparations. Still, the program began as scheduled. First, on February 15, 1977, three taxi runs of the prototype Orbiter *Enterprise* mounted on its 747 transporter tested the structural loads and ground-handling qualities of the combined vehicle. Then five captive-carry flights, accomplished with no crew in the Orbiter, recorded data on the aerodynamics and flying characteristics of the paired aircraft in a sequence of take-off, climb,

[5]Comments on a draft of this chapter by Mary Shafer, Dryden engineer, 13 July 1999, DFRC Historical Reference Collection; Dryden Fact Sheet, "Space Shuttle Mate-Demate Device," August 1995, DFRC Historical Reference Collection; Richard Day, interview with Richard Hallion, 24 February 1977, 62, DFRC Historical Reference Collection; David Scott to John Yardley, 29 July 1975, Milt Thompson Collection, DFRC Historical Reference Collection (quoted passage).

cruise, and landing. The first *Enterprise* free flight--a piloted glide launched from the 747 at 22,000 feet--occurred six months after the taxi tests, on August 12, 1977. During this pivotal test, the second of the four redundant computers shut down shortly after commander Fred Haise (of Apollo 13 fame) and pilot Gordon Fullerton pushed off from the airliner, but the remaining three digital sentries assumed control. The two astronauts maneuvered during their five minute twenty-two second descent to lakebed runway 17 and landed without further incident, although a mile beyond the agreed upon touchdown point. In flights two, three, and four *Enterprise* underwent additional evaluations of its aerodynamics, flight control systems, stability and control, handling qualities, angle-of-attack responses, and structural integrity.

Perhaps in the spirit of his earlier declaration to proceed regardless of impediments Deke Slayton decided to reduce these piloted flights from the planned eight to only five, even though the cockpit encountered significant braking ineffectiveness, brake "chatter," and missed landing targets on the first four. On the fifth and final approach and landing of the Orbiter prototype, flown on October 26, 1977, Fred Haise again took the controls, as he had on flight three. This time, however, the flight plan called for a touchdown not on Rogers Dry lakebed, but on the Edwards 15,000 foot concrete runway. Traveling at 245 knots when it uncoupled from the 747, *Enterprise* obeyed Haise's commands until approach and landing. Because Slayton mandated that this would be the last of the research flights, Haise felt some anxiety to redeem the reputation of the program and to land close to the targeted 5,000 foot marker. He had other reasons to be nervous; on this glide the tail cone fairing present in the initial approach and landing tests had been removed, changing the vehicle's aerodynamics. Moreover, Haise knew that among those witnessing the event was no less a dignitary than His Royal Highness Prince Charles, himself an experienced pilot. Reflecting his desire for pinpoint accuracy, as well as his state of mind, Haise worked the controls in an intense, high-gain fashion. He almost succeeded, running just a few hundred feet long. But in the heat of concentration he came in a bit too high and a little fast, so he applied the speed brake and then released it. Then, just eight seconds before landing, Haise used the control stick to modify the Orbiter's sink rate, resulting in 12

9

degree swings in the ship's elevons which, in turn, caused pitching oscillations. To combat these motions, Haise introduced some inputs to the computer system, but the response happened slowly due to a time lag in processing by the flight control system. These commands finally reached the control surfaces just as the wheels brushed the concrete, bouncing the Orbiter back into the air, at which time it rolled right and then fell into pilot induced oscillations for four seconds before the tires met the runway for the second and final time. Finally safe on the ground, Gordon Fullerton expressed surprise when observers described the extreme roll the *Enterprise* had just experienced. Obvious to those watching the approach, the danger was not so apparent to those inside the ship because the cockpit pivoted at the center of rotation. To the degree he realized the full peril of the situation, Haise fought the hazardous lateral movements with all his skill. Apparently, however, only when Fullerton yelled "Hey, let loose," did the commander take his hands off the controls, allowing the swaying to damp and the Orbiter to make a safe recovery at the last instant. Subsequent analysis revealed three gremlins which bedeviled Haise: time delay of the digital computers, the Orbiter's difficult handling qualities during landings, and rate limiting of the elevator actuators.[6]

While the Orbiter approach and landing tests did end with only five piloted flights, the Dryden Flight Research Center launched "a massive campaign" to comprehend and to solve the problem of pilot induced oscillation experienced in the *Enterprise*. Early in 1978, Milt Thompson, in his capacity as Director of Research Projects, drafted a flight plan designed to "obtain a current data base that will sharpen our awareness of all factors (subtle and obvious) that

---

[6]Dryden Fact Sheets, "Space Shuttles and the Dryden Flight Research Center," September 1995, 4, DFRC Historical Reference Collection; Wen Painter, interview with Michael Gorn, 15 May 1997, 42, DFRC Historical Reference Collection (first quoted passage); Duane McRuer, interview with Lane Wallace, 31 August 1995, 8, DFRC Historical Reference Collection; Approach and Landing Test Evaluation Team, "Space Shuttle Orbiter Approach and Landing Test Final Evaluation Report," Johnson Space Center, February 1978, 4-3 to 4-11, 4-89, DFRC Historical Reference Collection; Gordon Fullerton, interview with Lane Wallace, 7 September 1995, 5-7, DFRC Historical Reference Collection; Ronald "Joe" Wilson Diary, entry for 26 October 1977, DFRC Historical Reference Collection (second quoted passage); comments on a draft of this chapter by Mary Shafer, DFRC engineer, 13 July 1999, DFRC Historical Reference Collection.

might influence a low [lift-to-drag] Orbiter runway landing in demanding situations." This program not only involved new F-8 Digital Fly-By-Wire tests to determine how the delayed computer response to human input might be reduced or eliminated; but also the application of *Enterprise's* approach and landing data to simulators for the purpose of teaching pilots the appropriate gain for critical landing situations. Like David Scott and many others, Thompson argued that the Orbiter, guided by aerodynamic control surfaces during its descent through the atmosphere, should be developed based on "aircraft experience rather than spacecraft experience...." As such, Thompson assessed its performance by applying the standards of vehicles very familiar to him: the HL-10, the M2-F3, the X-24A, the F-8 Supercritical Wing and Digital-Fly-By-Wire, and others. He concluded that "lateral-directional stability and control margins are inadequate for hypersonic and supersonic flight [of the Shuttle Orbiter]. I, personally, would not approve the first orbital flight. [I]t just doesn't seem tidy," he warned with mock humor, "for NASA to produce a vehicle that has the potential of turning left when you want to go right."

In all likelihood, Thompson's mood did not brighten when the Johnson Space Center management decided to transfer the *Enterprise* to the Marshall Space Flight Center for ground vibration tests. This decision left the veteran pilot with no Orbiter to conduct the approach and landing flight research. Lacking the actual aircraft for this research, Thompson improvised with a substitute vehicle which was large, but *not* highly susceptible to PIOs. He also needed proven test pilots *not* familiar with this machine in order to eliminate the possibility that their past experiences might mask the aircraft's deficiencies. He chose (among other aircraft) an interceptor version of the SR-71 known as the YF-12A and selected Dryden pilots William Dana, Einar Enevoldson, and Fitzhugh Fulton to conduct the flight program. Flying just one time each, the three men conducted conventional "touch and go" landings, exhausting whole tanks of fuel in the repetitive process. Although neither of them experienced pilot induced oscillations, the data recorded by instrumentation and augmented by pilot impressions helped establish a baseline for approach and landing handling qualities both in the air and in the

11

simulator. Meanwhile, Ken Szalai's DFBW experts examined the recordings from Haise's wild ride in flight number five, as well as the results of 60 F-8 DFBW landings designed to simulate the Orbiter's characteristics. The five pilots who flew these touchdowns found that lags as short as 200 milliseconds between pilot input and discernible changes in the flight control surfaces made a profoundly detrimental impact on safe shuttle landings. In response, the DRFC engineers devised and the pilots flight tested a relatively simple software modification. Like a filter that dampened the kind of pilot motions likely to cause oscillations, the new electronics package suppressed high frequency longitudinal stick inputs without affecting the aircraft's handling qualities and without causing time delays. The improvement, however, came at a price; greater landing control resulted in a corresponding loss of stick responsiveness. Nonetheless, Johnson officials incorporated the software changes into the Orbiter in 1979. Because the suppression software restrained, but did not cure the Orbiter's latent PIO tendencies, Dryden continued its participation in Orbiter landing research. At the further request of Houston, DFRC computer specialists devised additional software modifications and simulator analyses to curb the Orbiter's propensity to pitch up on approach to touchdown. As the date neared for launch of STS-1 into orbit (April 12, 1981), Milt Thompson admitted the software revisions offered improvements in the Orbiter's safety. But, reflecting on the machine from the research pilot's viewpoint, he still felt uneasy about its fundamental properties as an aircraft, telling a colleague at Johnson during the last days of 1980 that pilot training and computer aids failed to alleviate persistent landing, as well as entry control deficiencies.

> I would improve the landing control system as soon as possible. Real handling quality improvements in the landing control system would eliminate the need for the PIO training. The PIO suppresser does not improve the handling qualities during landing. The PIO suppresser is simply a crutch which does not address the real problem. I do not feel the entry handling qualities are as good as they should be...[and] should be improved before the Shuttle becomes operational. To expand...on the entry handling qualities, every maneuvering...is a two-handed task due to the excessive stick forces. It is very easy to inadvertently make a pitch input when attempting to make a pure lateral input. This tendency to inadvertently couple a control input, particularly into the pitch axis, aggravates the mediocre longitudinal characteristics. Longitudinally, the orbiter is not as tight as it should be. The poor pitch trim characteristics compound the problem even

12

more. We plan to investigate some ideas for handling qualities improvements as soon as we have accomplished...evaluating the overall controllability. A pilot can do a completely acceptable job of flying the entry with adequate training. On the other hand, the handling qualities don't have to be mediocre. They can be improved and they should be for operational flights.[7]

Indeed, under Thompson's direction Dryden engineers continued in the 1980s to experiment with the Orbiter to make it a better airplane and improve to its landing qualities through the use of more sophisticated computer programs. But many other Shuttle projects vied for DFRF time and resources. During the year before the flight of STS-1 Dryden research pilots tested the heat-resistant tiles attached to the outer skin of the Orbiter. They flew an F-104 and an F-15 a total of 60 times at supersonic speeds to determine whether the ceramic squares deformed under the conditions of flight. These tests demanded some exacting maneuvers in order to replicate the prevailing airloads on the exterior of the Shuttle. The F-15, in particular, flew trajectories with an altitude and Mach number profile in keeping with the predicted STS-1 launch conditions. Using a new and easy-to-read guidance display, rather than scanning several instruments in the traditional manner, pilots succeeded in duplicating the Orbiter flight path with

[7]William Dana, interview with Peter Merlin, 14 November 1997, Dryden Flight Research Center, 18-20, DFRC Historical Reference Collection (first quoted passage); Handwritten briefing charts (Milt Thompson's handwriting), "DFRC Flight Investigation of Factors Affecting Orbiter Type Landings," n.d., and "Features of Candidate A/C for Landing Program," n.d., Milt Thompson Collection, DFRC Historical Reference Collection (second quoted passage); essay (by Milt Thompson) on Space Shuttle handling, no title, n.d., 4, 6, Milt Thompson Collection, DFRC Historical Reference Collection (third quoted passage); Annual Report of Research and Technology Accomplishments and Applications, FY 1978 (Draft), "Space Shuttle Support," 1-2, Milt Thompson Collection, DFRC Historical Reference Collection; comments on a draft of this chapter by Mary Shafer, DFRC engineer, 13 July 1999, DFRC Historical Reference Collection; Annual Report of Research and Technology Accomplishments and Applications, FY 1979, "Shuttle Orbital Flight Test Support," 25, Ken Szalai Collection, DFRC Historical Reference Collection; Bruce G. Powers, "Experience with an Adaptive Stick-Gain Algorithm to Reduce Pilot-Induced-Oscillation Tendencies," *AIAA 7th Atmospheric Flight Mechanics Conference, Danvers, Massachusetts, August 11-13, 1980*, AIAA Paper Number 80-1571, 2-3, DFRC Historical Reference Collection; John T. Gibbons to Distribution, "Approval of OPD 78-28-- Improvement of Orbiter Landing Characteristics," 26 April 1979, 1-2, Milt Thompson Collection, DFRC Historical Reference Collection; Isaac T. Gillam to Director, Johnson Space Center, 23 February 1981, Milt Thompson Collection, DFRC Historical Reference Collection; Milt Thompson to Robert Thompson, 17 December 1980, Milt Thompson Collection, DFRC Historical Reference Collection (block quote).

great accuracy. The data acquired by these methods suggested the need for improved bonding methods. During the 1980s researchers at Ames and Dryden tested new external insulating materials which challenged the supremacy of the famous ceramic covering. No one doubted the effectiveness of the thermal tiles, but their utility exacted a premium in cost; after each flight many needed to be replaced, a time-consuming process of sizing, shaping, and affixing them to the contours of the Orbiter. Two alternative materials underwent trials: felt reusable surface insulation (FRSI) and advanced flexible reusable surface insulation (AFRSI). Composed of heat treated aromatic polyimide coated with white silicone, both conformed like blankets to complex shapes. The results appeared promising; no failures occurred in either material, even at flights on an F-104 producing 40 percent more airload than a Shuttle launch. But these results happened with the insulation attached to flat surfaces. When AFRSI underwent exposure to actual loads during STS-6 (*Challenger*, launched on April 4, 1983) the portions adhering to curved surfaces broke down under the pressures of the air. Despite such research, the tiles continued in service, even though they remained a source of difficulty.[8]

Orbiter landing flight research--although of a much different kind--continued at Dryden into the 1990s. Representatives of the Johnson Spaceflight Center approached DFRF in 1992 to undertake tests of the Shuttle landing gear tires and wheels to determine whether the tire life of the Orbiter could be extended and whether safe landings could be achieved with ground wind

---

[8]The flight test trajectory guidance system flown on the Shuttle tile tests had broad implications for flight research by facilitating greater pilot accuracy in complicated maneuvers and increasing the amount of data collected per flight. E. L. Duke, M. R. Swann, E. K. Enevoldsen, and T. D. Wolf, "Experience with Flight Test Trajectory Guidance," *Journal of Guidance, Control, and Dynamics* (September-October 1983): 397-398; Bruce Powers and Shahan Sarrafian, "Simulation Studies of Alternate Longitudinal Control Systems for the Space Shuttle Orbiter in the Landing Regime," in *A Collection of Technical Papers presented at the AIAA Atmospheric Flight Mechanics Conference, 18-20 August 1986, Williamsburg, Virginia,* 1, 4, DFRC Historical Reference Collection; Dryden Fact Sheet, "Space Shuttles and the Dryden Flight Research Center," September 1995, 3, DFRC Historical Reference Collection; Bianca Trujillo, Robert Meyer, Jr., and Paul Sawko, NASA TM 86024, "In-Flight Load Testing of Advanced Shuttle Thermal Protection Systems" (Washington, D.C.: NASA, 1983), 1,4; Timothy Moes and Robert Meyer, NASA TM 4219, "In-Flight Investigation of Shuttle Tile Pressure Orifice Installations, (Washington, D.C.: NASA, 1990), 1, 6, DFRC Historical Reference Collection.

conditions as high as 20 knots per hour. Until these experiments, 15-knot winds had been the maximum tolerated, often resulting in landing delays during stronger winds. Also, the Orbiter's tires wore out almost instantaneously at the Kennedy Space Center, caused by an abrasive runway designed intentionally to allow safe landings on damp runways, on high speed landings, on high tire loadings, and on crosswind conditions. Kennedy's alligators, who inhabited the ditches alongside the runways, added extra urgency to pilot worries about tire blowouts. Although occurring late in the Shuttle program, these flight tests required two years of research, 155 flights, and the skills of no less an aviator than Gordon Fullerton. Fullerton not only served, in effect, as co-pilot during the 1977 Orbiter glide flights, but also piloted one Shuttle mission (STS-3), commanded a second (STS-51F), and then became a Dryden research pilot after leaving the astronaut corps. Like Fullerton, the aircraft on which the tests occurred won its spurs after a long time in the air. Assembled in 1962, the Ames-owned Convair (CV)-990, once a part of the American Airlines fleet, approximated the loaded weight of the Orbiter and landed at about the same speed (256 miles per hour). But it required extensive internal modifications at mid-fuselage in order to install a tire test fixture capable of simulating the Orbiter's vertical tire load and yaw angle. Actual Orbiter components were incorporated into the test device, which could accommodate Shuttle tires loaded to 140,000 pounds. For safety, the Convair's original landing gear remained fully extended in flight. Once modified, its empty weight rose from 115,000 to 177,000 pounds and its designation changed to the Landing Systems Research Aircraft (LSRA). In a typical approach during flight testing, Fullerton touched down, slowed the aircraft to the required speed, and extended the test landing gear to yield the planned vertical load, tire braking, and slip angle data. Flights occurred at the Dryden and (to test the roughened surface) at the Kennedy Space Center runways. The results added important knowledge to Shuttle operations. On the Dryden concrete runway, researchers demonstrated the safety of a 20-knot crosswind landing, while the instrumented data obtained from load cells located in three axes yielded refinements in the tire drag model for the Orbiter simulator. At Kennedy, the runway received a new, smoother surface based on studies of Orbiter tire wear conducted on the LSRA.[9]

But the DFRF's commitment to the STS did not end there. Since STS-1, Dryden supported this highly visible, national program as a landing site. Touchdowns one through four, in particular, absorbed a great deal of time and effort by Dryden personnel. At the approach of an incoming Orbiter, dozens of Dryden employees staffed the mission control room, prepared for post-landing Orbiter servicing and for physical exams for the astronauts, and handled the global media inquiries. They also maintained and operated one of the two 747 Shuttle carriers. The many landings also allowed Dryden engineers like Kenneth Iliff and others to analyze Shuttle flight data and to publish their findings. Gradually, however, these responsibilities and opportunities diminished. Between 1981 and 1996, 45 of the 76 Shuttle flights--nearly 60 percent--landed at Edwards. But the trend line sloped downwards when Edwards assumed the role of alternate landing location. During the 1980s, fully *80 percent* of all Shuttle touchdowns occurred at Dryden. During the early 1990s only about 45 percent ended their missions at DFRF; and from 1992 to 1996 just 30 percent. Nonetheless, its long-term association with the Shuttle and its valuable service to the nation rewarded Dryden with a measure of international prominence, with periodic bursts of laudatory press coverage, and with enhanced notoriety for its flight research mission. It also brought money. But with these undeniable advantages came a disadvantage, one like that heard in the hallways of the FRC during the 1960s: when a program the size of the X-15 ends, what happens to the center? Similarly, during the 1990s the gradual reduction in Shuttle activity at Dryden raised the same question, one not easily answered in a

_____

[9]Don Nolan, "CV-990 Expands Orbiter Crosswind Limits," Dryden *X-Press*, September 1994, 1, DFRC Historical Reference Collection; Dryden Fact Sheet, "Landing Systems Research Aircraft," November 1994, DFRC Historical Reference Collection; Cheryl Heathcock, "CV-990 Completes Orbiter Wheel and Tire Tests," Dryden *X-Press*, September 1995, 1, DFRC Historical Reference Collection; C.D. Michalopoulos and David Hamilton, "Orbiter Tire Traction and Wear," *AIAA Report Number 95-1256-CP*, 1995, 851, 859, DFRC Historical Reference Collection; Comments on a draft of this chapter by Mary Shafer, DFRC engineer, 13 July 1999, DFRC Historical Reference Collection; John Carter and Christopher Nagy, NASA TM 4703, "The NASA Landing Gear Test Airplane" (Washington, D.C.: NASA, 1995), 1-3, 5-7.

decade of tight federal budgets and one especially meaningful at a time when Dryden lacked bureaucratic autonomy. For reasons of institutional survival, the twin issues of a Shuttle follow-on and a return to center status assumed the highest importance to DFRF during the 1990s.[10]

## AN OLD FASHIONED PROGRAM

While Dryden found itself immersed in the complexities of Space Shuttle flight research and support, across the continent another NASA center awakened a low cost and venerable flight research project dormant for many years. Aircraft icing ranked among the great conundrums of flight safety, resulting in the loss of many lives and machines. Charles Lindbergh reported that wing icing almost brought down the *Spirit of St. Louis.* Northrop Alpha and Gamma aircraft carrying the mails during the winter of 1932 and 1933 were plagued by icing, yet all but one survived because of an experimental rubber de-icing fixture. The lost plane flew the same route without the protective device. Two commercial airliners--a Curtiss Condor in 1935 and a Douglas DC-2 in 1937--both succumbed to ice in the skies over Pennsylvania. Concerned for the survival of its crews and aircraft, the Army and the Navy asked the NACA to investigate this puzzling phenomenon. As early as 1928, Langley pilots William McAvoy and Thomas Carroll decided the existing first hand accounts offered nothing but contradictory evidence. By systematically flying a Vought VE-7 into freezing cloud formations above the laboratory the two men arrived at some of the earliest judgments about ice formation. They found clear, solid ice, often shaped like mushrooms and attached to the leading edge of airfoils, just below 32 degrees Fahrenheit. At significantly lower temperatures, so-called rime ice accumulated, characterized

---

[10]Dryden Fact Sheet, "Space Shuttles and the Dryden Flight Research Center," September 1995, 6, DFRC Historical Reference Collection; Dryden Fact Sheet, "Completed Space Shuttle Missions," May 1996, DFRC Historical Reference Collection; Richard Day, interview with Richard Hallion, 24 February 1977, 56-58, DFRC Historical Reference Collection.

by a lack of mushrooming, a white and opaque appearance, and a granular texture. Carroll and McAvoy speculated that the combination of the added weight of the ice coupled with the reduction in lift resulting from the irregular (non-aerodynamic) surfaces contributed to many winter-time crashes.

Like others after them, the two pilots found solutions hard to come by. Their own flight research in the VE-7 demonstrated that slick substances like wax or paraffin attracted more ice than bare metal; that heating a wing with engine exhaust reduced ice at the leading edge, but did nothing to prevent its accumulation farther back along the chord; and that solutions designed to mix with the water droplets and cause a lower freezing temperature were effective initially but soon washed off of the aircraft by rainwater. The only sure remedy in 1929 lay in avoiding cold, moisture-filled clouds. Two years later, however, a pair of Langley researchers issued a far more optimistic report. The problem became increasingly acute as new aircraft achieved greater altitude and range, thus increasing susceptibility to the ill-effects of freezing conditions. The situation attracted the attention of one of the laboratory's great minds, Norwegian-born physicist Theodore Theodorsen. After Theodorsen studied the icing phenomenon with colleague William Clay, the two men advocated a wing heating system using a mixture of exhaust vapor and alcohol. But, again, the preventive effect was not flawless; while airfoil heating required only one-tenth of the entire engine gases, researchers still awaited the discovery of a method to distribute the hot air *evenly* over the full span of the leading edge during flight. Aircraft icing retained its reputation as an intractable problem.[11]

---

[11]Willson Hunter to Executive Engineer, 26 May 1945, National Archives and Records Administration (NARA), Record Group (RG) 255, Box 48, Folder 61, "Icing Research," 1944-1945, copy in DFRC Historical Reference Collection. (The Dryden Flight Research Center Historical Reference Collection contains copies of all of the National Archives Records Group (RG) 255 documents mentioned in this chapter). Glenn E. Bugos, "Lew Rodert, Epistemological Liaison, and Thermal De-Icing at Ames," in Pamela Mack, ed. *From Engineering Science to Big Science: The NACA and NASA Collier Trophy Research Project Winners* (Washington, D.C.: NASA SP-4219, 1998), 29-58; Thomas Carroll and William McAvoy, NACA TN 313, "The Formation of Ice Upon Airplanes in Flight" (Washington, D.C.: NACA,1929), 1-13; Theodore Theodorsen and William Clay, NACA TR 403, "Ice Prevention on Aircraft by Means of Engine Exhaust Heat and a Technical Study of Heat Transmission From a Clark Y Airfoil"

After several partial attempts and limited successes, it became apparent that to make progress in the understanding of ice, it needed to be pursued as a specialized study. Engineer Lewis Rodert, a wiry, relentless, and somewhat difficult midwesterner who graduated from the University of Minnesota in 1930, joined the Langley Laboratory in 1936 and plunged into the void. Never intimidated by the subject's complexities, he began by exploring ways to protect propeller blades from the build-up of ice. Like others after him, Rodert realized icing was not a subject likely to yield its full complexities in a laboratory, so he turned instinctively to flight research as his main instrument of discovery. Rodert began his investigations on an Army XC-31 cargo aircraft with a 12.5 foot diameter propeller. Rodert employed a pumping system to move an ice-repellent cocktail (85 percent alcohol, 15 percent glycerin or ethylene glycol) to the propeller hub, from which centrifugal force propelled it to the root cup and from there by tubes to the blades' leading edges. While this research did not prevent the accumulation of blade ice, the fluid and the flow over the propeller showed promise. Rodert then tried his hand at the application of exhaust heat to icing. He and his assistants mounted a model airfoil, with a chord of three feet and a span of four feet, between the wings of a Navy XBM biplane and flew the machine at 100 miles per hour. Rodert discovered that a tube running inside of the wing's leading edge removed about one-third of the heat available from the exhaust gases. He also derived a formula (involving airplane speed, chord, and air temperature) for calculating the heat necessary for ice prevention.[12]

Rodert's research soon acquired a critical mass of success. Encouraged by his early work, the Army Air Corps agreed to sponsor new investigations into aircraft icing. Accordingly, in early 1940 Rodert asked his superiors to form an ice research unit. Already immersed in war research, the Langley leadership agreed, but decided to assemble the group at the new Ames

(Washington, D.C.: NACA, 1931), part one, 91, and part two, 111-112; Hansen, *Engineer in Charge*, 421.

[12]Bugos, "Lew Rodert," 29-30; Lewis Rodert, NACA TN 727, "A Flight Investigation of the Distribution of Ice-Inhibiting Fluids on a Propeller Blade" (Washington, D.C.: NACA, 1939), 1-2, 6; Lewis Rodert and Alun Jones, NACA TN 783, "A Flight Investigation of Exhaust-Heat De-Icing" (Washington, D.C.: NACA, 1940), 1-4, 11.

Laboratory in California. Pilot William McAvoy joined Rodert on the West Coast at the start of 1941. A note of urgency entered the Rodert-McAvoy collaboration just after their arrival; the U.S. Weather Bureau disclosed that it had on file between 800 and 1,000 pilot reports of aircraft icing and made them all available to the NACA researchers. Luckily, NACA headquarters recognized the momentum gathering around Rodert's research; Lockheed Aircraft received a purchase order from Hampton for a new Model 12A with heated wings built to Rodert's specifications. It began to ply the skies in February 1941 in an ice belt located about 40 miles north east of Sacramento. At first, nothing went smoothly. A bolt of lightening struck the new aircraft, resulting in time-consuming inspections of the engine bearings and delays due to repair of the radio compass, essential to the all-instrument flying required in icing research. With the winter melting away, further flights shifted to the upper Midwest. For about a week in mid-April the heated wing received the hard test of Northern Minnesota and Wisconsin weather. It performed in a "completely satisfactory" manner. More extensive flight research occurred the following winter and Rodert published the findings in a NACA paper entitled "A Flight Investigation of the Thermal Properties of an Exhaust Related Wing De-Icing System on a Lockheed 12A Airplane." The results showed that with the Rodert heating system, the forward 20 percent of the wing rose 70 degrees over the air stream temperature. Rearward of this portion the warming effect diminished gradually to only a 10 degree rise. This benefit occurred at the cost of an increase in weight ranging from one-half to one and one half percent of the aircraft's total.

But not all of the hazards associated with this phenomenon had been understood and conquered. On a flight from North Dakota at the very end of 1942 ice accumulated steadily on the vertical tail surfaces forcing pilot Lawrence Clousing to increase power in order to counteract the noticeable effects of drag. Clousing decided to pull out of the overcast in which the 12A flew, losing 23 to 30 miles per hour as he climbed slowly. Then, as often happens in flight research, the unexpected struck: "a rather violent clockwise rotation was imparted to the wheel control, followed by a yawing movement to the right, and dropping of the right wing." Moving

20

the ailerons had no effect. Pushing the nose down resulted in pilot recovery, allowing him to resume the climb with no further incidents. This classic, if frightening case of a stall triggered by loss of aerodynamic effectiveness due to vertical tail ice represented a cautionary event for the engineers and pilots engaged in these hazardous missions. It also demonstrated the dangers still awaiting military aviators, as well as the flying public.[13]

Although Rodert's heating system constituted only a partial solution to the vexing icing puzzle, George Lewis and Jerome Hunsaker urged Army Air Forces General Hap Arnold to consider it for all Army Air Forces aircraft. Never slow to make up his mind, Arnold decided in August 1942 that by the following winter every B-17E and B-24 flying the North Atlantic would have the equipment aboard. The Ames group under Rodert received instructions to advise immediately both Convair and Boeing on the appropriate modifications. Furthermore, the AAF not only sent a North American O-47A to Ames; it also provided a virtual air armada, including an XB-24F, an XB-26D, an XC-53A, an X-60B, a B-25, and a B-17F. These machines were flown to Moffett Field to participate in the Aircraft Icing Research Project, carried out in winter 1942-1943 by the NACA in cooperation with the Army Air Forces and the Minneapolis office of the Weather Bureau. It constituted an intense program of flying under the most extreme and varied icing conditions in order to gather as many ice measurements and as many flight records

---

[13]George Lewis to F.W. Reichelderfer, 3 February 1941, NARA RG 255, Box 48, "Icing Research," 1939-1941, DFRC Historical Reference Collection; Smith DeFrance to NACA, 6 February 1941, NARA RG 255, Box 48, "Icing Research," 1939-1941, DFRC Historical Reference Collection; Robert Littell to George Lewis, 25 February 1941, NARA RG 255, Box 48, "Icing Research," 1939-1941, DFRC Historical Reference Collection; Henry Reid to NACA, 29 February 1941, NARA RG 255, Box 48, "Icing Research," 1939-1941, DFRC Historical Reference Collection; Smith DeFrance to George Lewis, 9 April 1941, NARA RG 255, Box 48, "Icing Research," 1939-1941, DFRC Historical Reference Collection; Lewis Rodert to Engineer in Charge, 25 April 1941, NARA RG 255, Box 48, "Icing Research," 1939-1941, DFRC Historical Reference Collection (quoted passage); George Lewis to the Chief of the Navy Bureau of Aeronautics, 8 November 1941, NARA RG 255, Box 48, "Icing Research," 1939-1941, DFRC Historical Reference Collection; Lewis Rodert and Lawrence Clousing , NACA Wartime Report A-45, "A Flight Investigation of the Thermal Properties of an Exhaust-Heated Wing De-Icing System on a Lockheed 12A Airplane" (Washington, D.C.: NACA, 1941), 7, 13; Lawrence Clousing and William McAvoy to Engineer in Charge, 29 December 1941, NARA RG 255, Box 48, "Icing Research," 1942, DFRC Historical Reference Collection.

as possible. Once the investigation ended, the Weather Bureau hoped to use the experience to improve its icing forecasts. Exposed to opportunities such as these, Rodert advanced quickly during World War II. He rose from junior to senior aeronautical engineer, and finally, in his mid-thirties, became chief of the Ames Ice Research Project, in which he directed a staff of more than 50. During the cold months of 1944 and 1945 his team flew from their Minneapolis base an AAF C-46 cargo airplane outfitted as an icing laboratory, which added to the understanding of prevention and accumulation on large transports and airliners.

By the end of the war the thermal system developed during successive winters demonstrated, in the words of one observer, "the complete protection of the wings, tail surfaces, and windshield...irrespective of the icing conditions encountered." While this description may have exaggerated the performance of Rodert's contributions, he and his colleagues could surely claim credit for saving many machines and lives during the war. Nominated by the NACA Executive Secretary John F. Victory, Rodert won the 1947 Collier Trophy for his labors. But by this time, Lewis Rodert had left California to direct flight research at the new NACA engine laboratory in Cleveland, Ohio (See chapter 4). The scientists and engineers who worked for him understood his decision. The future of icing work lay in the protection of turbine powerplants, highly vulnerable to the ingestion of ice particles into delicate engine components like compressors. In the pursuit of this research, Cleveland offered not just the advantages of colder winters, but also a new piece of equipment. Shortly after the Flight Research Division opened, the Icing Research Wind Tunnel (IRT) roared into action in 1944. Because of these factors, George Lewis and his headquarters staff decided in September 1946 to transfer all of the Ames icing work to Ohio by the summer of 1947.[14]

[14]George Lewis to Smith DeFrance, 28 August 1942, NARA RG 255, Box 48, "Icing Research," 1942, DFRC Historical Reference Collection; Benjamin Chidlaw to George Lewis, 31 August 1942, NARA RG 255, Box 48, "Icing Research," 1942, DFRC Historical Reference Collection; Lewis Rodert to Engineer-in-Charge, 7 October 1942, NARA RG 255, Box 48, "Icing Research," 1942, DFRC Historical Reference Collection; William Littlewood to George Lewis, 27 October 1942, NARA RG 255, Box 48, "Icing Research," 1942, DFRC Historical Reference Collection; Organization Chart, January 1943, NARA RG 255, Box 48, "Icing research," 1943,

Yet, even the drive and enthusiasm of Lew Rodert failed to launch the Cleveland icing research program with the speed demanded by aircraft industries and by NACA headquarters. Manufacturers just installing jet propulsion systems on their airframes clamored for research about their susceptibility to freezing conditions. In part, Rodert faced delays resulting from the vagaries of integrating the Ames personnel and equipment into the new laboratory. In addition, the icing tunnel experienced problems adapting to the demands of the researchers who expected it to simulate the type of freezing water droplets encountered in all different flight regimes and in all weather conditions. Because no one had ever designed a nozzle capable of producing exactly the kind of freezing spray encountered in nature, a significant learning period ensued before the right type of droplets and ice could be produced reliably indoors. In the meantime, engineers at Douglas Aircraft in particular agitated for data about measures necessary to protect their new turbojet engines from ingesting ice and stalling. At the same time, Harrison Chandler, a recent visitor to the laboratory from NACA headquarters, blasted the staff's slowness in publishing technical literature for Douglas and the other aircraft firms. He even told acting Director of Aeronautical Research John Crowley that "the lengthy delays in the preparation of reports are typical of the situation that has existed throughout the conduct of icing research at the [Cleveland] Laboratory." As a consequence, the icing staff shifted into overdrive, laboring virtually around the clock to prepare and fly their B-24 bomber testbed to unscramble the turbine engine dilemma. One early technique involved installing screens over the jet inlets, but the power required to heat them exceeded the capacity of most aircraft. To conserve energy, retractable screens seemed to offer promise. Ultimately, warming the turbine blades themselves proved the most effective method.

The situation further righted itself as the engineering staff worked to develop

DFRC Historical Reference Collection; Harrison Chandler, "NACA Conducts Research to Protect Airplanes Against Icing," 23 December 1944, NARA RG 255, Box 48, "Icing Records," 1944-1945, DFRC Archives (quoted passage); anon., *Icing Research Tunnel History*, NASA Lewis Research Center, September 1944, 1-6, DFRC Historical Reference Collection; Bugos. "Lew Rodert," 29, 51-53; Harrison Chandler to John Crowley, 16 September 1946, NARA RG 255, Box 49, "Icing Research," DFRC Historical Reference Collection.

instrumentation able to record the process of droplet formation and propagation in the atmosphere. They received help from an unexpected quarter when the Army Air Forces persuaded scientist Irving Langmuir of General Electric to advise on this complicated problem. A brilliant polymath willing to participate in varied projects, Langmuir made two essential contributions: a rotating, multi-cylinder device capable of determining the size and the water content of droplets in icing clouds, and a mathematical formula to interpret the results. When assembled, the Langmuir recorder resembled a small, six-tiered wedding cake which widened toward the bottom and whose individual stages moved in unison as they rotated in the atmosphere, gathering droplet samples. Porter Perkins, one of the Cleveland lab's bright young engineering minds who arrived just as the icing tunnel opened, pioneered the application of this instrument to flight research. Such progress was hastened by several factors. The NACA Sub-Committee on De-Icing met at Cleveland in April 1947 and passed a resolution approving the laboratory's plans to investigate the physics of icing clouds. To guide this research the lab appointed Irving Pinkel of the physics section to lead the icing research team. In addition, meteorologist William Lewis of the U.S. Weather Bureau was assigned to Ames (where he flew on the C-46) and then to Cleveland. Lewis and his fellow researchers actually drew their own weather maps each morning based on overnight teletype reports. With collaborators, William Lewis produced four NACA technical notes by mid 1949 which described the data collected over several winters using the Langmuir machine in extensive flight research. Collectively, Lewis' work constituted an early attempt to classify the varieties of water content and droplet size and determine their effect on aircraft ice formation, with the eventual objective of influencing the designs of anti-icing machinery.[15]

---

[15]John Crowley to Cleveland Laboratory, 21 January 1947, NARA RG 255, Box 49, Folder 61: "Icing Research," DFRC Historical Reference Collection; Harrison Chandler to John Crowley, 4 February 1947, NARA RG 255, Box 49, Folder 61: "Icing Research," DFRC Historical Reference Collection (quoted passage); anon., *Icing Research Tunnel History*, 7; Porter Perkins and William Rieke, interview (by telephone) with Michael Gorn, 2 June 1998, DFRC Historical Reference Collection; Bugos, "Lew Rodert," 53-54; Subcommittee on De-Icing Problems to the Committee on Operating Problems, NACA, 6 June 1947, NARA RG 255, Box 49, File 61:

During the 1950s, the experiments at Cleveland (by now known as the Lewis Laboratory in honor of the NACA's late Director of Research George Lewis) followed the patterns set out just after the war. The icing research took three main forms: measuring meteorological parameters like droplet size, developing flight instrumentation to record the characteristics of freezing water in the atmosphere and on flying machines, and capturing in-flight data on the accumulation of ice on aircraft. Several projects characterized these avenues of research. Pilot Joseph Walker teamed with William Lewis' Weather Bureau colleague Dwight Kline to investigate icing encountered in low altitude stratiform clouds. Walker and Kline represented an uncommon pairing. Walker represented the archetype flight research pilot, one who later distinguished himself flying the X-15 research airplane at the Flight Research Center. Kline, according to one project engineer who flew on virtually every parameter icing mission, "always got sick, once even when taxiing out." Flying bomber aircraft in 22 flights, this odd couple measured the droplet size and water contents, as well as the areas in the cloud structure carrying different kinds of precipitation.

Complementing Langmuir's seminal contribution, the NACA Lewis engineers developed some important new icing instruments of their own. One of the most significant was designed by Porter Perkins and two colleagues who fabricated a machine for installation on commercial aircraft which measured the frequency and intensity of ice encounters. Aside from promising an immense statistical database, it also offered information useful to the invention of future ice prevention equipment. Called a pressure-type icing meter, it weighed just 18 pounds, offered a continuous record, required no maintenance, and operated automatically in icing conditions.

---

"Icing Research," DFRC Historical Reference Collection; William Lewis, NACA TN 1393, "A Flight Investigation of the Meteorological Conditions Conducive to the Formation of Ice on Airplanes" (Washington, D.C.: NACA, 1947), 1-3; William Lewis and Dwight Kline, NACA TN 1424, "A Further Investigation of the Meteorological Conditions Conducive to Aircraft Icing" (Washington, D.C.: NACA, 1947), 1-2; Alun Jones and William Lewis, NACA TN 1855, "Recommended Values of Meteorological Factors to be Considered in the Design of Aircraft Ice-Prevention Equipment" (Washington, D.C.: NACA, 1949), 1-5; William Lewis and Walter Hoecker, NACA TR 1904, "Observations of Icing Conditions Encountered in Flight During 1948" (Washington, D.C.: NACA, 1949), 1-4.

Gradually, almost every American carrier equipped at least one of its airliners with the Perkins ice meter. It eventually flew on about 50 passenger aircraft and also on a good many Air Force F-89 fighters. Meantime, after ten years of use in the field under almost every conceivable icing situation, the Langmuir multi-cylinder ice collection instrument was introduced to the worldwide aeronautics community in a lengthy NACA Report published in 1955 and written by William Lewis, Porter Perkins, and other members of the Lewis Laboratory staff. Paradoxically, its publication coincided with the decline of icing as a field of flight research. Powerful turbine engines in airliners and in military vehicles generated so much heat and traveled so quickly through patches of harsh weather that, armed with thermal de-icing equipment on the major flight surfaces and exposed engine parts, they appeared to be all but invulnerable to the hazards of ice. Moreover, with the creation of NASA in 1958, icing research joined many other aeronautical projects eliminated in the rush to conquer space. In any case, the war against one of the major enemies of safe and predictable flight seemed to be won and Lewis Rodert did not mind declaring victory.

> Weather conditions cause ice to form on airplanes during flight. Such vital airplane components as the wings, propeller, engine carburetor, windshield and radio antennas are seriously affected by ice formations. The increase in drag, the loss in propeller thrust, the losses in engine power, in vision, and in communication and other effects of icing would be serious enough in fair weather; but icing does not occur when the sun is shining. It occurs when the pilot is flying on instruments, in clouds, often in turbulent clouds, and at night and when atmospheric static is causing a further reduction in radio communication. For these reasons a strong and persistent request was made that the NACA find a solution to the icing problem....The investigations by the NACA of the thermal anti-icing system and its continued development and application by the aviation industry has enabled the commercial operator "to get his load of passengers through" and the military operator to plan his operations irrespective of possible icing conditions.[16]

---

[16]Porter Perkins, "Objectives of Flight Icing Research at NACA Lewis," n.d., DFRC Historical Reference Collection; Porter Perkins and William Rieke, interview with Michael Gorn, 1 June 1998, DFRC Historical Reference Collection (first quoted passage); Dwight Kline and Joseph Walker, NACA TN 2306, "Meteorological Analysis of Icing Conditions Encountered in Low-Altitude Stratiform Clouds" (Washington, D.C.: NACA, 1951), 1-3; Porter Perkins, Stuart McCullough, and Ralph Lewis, NACA RM E51E16, "A Simplified Instrument for Recording and Indicating Frequency and Intensity of Icing Conditions in Flight" (Washington, D.C.: NACA, 1951), 1-11; R.J. Brun, W. Lewis, P.J. Perkins, J.S. Serafini, NACA TR 1215,

The declaration of victory proved premature. Twenty-five years after the Lewis Research Center abandoned icing research, it rose like a latter-day phoenix from the ashes of aircraft accidents. Actually, several contributing factors caused the revival. After President Jimmy Carter signed the Airline Deregulation Act in October 1978, small commuter carriers sprang up like wildflowers to cater to underserved regional markets. Never had commercial aviation witnessed this phenomenon. These companies, almost all new, needed to find pilots and aircraft without delay. For equipment they turned to European, Japanese, and even Latin American manufacturers since few U.S. firms produced machines designed to carry 20 to 30 passengers at a time. As a result, many of these aircraft received icing certification in their countries of origin. Pilots, on the other hand, who often were recruited from the ranks of recreational fliers, not infrequently found themselves in the left seat on their first assignment. Moreover, because such a long time elapsed between the 1950s, when icing seemed an important safety factor, and the 1970s, when no one even considered it, pilot knowledge of the phenomenon had all but disappeared. Finally, in the race to gain a share of this virgin market small airlines felt compelled to squeeze profit from every quarter, and they purchased flight vehicles with one quality uppermost: the capacity to operate cheaply. This insistence on economy translated into machines which met the bare FAA standards for icing (and, indeed, other safety standards as well) while delivering the lowest cost per passenger mile. Efficient consumption of fuel topped the list of desirable airplane attributes, often attained by slender, sharp-edged airfoils. Unfortunately, such designs invited ice accumulation, in contrast to the thick, blunt-edged wings on the big commercial jets. The turboprops also lacked the power of the airliners to tear through spots of inclement weather speedily; when the slower, smaller machines hit a pocket of intense freezing they often had no option but to fly through it for some time, thus accumulating ice.

---

"Impingement of Cloud Droplets on a Cylinder and Procedure for Measuring Liquid-Water Content and Droplet Sizes in Supercooled Clouds by Rotating Multicylinder Method" Washington, D.C.: NACA, 1955), 1-2, 18; Lewis Rodert and W.T. Olson, "Research on the Icing Problem in NACA," n.d., attached to routing sheet from Lewis Laboratory to the NACA, 10 February 1950, NARA RG 255, Box 49, Folder 61: Icing Research, DFRC Historical Reference Collection (block quote).

Finally, a sad truth about airline accident reporting allowed these conditions to prevail. When 20 to 30 individuals died in a commuter crash the press devoted far less time and space to the event than when a jumbo jet fell from the sky.[17]

A number of persons recognized the inherent danger in these conditions. One, a former pilot at Lewis, resurrected the subject almost single-handedly. Jack Enders, who left Cleveland to become Chief of the Aviation Safety Office at NASA Headquarters started a one-man campaign in 1978 to resume icing research for general and corporate aviation. He realized in a world in which few feeder airlines existed, the odds of icing fatalities were slim. But in light of the burgeoning number of commuter departures each day, even a tiny fraction of affected flights would result in many deaths. Before succeeding, Enders' salesmanship encountered the passage of time (four years), many crashes and fatalities, and strong opposition from Langley Research Center to win the role for itself. Finally, in the summer of 1982 a DeHavilland Twin Otter aircraft was delivered from Langley and Porter Perkins, one of the original Lewis icing researchers who retired in 1980, returned to Lewis Research Center on contract. Former naval aviator William Rieke joined Perkins to form the nucleus of a small, select team, which also included research pilot Richard Ranaudo. During the first two years researchers concentrated on gathering parameter icing data with such new techniques as laser probes and computer-aided averaging applied to the traditional multi-cylinder collection techniques. Indeed, these two years of flight research proved the original Langmuir instrument no less accurate than the modern equipment. Another line of inquiry, called the LEWICE (Lewis Ice) program, evaluated the physics of icing, while a third project employed computer modeling to detect the presence of rime ice, a less hazardous type than the more commonly studied glazing ice. Despite these endeavors, until the late 1980s the airlines and the major manufacturers expressed little interest in the resuscitated program. Builders of commuter aircraft, whose products might be tested and

[17]Porter Perkins and William Rieke, interview with Michael Gorn, 1 June 1998, DFRC Historical Reference Collection; Porter Perkins and William Rieke, "Aircraft Icing Problems--After 50 Years," *AIAA 31st Aerospace Sciences Meeting and Exhibit, Reno, Nevada, 11-14 January, 1993*, 1-8.

found wanting, showed even less sympathy. Even the disastrous loss of 69 lives aboard an American Eagle flight near Roselawn. Indiana (due to an encounter with large ice droplets), failed to break the inertia. The turning point occurred when two engineers from the FAA certification branch visited the Cleveland laboratory with news of a troubling tail stall problem. Perkins and Rieke identified tailplane icing as the cause. They then organized a conference in November 1991 to discuss current knowledge and possible new research. Now the regional airlines took notice, sending representatives who rubbed shoulders with engineers and scientists from 13 countries. This meeting gave the renewed icing project the boost it needed. Open discussion (and publication) ensued, concentrating on the many fatal accidents of the 1980s and how they might be prevented in the future. *Airline Pilot*, the magazine of the Airline Pilots Association, ran a lead story on turboprop icing, as did *Accident Prevention*, the journal of the Flight Safety Foundation (headed by none other than Jack Enders after his retirement from NASA). Soon afterwards, one commuter pilot wrote to the Lewis ice researchers thanking them for saving his life and the lives of his passengers. During one of his flights the aircraft, flying through icing conditions, suddenly nosed down, a possible sign of a tail stall. But warned in advance by Perkins and Rieke to avoid the textbook maneuver (put down the flaps), he instead followed their advice to *retract* the flaps. In short order, many of the old technical notes and reports describing the meteorological conditions of icing, the most ice-prone parts of clouds, and the importance of droplet size all became relevant once more. Aboard the Twin Otter, Rieke and the Lewis engineers and scientists resumed an aggressive flight research program of cloud physics, probing deeper into severe icing conditions than anyone dared before in the hunt for dangerously large droplets. These results began to take their place in the literature beside the venerable studies of the 1940s and 1950s. For the first time, training films and programs on turboprop icing, sponsored by the Lewis researchers, became available to pilots and lifelike simulations of icing conditions, never before undertaken, became a subject of urgent discussion.

Ultimately, the renewal of icing research saved countless lives.[18]


## AN INDEPENDENT ENTITY


During the years in which the icing program underwent an extraordinary revival, the Dryden Flight Research Facility conducted many programs of keen interest to the military services and to the aircraft industry. Moreover, despite its merger with Ames, the DFRF's physical plant at least maintained, and in many instances improved its quality. The Flight Loads Research Facility experienced a costly modernization. The Space Shuttle Facility Area continued to operate with full support. A multi-million dollar Data Analysis Facility and a $15 million Integrated Test Facility both became realities. Additionally, in the actual practice of flight research, Dryden continued to operate without any outward slackening in the quantity, the scope, or the imagination of its projects. For example, just before the consolidation with Ames, Dryden flew yet another of Richard Whitcomb's theories. Flight research proved that small, almost vertical fins attached to the wing tips of airliners channeled forward the vortices of air normally eddying at the end of the airfoil, thus reducing drag and decreasing fuel consumption. Passengers aboard such airliners as the McDonnell-Douglas (MD)-11 soon noticed these little shapes at the end of the plane's wings. While some of the DRFC engineers and pilots tested the

---

[18]Porter Perkins and William Rieke, interview with Michael Gorn, 1 June 1998, DFRC Historical Reference Collection; Richard Ranaudo, Kevin Mikkelsen, Robert McKnight, and Porter Perkins, NASA TM 83564, "Performance Degradation of a Typical Twin Engine Commuter Type Aircraft in Measured Natural Icing Conditions" (Washington, D.C.: NASA, 1984), 1-6; Kevin Mikkelsen, Robert McKnight, Richard Ranaudo, and Porter Perkins, NASA TM 86906, "Icing Flight Research: Aerodynamic Effects of Ice and Shape Documentation With Stereo Photography" (Washington, D.C.: NASA, 1985), 1-6; John Reinman, Robert Shaw, and Richard Ranaudo, NASA TM 101989, "NASA's Program on Icing Research and Technology" (Washington, D.C.: NASA, 1989), 5-9; Jan Steenblik, "Turboprop Tailplane Icing," *Airline Pilot* (January 1992): 30-33; Porter Perkins and William Rieke, "Tailplane Icing and Aircraft Performance Degradation," *Accident Prevention* (February 1992): 1-6.

winglets, as they came to be called, others fabricated and flew the AD (Ames/Dryden)-1 oblique wing. This vehicle, designed by Robert T. Jones of swept-wing fame, featured a one piece airfoil that swung laterally up to 60 degrees from a fixed point on top of the fuselage. Jones predicted significant fuel economies for commercial aircraft using his concept. Laminar flow research on a Lockheed JetStar and later aboard an Air Force F-16XL experimented with wing modifications designed to propagate the smoothest possible stream of air in flight. The Highly Maneuverable Aircraft Technology (HiMAT) aircraft--a half size remotely piloted vehicle--offered great promise for the next generation of fighters by demonstrating turns twice as tight as existing warbirds. Carrying the art of fast maneuver still further, during the 1980s and early 1990s the forward-swept wing X-29 tested the limits of computer controls on an aircraft whose very survival depended on its digital brain but whose very instability rendered it beyond compare in maneuvering characteristics. Another demonstrator, the F/A-18 High Angle-of-Attack Research Vehicle (HARV) explored the limits of one aspect of maneuver--high angle of attack--using thrust vectoring paddles and nose strakes to control the machine at the steepest angles. Finally, an international project known as the X-31 research aircraft combined the high-angle-of-attack and extreme agility of the HARV and the X-29, but with the express purpose of improving the maneuverability of fighter aircraft. Conceived and partially fabricated in Germany, the thrust-vectoring X-31's flight research program at Dryden lasted from 1992 to 1995 and demonstrated such extraordinary capabilities as stabilized flight *and* 180 degree turns at 70 degrees angle-of-attack.[19]

Thus, during the 1980s and early 1990s Dryden's flight research program prospered in spite of (rather than because of) the awkward institutional relationship imposed in 1981. The wedding between Ames and Dryden failed to mature into a satisfying marriage for either party. Dryden experienced a change of command in mid-1984 when former Central Intelligence Agency pilot Martin Knutson replaced John Manke as Ames Director of Flight Operations and as

---

[19]Briefing, "Dryden Experimental Facilities," 13 July 1982, Ken Szalai Collection, DFRC Historical Reference Collection; Wallace, *Flights of Discovery*, 93-110.

Dryden Site Manager. Unlike Manke, who was associated with Dryden, Knutson had been employed by Ames since 1971. A man who operated quietly and gave wide latitude to his subordinates, some thought Knutson seemed out of place at Dryden and unfamiliar with its ways. Still, he adapted well and developed a genuine sympathy for Dryden's methods and achievements. Moreover, in his difficult role of Ames representative, Knutson astutely avoided sweeping reforms and instead concentrated on steady stewardship. Subjected to the daily stresses of balancing the needs of Dryden with the requirements of Ames, Knutson and his deputy Ted Ayers made every effort to impress on NASA Headquarters the necessity of redefining the relationship between Moffett Field and the DFRF.

Knutson and Ayers were joined in their campaign for Dryden autonomy by Ken Szalai, a man who gained notice during the F-8 DFBW project. Szalai rose in the DFRF hierarchy during the 1980s. During much of the decade he held the pivotal position of Chief of Research Engineering, a division that embraced five branches: aeronautics, dynamics and control, facilities engineering, facilities management, and aerostructures. These years afforded him the opportunity to acquire experience as a cross-disciplinary manager and leader. In a move perhaps designed to groom him for wider responsibilities, in 1989 he served briefly as Acting Associate Director of Ames. In this capacity he learned to appreciate the Ames viewpoint and to learn the breadth of duties exercised by center director Dr. Dale Compton. But Szalai also recognized that DFRF lacked adequate administrative support, that NASA headquarters showed decreased interest in Dryden's activities and needs, and that morale at Dryden felt the effects of the institutional impasse. Evidently, Szalai found some sympathetic ears in Washington. When Martin Knutson announced his decision to return to Ames as director of Flight Operations in December 1990, Ken Szalai assumed his position, but not as Dryden Site Manager; rather, he became Deputy Director of Ames for Dryden and, at the same time, Director of the DFRF. No longer one of many Ames directorates, Dryden now held the rank of an autonomous facility and, significantly, Szalai won control over such administrative machinery as budgeting and policy-making. Still, the bureaucratic apparatus remained unworkable, prompting Szalai to describe his

32

new role as, "probably the most difficult job I've ever had...."[20]

Not surprisingly, even before embarking on this position, and certainly once he held it, Ken Szalai argued privately for Dryden's total independence from Ames. First, no real collaborative relationship had evolved between the two organizations; the theoretical and experimental aeronautics projects conducted at Ames were not conceived or planned with the intention of subsequent flight research at Dryden. Each institution followed its own research path with little connection to the other. Second, no significant sharing of equipment or staffs ever materialized. Among the several high angle-of-attack programs pursued at Dryden, only one Ames civil servant participated full time, and he reported each day to his office on Moffett Field. A regular air shuttle failed to close the distance between the two operations. Finally, since no one advanced the argument that the consolidation yielded savings in dollars and cents, Szalai emphasized the loss of a more precious asset.

> What is indisputable is that much valuable time is being used by senior managers to implement the 1981 decision, to travel between the sites, to solve intersite problems, to coordinate administrative and financial activities, and to attempt to advocate and formulate Center wide programs over two geographically widely separated sites. Also, it is clear that the principal DFRF mission and capabilities are significantly different than the Ames-Moffett mission. A great deal of effort has been expended to overcome this gap, by managers, technical staff, and administrative staff. There is no obvious commensurate return on this investment. It is time to admit that the 1981 reorganization did not produce the desired resource savings and has placed an excessive burden on Ames management at both sites which is hurting the Agency's aeronautical research program.

---

[20]Ames New Release, "New NASA Ames Research Center Director of Flight Operations, Ames Dryden Site Manager Named," May 1984, DFRC Historical Reference Collection; Ames Biographical Data, "Martin A. Knutson," December 1990, DFRC Historical Reference Collection; Ted Ayers to J.D. Hunley, (approximately) 9 March 1999, DFRC Historical Reference Collection; Dryden biographical Data, "Kenneth J. Szalai," March 1994, DFRC Historical Reference Collection; Organization Chart, Research Engineering Division, October 1985, DFRC Historical Reference Collection; Martin Knutson, "Knutson Bids a Fond Farewell to Dryden Employees," Dryden X-press, 30 November 1990, DFRC Historical Reference Collection; Ken Szalai to Bill (Ballhaus?) and Dale (Compton?), 7 May 1989, with attached comparisons of Dryden claims versus Szalai's observations at Ames, Milt Thompson Collection, DFRC Historical Reference Collection; Ames News Sheet, "New Ames Deputy Director to Head Ames-Dryden," 5 November 1990, DFRC Historical Reference Collection; anon., "New Dryden Director: Challenges Ahead," Dryden X-press, 30 November 1990, 1, DFRC Historical Reference Collection (quoted passage).

It       is time to reestablish Ames and Dryden as separately managed facilities in [Headquarters NASA] OAST.[21]

What Szalai expressed privately at Dryden soon became something he said aloud at headquarters. At the invitation of Arnold Aldrich, Headquarters Associate Administrator for Aeronautics, the two men reviewed the situation face-to-face and subsequently corresponded. Szalai admitted candidly that he accepted the "current arrangement ...[due to] loyalty to NASA. It is," he added, probably with himself in mind, "a stressful situation for several people at DFRF; this probably filters down to the rank and file inadvertently." He advised Aldrich to separate the two entities completely and, on a positive note, expressed the conviction that it could be done amicably. Szalai even suggested Aldrich contact Marty Knutson for elaboration. But Ken Szalai did not beat this drum on his own. Inspired by a letter from Milt Thompson, Walt Williams also weighed in. In October 1989, Williams wrote to NASA Administrator Richard Truly--well known at Dryden as crew member on the *Enterprise's* approach and landing tests and later as a Shuttle pilot--and appealed to him to rescue Dryden from the burden of consolidation. Inside the Dryden community, none expressed the case for independence more fully or forcefully than Thompson. His argument turned on the decline in morale at Dryden since 1981, as well as on psychological factors. Thompson felt the timing of the unification announcement doomed it from the start. Dryden employees thought demotion to facility status an ungrateful act of headquarters bureaucrats who, at the moment of the Shuttle's first orbital flight, conveniently forgot all of the DFRC contributions to the STS, over and above the Orbiter flight research program. Had the center been placed under direct headquarters supervision, Thompson thought the staff might have accepted it; but being *removed* from direct communication with Washington, D.C., by the imposition of a new reporting layer represented "a terrible blow to the Dryden ego," entailing a loss of "status...[and]...stature."

After nine years of consolidation, none of the older [Dryden] employees have really

---

[21]Ken Szalai to Bill (Ballhaus?) and Dale (Compton?), 5 July 1989, with attached essay, "Should NASA Spin-Off Dryden As A Separate Entity?" Milt Thompson Collection, DFRC Historical Reference Collection.

accepted consolidation. They have no sense of belonging to the Ames Research Center. They have no loyalty to Ames. Dryden employees are still proud of and loyal to NASA, but they still hope for deconsolidation. They have seen no benefits of consolidation. On the contrary, they have witnessed a decline in the quality of life at Dryden. Ames management is seen as another superfluous layer of management that unlawfully taxes Dryden funding in a somewhat arbitrary manner. Dryden has for example been assessed to support a number of activities at Ames-Moffett due to shortfalls in funding, but this never seems to work in reverse. Dryden shortfalls are Dryden problems. The younger Dryden employees, those hired after consolidation, are not as emotionally effected [sic] by consolidation, and yet they wonder why Dryden lost its center status after thirty five years of independence. There is no good answer. There was no obvious benefit of consolidation. The two sites represent two different cultures.[22]

Daniel Goldin--Truly's successor as NASA Administrator--announced the establishment of Dryden as an independent entity on March 1, 1994. Perhaps Goldin heard the arguments of Szalai, Thompson, and Williams; or perhaps he learned about Dale Compton's apparent viewpoint that Dryden should return to fully autonomous status. In any event, Goldin initiated a six month transition period after which Ken Szalai would assume the role of Dryden Director. "This change," said Goldin speaking by satellite from Washington to NASA employees across the country, "reflects the commitment on the part of NASA to reduce layers of management and empower operating organizations to carry out their missions with maximum benefit to the country." Szalai greeted the news with satisfaction, saying the end of the 13-year interlude suggested NASA once again "trusts us" with the flight research mission. He then acted quickly to restore a self-governing organizational structure consisting of five major functional areas: Research Facilities, Research Engineering, Flight Operations, Aerospace Projects, and Intercenter Aircraft Operations. The center issued a Basic Operations Manual (BOM) in February 1995--its first independent one in many years-- emphasizing safe operations and the lessons learned "through tears, sweat, and worse...." Finally, to re-establish the center's historic

---

[22]Thompson miscalculated the number of years of Dryden independence. Dryden lost center status after 27, not 35 years of independence, separating from Langley in 1954 and consolidating with Ames in 1981. Ken Szalai to (?) Geastman, n.d., Milt Thompson Collection, DFRC Historical Reference Collection (first quoted passage); Milt Thompson essay on Dryden/Ames consolidation, n.d., Milt Thompson Collection, DFRC Historical Reference Collection (second quoted passage and block quote).

role, Szalai declared a year of celebrations to commemorate Dryden's 50th anniversary in 1996.[23]

But Szalai wanted to look ahead, as well as to the past. Starting in 1993, he and his inner circle recognized that the end of the Cold War threatened to reduce much of the military testing that comprised such a large part of Dryden's flight research. Szalai felt the situation demanded a serious assessment, so he created a new position--the Assistant for Strategic Planning and New Program Development--and asked Robert Meyer to be its first incumbent. Then, in cooperation with other senior advisors, Meyer launched a review of the existing programs and began to chart a new course for the future. What roles should Dryden carve out for itself in light of this impending void? His preliminary findings were aired at an off-site strategy meeting in November 1993 where most of the top Dryden officials participated. This frank and pivotal discussion centered on new avenues of research in which Dryden should involve itself. During the weeks immediately following, Meyer, working closely with Szalai, translated these ideas into a formal briefing, presented to the other center directors and headquarters aeronautics representatives in January 1994. Meyer reported to his listeners a disturbing assessment: most of Dryden's existing programs, including high performance research, faced curtailment by mid 1995. Perhaps more alarming, no plans existed to narrow the yawning gap between these losses and the anticipated levels of support for civil and hypersonic projects. Szalai and Meyer wanted to close the gap. They identified the essential resources of Dryden (the lakebed and long runways on Edwards Air Force Base, the good weather, the open airspace, the ground facilities,

---

[23]"An Important Briefing," 4 June 1993, Milt Thompson Collection, DFRC Historical Reference Collection; NASA News, "NASA Administrator Announces Management Changes," 6 January 1994, DFRC Historical Reference Collection (first quoted passage); Jim Skeen, "Dryden Instructed to Split from Ames," *L.A. Daily News*, 8 January 1994, DFRC Historical Reference Collection (second quoted passage); Sharon Moeser, "NASA Facility Will Gain Its Independence," *L.A. Times*, 7 January 1994, DFRC Historical Reference Collection; Nancy Lovato, "Up Front with Dryden Director Ken Szalai," Dryden *X-Press*, March 1994, 1, DFRC Historical Reference Collection; Organization Chart, DFRC, October 1995, DFRC Historical Reference Collection; Basic Operations Manual, DFRC, March 1995, (introductory letter by Ken Szalai, inside cover), 17 March 1995, DFRC Historical Reference Collection; Ken Szalai, essay on Basic Operations Manual entitled "Back to Basics," n.d., DFRC Historical Reference Collection (third quoted passage).

and the competencies of the staff) and suggested the types of activities which might be attracted by such conditions  Five areas of concentration--subsonics, high speed, high performance, hypersonics, and research and technology--constituted the traditional Dryden pursuits.  But the main event of the briefing involved the unveiling of the principal new initiative, one called Access to Space.  It offered many attractive features: an affinity with most of Dryden's resources; an opportunity to participate in the fulfillment of one of NASA Administrator Daniel Goldin's most cherished objectives, lowering the cost of sending payloads into orbit; a chance to become associated with the fields of communications and satellites, both growing areas of technology experiencing expansion and cutting edge research; and it allowed the Dryden director to open lines of discussion with the satellite and communications industries, as well as with NASA's launch vehicle experts at the Marshall Space Flight Center (MSFC).  Moreover, Access to Space had the ring of familiarity and the weight of precedent.  Two of the most outstanding Dryden programs of the past--the lifting bodies and the X-15--opened space travel to the nation. Finally, the new initiative was timely; contemporary discussions in Washington about funding new X-vehicles for space launch made the pursuit of Access to Space all the more appealing to Dryden's leaders.[24]

## ACCESS TO SPACE

---

[24]Toward the end of November 1994 Robert Meyer's strategic initiative received the assistance of John McCarthy, a highly placed headquarters official with extensive long-range planning experience.  Ultimately, Meyer and McCarthy added two more high visibility initiatives to Access to Space: Environmental Research and Sensor Technology (ERAST) which consisted partly of remotely piloted aircraft; and Unpiloted Air Vehicles (UAVs).  Like Access to Space, both seemed to have a close kinship with Dryden resources.  But in a center whose reputation, fairly or not, hung on the role of the pilot, both initiatives engendered much controversy.  See Robert Meyer, interview with Michael Gorn (by telephone), 18 March 1999, DFRC Historical Reference Collection; Robert Meyer, notes of a meeting with Michael Gorn, 30 March 1999, DFRC Historical Reference Collection; Robert Meyer, interview with Michael Gorn (by telephone), 28 July 1999, DFRC Historical Reference Collection; Briefing, "Dryden Strategic Planning Brief for Associate Administrator and Division Directors," 21 January 1994, DFRC Historical Reference Collection.

At the time Ken Szalai assumed his duties as Dryden Director, he forecasted a resurgence in the experimental (X-series) airplanes, perhaps the signature programs of the center's 50 year history. Before long, his prediction seemed to come true. After the destruction of the Space Shuttle *Challenger* and its crew in January 1986 the prospect of alternate means of transportation into space found receptive audiences in Washington, D.C. Gradually, a consensus emerged that the Shuttle's technology had become outmoded and that the European consortium Arianespace offered stiff competition to American space launch. At the same time, the end of the Cold War imposed on Congress a mandate to pare down the federal budget. Republican President George Bush and his administration weighed these realities and sought alternatives to the STS. In his role as chairman of the National Space Council from 1989 to 1992, Vice President Dan Quayle endorsed a number of studies devoted to this question, most of which agreed on the necessity of updating Shuttle access to space. Near the end of his term the President selected Dan Goldin, Vice President of satellite systems at TRW, to direct NASA. To the surprise of many, he retained his position despite the victory of Democratic President Bill Clinton. Goldin may have survived because of his proven desire to implement reforms. In an age of smaller budgets, the new Administrator decided the well-being of NASA depended on the success of a few high-profile, high-cost programs, augmented by many projects fashioned under Goldin's formula of "cheaper, faster, better." Accordingly, in January 1993 (the month of the new President's inauguration) he directed his agency to initiate a top to bottom review of its long-range objectives, resulting in a report entitled *Access to Space*. Unlike its predecessors, this document, prepared by a commission appointed by Goldin, proposed an entirely different system to lift human beings beyond the atmosphere; a *fully reusable* launch vehicle (RLV) propelled neither by jettisoned boosters nor by an expendable fuel tank. Referred to by the shorthand designation Single Stage to Orbit (SSTO), it later became known as Venture Star, a full-scale follow-on to

38

the X-33 technology demonstrator.[25]

The issuance of a Request for Proposal for X-33 Phase I--open for just two months under the Goldin Rule of speedy competitions--yielded three bidders: Lockheed Martin, McDonnell Douglas Aerospace, and Shuttle designer Rockwell International. The competition began in March 1995, at the end of which time the firms agreed to present the government with concept definitions and technical designs. After reviewing them, NASA then faced the critical question of whether any one, or any combination of the three, should advance to Phase II, consisting of the construction and flight of a prototype machine. A contract for $941 awaited the winner. While a 15 month review of the three plans ensued (in which Dryden representatives participated in the negotiations with the three companies), NASA parceled out the respective responsibilities to its centers. The Marshall Space Flight Center became the lead center for the X-33 program and even though Dryden did not win a starring role in the drama, it stood a chance to make a significant contribution to a program of national stature. In supporting Marshall, Dryden assumed such duties as:[26]

1. Advising MSFC about vehicle safety.

2. Planning the flight research program, starting with a staff of about 25 and building to

[25]Nancy Lovato, "Up Front with Dryden Director Ken Szalai," Dryden *X-Press*, 2, DFRC Historical Reference Collection; Andrew Butrica, "X-33 Fact Sheet #1: Part I: The Policy Origins of the X-33," 7 December 1997, 1-8, The X-33 Home Page on the World Wide Web (http://www1.msfc.nasa.gov/NEWSROOM/background/facts/x33.htm), DFRC Historical Reference Collection; Bill Sweetman, "VentureStar: 21st Century Space Shuttle," *Popular Science* (October 1996): 43-47, DFRC Historical Reference Collection; anon., "RLV Overview: About the Reusable Launch Vehicle Technology Program," 6 December 1995, 1, The X-33 Home Page on the World Wide Web (See above, this footnote, for web site address), DFRC Historical Reference Collection.

[26]The Air Force, rather than any NASA center, assumed range safety responsibilities. NASA News Release, "X-33, X-34 Contractors Selected for Negotiations," 8 March 1995, DFRC Historical Reference Collection; NASA News Release, "X-33 Cooperative Agreements Signed," 29 March 1995, DFRC Historical Reference Collection; draft NASA News Release, "NASA Langley Plays Major Role in New X-33, X-34 Programs," n.d., DFRC Historical Reference Collection; anon., "X-33 RTQ" (Response to Questions), n.d., DFRC Historical Reference Collection; Gray Creech, "Dryden Plays Major Role in X-33," Dryden *X-Press*, August 1996, 1,8, DFRC Historical Reference Collection; NASA News Release, "X-33 Phase I Cooperative Agreement Notice Issued," 2 April 1996, DFRC Historical Reference Collection.

approximately 70.

3. Designing the range and overseeing the construction necessary to extend communications, as well as radar and Global Positioning System (GPS) tracking.

4. Assisting Marshall in the development of flight control components and real time computer simulations.

5. Supporting the X-33 launches and cooperating with Edwards Air Force Base in making recommendations for take-off and landing locations.

6. Undertaking tests of X-33 propulsion systems aboard the SR-71 aircraft.

7. Factoring X-33 support into DFRC program planning, thus preserving all of Dryden's other flight research projects.

Despite the truly formidable technical hurdles faced by the three contractors vying for the X-33, some critics believed the structure of the program presented equally high obstacles. Manufacturers had been asked to develop an entirely new system able to leap into space without multiple stages, to do so quickly, and to achieve the objective with relatively little government backing. Others expressed concern about the short flight research schedule for a machine so radically different, as well as the decision to build just one prototype. But the contractors proceeded nonetheless. Early wind tunnel data on the three likely planforms (gleaned from an intermediate review of the Lockheed, McDonnell Douglas, and Rockwell concepts) allowed the Dryden simulations engineers to implement representations of the X-33 design and the DFRC research staff to begin preliminary evaluations on the flight control systems and on the aerodynamic characteristics of the vehicles. These researchers produced a piloted simulation and collaborated on the software integration of the X-33 avionics suite. Meanwhile, a full-scale segment of a graphite-composite structural component underwent extensive tests at Langley. Finally, on July 2, 1996, Lockheed Martin won the contest to build the X-33 RLV. The company relied on three essential technical ingredients, two of which possessed long histories: a wedge shaped airframe borrowed straight from Dryden's lifting body flight research programs; a linear aerospike engine, a cooperative venture between Rocketdyne and the USAF dating back to

the 1960s but never actually flown; and metal thermal protection systems. While the Lockheed
Martin Skunk Works in Palmdale, California, constructed the airframe, its subcontractor
Rocketdyne/Boeing in nearby Canoga Park would fabricate the engines. Together they agreed,
with the other subcontractors, to deliver the X-33 (a half-size prototype of the Venture Star)
measuring 67 feet in length and 68 feet across the widest (tail section) of the wedge, weighing
64,000 pounds empty, and containing a five-by-ten-foot cargo bay. Dan Goldin heralded the
announcement as the first step in a process calculated to drive down space payload costs from
$10,000 per pound on current vehicles to $1,000 on the new RLV.[27]

Dryden's participation in this unique project intensified with the passage of time.
Although the agreement signed with Lockheed relegated DFRC to little more than a
subcontracting role, Ken Szalai, eager to advance his Access to Space initiative, decided to
squeeze all he could from the opportunity. He and his top assistants recognized that this
undertaking, initiated just months after Dryden ended its 13 year amalgamation with Ames,
offered a chance for DFRC to reclaim some of the national recognition achieved in such past
programs as the X-15 and the Space Shuttle. Moreover, should Venture Star become a reality, it
would elevate Dryden once more to the role of a spaceport. Thus, not content merely to

---

[27]For a full discussion of the origins and features of the three competing X-33 designs, see
Kenneth Iliff, "30 Minute Draft" of Reusable Launch Vehicle History, 2 March 1999, DFRC
Historical Reference Collection and Kenneth Iliff, interview with Michael Gorn, 22 March 1999,
DFRC Historical Reference Collection. Anon., "X-33 Advanced Technology Demonstrator,"
Marshall Space Flight Center Fact Sheets, as of July 27, 1999, The X-33 Home Page on the
World Wide Web (http://www1.msfc.nasa.gov/NEWSROOM/background/facts/x33.htm),
DFRC Historical Reference Collection; anon., "X-33 Program Risks Unnecessarily High,"
*Aviation Week and Space Technology* (29 January 1996): 74, DFRC Historical Reference
Collection; anon., Briefing Charts, "X-33," (two pages showing X-33 workyears,
accomplishments, and major issues), February 8, 1996, DFRC Historical Reference Collection;
NASA News Release, "Lockheed Martin Selected to Build X-33," 2 July 1996, DFRC Historical
Reference Collection; Lockheed Martin Information Release, "Lockheed Martin VentureStar
Wins X-33 Competition: Program Valued at More than $1 Billion Through 2000," July 1996,
DFRC Historical Reference Collection; anon., Specifications Sheet, "X-33 Advanced
Technology Demonstrator," n.d., DFRC Historical Reference Collection; Gray Creech, "Dryden
Plays Major Role in X-33," Dryden *X-Press*, August 1996, 1, DFRC Historical Reference
Collection.

participate, Szalai, project manager Gary Trippensee, and their colleagues at X-33 technical meetings actually volunteered *additional* services which Dryden could render to the contractors and to the other centers and through which it might widen its fairly circumscribed role. Dryden engineers collaborated with their Lockheed counterparts on sensor development and supplied the contractor with system configuration and operational know-how. Originally, DFRC assumed responsibility for the Shuttle Carrier Aircraft (SCA) during its transfiguration into the X-33 SCA, including flight research on the mated X-33/747, oversight of maintenance and flight operations, and envelope expansion tests. However, planners later abandoned air conveyance of the X-33 in favor of ground transportation. Finally, in weekly telephone conversations Dryden researchers advised Marshall representatives about the design of flight control laws applicable in case of failure.[28]

But even without a single voluntary act, Dryden assumed some significant tasks for an operation its size. Its staff cooperated with Lockheed in assessing the aerodynamic model (based on some 3,500 hours of wind tunnel data) in order to predict flying characteristics, energy management, and to prepare appropriate flight control systems. At the same time, the linear aerospike engine made its debut at Dryden. The tests conducted above Edwards Air Force Base on this unusual powerplant had an important bearing on the failure or the success of the entire program. Unlike conventional rockets, this one expelled the thrust and exhaust from liquid hydrogen and liquid oxygen--propellants with the highest known performance--not from bell-like nozzles attached to a central combustion chamber, but from 20 small combustors arrayed in parallel rows and trained to fire on curved, rectangular plates. On nozzles of a fixed shape, like those used on the Shuttle, their expansion ratios represented a compromise configuration designed to work most effectively on a critical part of the launch. But in the linear aerospike, because the combustors' flow was not constrained like the standard nozzle, it could adjust its

---

[28]Robert Meyer, interview with Michael Gorn, 30 March 1999, DFRC Historical Reference Collection; Briefing Charts, "X-33: Director's Weekly Update," 14 February 1997, DFRC Historical Reference Collection; Gray Creech, "Dryden Plays Major Role in X-33," Dryden *X-Press*, August 1996, 1, DFRC Historical Reference Collection.

angle as it rose through the atmosphere, gradually altering thrust as it sensed lessening air pressure at increasing elevations. If workable, this efficient system promised large savings in fuel consumption and weight, added advantages to an engine whose virtues also included simple design, high thrust, light weight, durable construction, and a history of thorough ground testing. The first Dryden contribution to this engine actually started during spring 1996, when DFRC technicians initiated the process of mating a ten percent scale X-33 model equipped with the aerospike to the rear of SR-71 Blackbird Number 844. Thus began the LASRE, or Linear Aerospike SR-71 Experiment, a flight research project conducted on the powerplant at Dryden. After a series of successful ground tests, the model was mounted on the back of the Blackbird in August 1997. Then, on October 31, the SR-71 testbed (on its own power, with the aerospike off) flew for nearly two hours as high as 33,000 feet and as fast as Mach 1.2 to collect data on the vehicle's aerodynamics, stability and control qualities, and structural integrity in flight while carrying the aerospike model. The first cold flow flying tests (cycling gasses through the engine during flight) occurred aboard the SR-71 on March 4, 1998, followed by three more during spring and summer. Leaks of liquid oxygen materialized during the latter three trials. Meantime, two engine hot firings were completed on the ground. Combining the data from the four cryogenic flights with the engine ground tests yielded sufficient information to extrapolate the behavior of the linear aerospike powerplant operating with hot gasses in flight. Because of the adequacy of these results, and because of concerns about the liquid oxygen leaks, LASRE project engineers decided not to attempt an actual hot firing on the SR-71. They concluded flight operations in November 1998.[29]

---

[29]Anon., "Linear Aerospike Engine--Propulsion for the X-33 Vehicle," Marshall Space Flight Center Fact Sheets, as of July 27, 1999, The X-33 Home Page on the World Wide Web (http://www1.msfc.nasa.gov/NEWSROOM/background/facts/aerospike.htm), DFRC Historical Reference Collection; Summary Paper, "X-33 Year in Review: Accomplishments and Challenges," 1996, DFRC Historical Reference Collection; Kathy Sawyer, "Bargain-Hunting NASA Picks Blast From Past, *Washington Post*, 3 February 1997, A03, DFRC Historical Reference Collection; Warren Leary, "Novel Rocket to Power Shuttle Successor," *New York Times*, 30 July 1996, C1, C8, DFRC Historical Reference Collection; Cheryl Agin-Heathcock, "Linear Aerospike Engine Fitted to Dryden's SR-71 #844," Dryden *X-Press*, April 1996, DFRC

While Dryden engineers put the LASRE through its paces, industry and NASA leaders met at DFRC in November and December 1996 for the Preliminary X-33 Design Review. They concluded the program had advanced sufficiently to permit more detailed design and fabrication, including an area of eventual importance to DFRC, the ground and launch facilities. Accordingly, during 1997 Dryden officials hosted Sverdrup Corporation representatives who initiated a survey of the base and the surrounding desert for a launch site. Eventually, a six-person panel consisting of Edwards and DFRC personnel examined seven alternatives recommended by the contractor. Among these, a 25-acre location a few hundred yards north of Haystack Butte, on the eastern side of the base roughly midway between Highways 58 and 395, emerged as the front-runner. This property seemed promising because the projected X-33 launch path would offer the least disruption to the main base and to the local population while still being only 30 miles by road from Dryden. Once the Air Force accepted this site formally, a 30 day period of public review of the Environmental Impact Statement (EIS) ensued. The EIS described the project and its likely consequences to people and to the terrain. The plan passed muster at these hearings and Headquarters NASA then affirmed the Haystack Butte option. Sverdrup broke ground for the $30 million project on November 14, 1997, and over the next year constructed the launch pad, the X-33 rolling shelter, the fuel storage tanks, a water storage tank for a sound suppression system, and a concrete flame trench. During the initial (suborbital) flights planned for the year 2000 the X-33 was scheduled to land at Michael Army Air Field on Dugway Proving Ground in Utah.[30]

---

Historical Reference Collection; Briefing Charts, "X-33 Status as of 20 June 1997," DFRC Historical Reference Collection; "Linear Aerospike SR-71 Experiment Talking Points" and "LASRE Project Information Summary," 4 October 1997, DFRC Historical Reference Collection; Dryden Press Release, "Linear Aerospike SR-71 Experiment Completes First Cold Flow Flight," 5 March 1998, DFRC Historical Reference Collection; Draft DFRC Fact Sheet on the LASRE Project, July 1999, 1-2, DFRC Historical Reference Collection.
[30]Dryden News Release, "X-33 Launch Facility Site Survey Underway," 6 March 1997, DFRC Historical Reference Collection; Gray Creech, "1st Phase Ends for X-33 Launch Site Survey," Dryden *X-Press*, 21 March 1997, DFRC Historical Reference Collection; anon., "Preferred X-33 Sites Chosen," Dryden *X-Press*, 3 October 1997, 1, DFRC Historical Reference Collection; Briefing Charts, "X-33 Director's Weekly Update," 14 February 1997, DFRC Historical

The prototype demonstration phase of the X-33 received its birth certificate in 1997. Over the course of the year, delegates from public and private institutions presented 51 detailed briefings about subsystems and components. Then, at the end of October, Edwards Air Force Base sponsored a meeting crucial to the fate of the program. About 600 individuals met for five days for a Critical Design Review and in the end gave approval to fabricating the final pieces and to assembling the X-33 technology demonstrator. Meanwhile, Dryden published important technical findings and initiated a second X-33 flight research project. The first paper announced the center's completion of a flush air data sensing (FADS) system, a series of pressure orifices implanted on the X-33's nose to record air flow in flight. A second paper reported the results of an ambitious undertaking by DFRC scientists and engineers to review flight data from six lifting body and the Shuttle Orbiter flight programs; to compare this historic information to X-33 wind tunnel findings; and to arrive at models of uncertainty for the new vehicle's subsonic and supersonic flights. Finally, the X-33 thermal protection system (TPS) underwent a rigorous flight research program at DFRC, similar to the ones conducted on the Shuttle during the mid-1980s. Affixed to an F-15B test fixture located underneath the aircraft, three materials were subjected to shear and shock loads: metallic Inconel tiles, soft advanced flexible reusable surface insulation tiles, and sealing materials. During several strenuous flights the pilot maneuvered the F-15B at speeds up to Mach 1.4 and altitudes of 33,000 feet. Subsequent examinations of the heat-resistant materials showed that no wear or damage resulted from air loads below the speed of sound nor from shock waves encountered through the transonic range. Roy Bryant, the F-15B project manager, expressed satisfaction at the speedy and frugal manner in which his colleagues delivered data to the X-33 designers. Timing was important; even as the prototype took shape on the Skunkworks floor, critical programming decisions remained to be made. Yet, if the all went

Reference Collection; NASA News Release, "X-33 Launch Facility Ground Breaking Held," 14 November 1997, DFRC Historical Reference Collection; Briefing Charts, "Striving for Affordable Access to Space," 17 October 1996, DFRC Historical Reference Collection; NASA News Release, "X-33 Program Completes Operations Review," 18 December 1996, DFRC Historical Reference Collection; Briefing Charts, "X-33 Preliminary Design Review (PDR)," 4-8 November 1996, DFRC Historical Reference Collection.

as planned, when the small lifting body arrived at Dryden it would join the historic company of experimental aircraft which made their maiden flights over Rogers Dry Lake and its barren environs.[31]

## NEW RESPONSIBILITIES

Fresh from a long period of institutional anonymity and eager to restore its former prominence, the Dryden Flight Research Center seized on the X-33 program as an opportunity to affiliate itself with an undertaking potentially equal to the marquee endeavors of the past: the X-1 and the D-558 (1940s and early 1950s); the X-15 ( mid 1950s to the late1960s); and projects supporting the Space Shuttle (1970s to the 1990s). But the sense of renewal brought about first by the center's return to independence in 1994, followed the next year by the inauguration of the X-33 program, gained additional momentum in 1996 from an entirely unlikely quarter. During a headquarters NASA Zero Base Review (ZBR) of the agency's roles and missions, infrastructure reductions, and bureaucratic streamlining, those alert to savings raised the specter of aircraft consolidation under Dryden, last suggested 25 years earlier during the demise of the Apollo program. (See Chapter 7). This time, however, headquarters issued its restructuring decision without consultation. Thus, on January 23, 1996, Associate Deputy Administrator General John

---

[31]NASA Ames contributed to the X-33 thermal protection system tests by conducting wind tunnel research and by offering advice relative to the TPS flight research program. Anon., "X-33 Program Completes CDR," Dryden *X-Press*, 21 November 1997, 1, DFRC Historical Reference Collection; Stephen Whitmore, Brent Cobleigh, and Edward Haering, NASA TM 206540, "Design and Calibration of the X-33 Flush Airdata Sensing (FADS) System" (Washington, D.C.: NASA, 1998), 1; Brent Cobleigh, NASA Technical Paper 206544, "Development of the X-33 Aerodynamic Uncertainty Model" (Washington, D.C.: NASA, 1998), 1; Dryden Press Release, "X-33 Thermal Protection System Materials Fly on F-15B," 18 May 1998, DFRC Historical Reference Collection; NASA News Release, "X-33 Thermal Protection System Tests Complete," 30 June 1998, DFRC Historical Reference Collection.

Dailey informed Ken Szalai and the Directors of Ames, Lewis, and Langley that a headquarters evaluation of aircraft consolidation confirmed its capacity to reduce costs. Accordingly, NASA reconstituted its fiscal year 1997 budget to reflect the projected savings and Dailey directed Associate Administrator for Aeronautics Robert Whitehead to centralize all of the agency's flight research *and* its platform research aircraft at the Dryden Flight Research Center. Dailey's timetable allowed no discussion; the other aeronautics center directors were given just a week to tell DFRC's Gary Krier, the manager of the consolidation project, their points of contact for the movement of people and machines. Robert Whitehead made plain the headquarters' desire for fast action in order to "demonstrate our commitment to...this decision...." He assigned Dryden the responsibility of drafting an implementation plan which set the "earliest possible" dates of conveyance, thus permitting the affected civil servants the maximum time to decide whether to follow the aircraft or choose other options. During September 1996, however, this guidance underwent revision, not due to any change of heart by headquarters, but by direction of the Congress of the United States. Evidently, a senator who wished to protect the air fleet stationed at Wallops Island joined forces with members of the House of Representatives from Cleveland and from Hampton wishing to protect the flying assets of NASA Lewis and Langley, respectively. As a consequence, provisions added to HR 103-812--the bill which included NASA's appropriations--prevented the permanent transfer of the space agency's aircraft east of the Mississippi River, effectively leaving just Dryden and Ames in the amalgamation process.[32]

All of the aircraft arrived from Moffett Field during 1997 and early 1998. A majority operated as platforms in Ames' 25 year old Airborne Science program, a high-prestige and high-visibility endeavor designed to monitor and to conduct flying experiments related to the earth's atmosphere and ecosystems, to record celestial observations, and to undertake sensor

---

[32]John Dailey to Directors of Dryden Ames, Langley, and Lewis Research Centers, 23 January 1996, DFRC Historical Reference Collection; Robert Whitehead to Directors of Dryden, Ames, Langley, and Lewis Research Centers, n.d., DFRC Historical Reference Collection (quoted passages); Mitzi Peterson to Ken Szalai, Kevin Petersen, Chuck Brown, Dwain Deets, Gary Krier, Tom McMurtry, Bob Meyer, and Joe Ramos, 24 September 1996, DFRC Historical Reference Collection.

development and satellite sensor verification. On a more utilitarian note, this project also assisted firefighters penetrate smoke and pinpoint hot spots during forest fires. Among the Ames aircraft involved in the transfer, a Lockheed C-130B Hercules landed at Dryden in June 1997. It constituted one of Ames' two Earth Resources and Applications Laboratories. The other one, a McDonnell Douglas DC-8 Super 72, arrived at Edwards during the last days of 1997. The month before, two high-altitude Lockheed ER-2 Earth Resources Survey Aircraft (updated, larger versions of the famous U-2) touched down at DFRC. The DC-8 and ER-2s served as the observation aircraft from which NASA scientists surveyed Antarctica and the Arctic from 1989 to 1992 and arrived at the conclusion that chlorofluorocarbons actively depleted the earth's ozone layer. Finally, Dryden received a smaller airborne sensing laboratory in February 1998, contained aboard a Learjet 24. Collectively named the Airborne Science Program, the aircraft and personnel began operation in DFRC Building 1623 early in 1998. The total complement numbered about 92, including 68 contract employees to support the ER-2s, the DC-8, and to provide overall assistance; and 24 civil servants to staff the program and serve as flight crew.[33]

At first, the new function took root in Dryden's Aerospace Projects Directorate under Deputy Director for Airborne Science Gary Shelton, the former Ames Assistant Director of the Earth Sciences Division. Shelton served previously at NASA headquarters as chief of the Airborne Science Office in the Mission to Planet Earth program. Although the platform aircraft did not represent an entirely new role for Dryden--the later X-15 flights conducted hundreds of

---

[33]Two general purpose aircraft from Ames also joined the Dryden inventory: a Lockheed YO-3A and a Beechcraft Model 200 Super King Air. Anon., *Ames Research Center: The Future Begins Here*, n.d. (about 1993), "Science and Earth Science," 18-19, DFRC Historical Reference Collection; Inventory of Ames Aircraft, compiled by Robert Burns, 3 March 1992, DFRC Historical Reference Collection; "Aircraft Transfers from Ames Research Center," 31 July 1998, compiled by DFRC Historian Dr. J.D. Hunley, DFRC Historical Reference Collection; J.D. Hunley, NASA Facts, "Dryden Historical Milestones," April 1998, 11, DFRC Historical Reference Collection; "Personnel Support Airborne Science at DFRC as of July 29, 1998," compiled by DFRC Historian Dr. J.D. Hunley, DFRC Historical Reference Collection; anon., "Around Center," Dryden *X-Press*," 4, 16 January 1998, DFRC Historical Reference Collection; Dryden News Release, "Airborne Science Flights Begin from NASA Dryden," n.d., DFRC Historical Reference Collection.

high altitude experiments related to exposure to or sampling in the upper atmosphere--a systematic, long-term program aboard specifically configured aircraft distinguished it from past undertakings at DFRC. The work began in January and February 1998 when the two ER-2s completed flights for entirely different clients. Johnson Space Center requested the collection of high altitude particulate matter while Harvard University's Anderson Group booked time to test a new instrument, an atmospheric thermal radiometer designed to read heat radiation more accurately than existing sensors on National Oceanic and Atmospheric Administration (NOAA) satellites. The DC-8, laid up temporarily for maintenance and improvements in its satellite communications system, went into action in April. Researchers at the Jet Propulsion Laboratory (JPL) employed the laboratory to test the imaging and data collection qualities of a sensing instrument known as the Airborne Synthetic Aperture Radar. Flying over the Pacific Northwest, the Gulf Coast, and the Missouri River, the JPL scientists used their device to penetrate forest foliage and cloud cover, obscuring such topographic and geologic features as soils, glaciers, ocean currents, and vegetation.[34]

## AIR AND SPACE

Thus, in its first months of operation, Airborne Science exposed the Dryden staff to a fresh set of contacts at Johnson, at Harvard, and at JPL. These connections promised to multiply with the passage of time, spreading DFRC's flight research practices and results to unforeseeable corners of science and engineering, and, in turn, infusing the center's traditional role with new ideas and new allies. In the broader sense, at the close of the century flight research seemed to

---

[34]Alan Brown, "Shelton Leads Airborne Science Program," Dryden *X-Press*, 3 April 1998, DFRC Historical Reference Collection; Alan Brown, "Dryden Launches Airborne Science Program," Dryden *X-Press*, 20 March 1998, DFRC Historical Reference Collection; Alan Brown, "DC-8 Studies the Earth," 1 May 1998, DFRC Historical Reference Collection; Dryden News Release, Fred Brown, "Airborne Science Flights Begin from NASA Dryden," n.d., DFRC Historical Reference Collection.

experience at least a partial revival, suggested by Dryden's return to institutional independence, by its participation in a major space project predicated on lifting body techniques, and by its assumption of a portion of the historic Ames flight research mission. This modest renewal also embraced a growing recognition of the kinship between aircraft and spacecraft, demonstrated historically in such flight research projects as the X-15, the lifting bodies, the Shuttle Orbiter, and more recently the X-33. Flight research at NASA continued to serve such traditional constituencies as the military services, the aircraft manufacturers, the air carriers, flight safety organizations, and university researchers. But at the same time, Dryden and the other centers showed a willingness to preserve the worthwhile tradition of conceiving much of their own aeronautical research agenda.[35]

---

[35]See earlier in this chapter, Milt Thompson's evaluation of the flying deficiencies of the Space Shuttle Orbiter in the atmosphere.

# REPORT DOCUMENTATION PAGE

Form Approved
OMB No. 0704-0188

| 1. AGENCY USE ONLY *(Leave blank)* | 2. REPORT DATE | 3. REPORT TYPE AND DATES COVERED |
|---|---|---|
| | 09/11/2000 | Contractor Final Report |

| 4. TITLE AND SUBTITLE | 5. FUNDING NUMBERS |
|---|---|
| Separating the Real from the Imagined: Flight Research at the NACA and NASA, 1915-2000 | NASA Purchase Order No. E-2021D |

**6. AUTHOR(S)**

Michael H. Gorn

| 7. PERFORMING ORGANIZATION NAME(S) AND ADDRESS(ES) | 8. PERFORMING ORGANIZATION REPORT NUMBER |
|---|---|
| 1848 Camino Vera Cruz<br>Camarillo, CA 93010 | |

| 9. SPONSORING/MONITORING AGENCY NAME(S) AND ADDRESS(ES) | 10. SPONSORING/MONITORING AGENCY REPORT NUMBER |
|---|---|
| NASA History Division<br>NASA Headquarters<br>Code ZH<br>Washington, DC 20546 | N/A |

**11. SUPPLEMENTARY NOTES**

N/A

| 12a. DISTRIBUTION AVAILABILITY STATEMENT | 12b. DISTRIBUTION CODE |
|---|---|
| Unlimited Distribution, Publicly Available | |

**13. ABSTRACT** *(Maximum 200 words)*

One of the most important, but under-appreciated, aspects of the NACA/NASA mission is its aeronautical R&D efforts. Within a short time of the first flight of the Wright brothers in 1903, the United States government recognized the importance of fostering development in the new and critical field of aeronautics. NASA's predecessor, the National Advisory Committee for Aeronautics (NACA), was chartered by Congress in 1915 specifically "to supervise and direct the scientific study of the problems of flight, with a view to their practical solution." This became an enormously important government research and development activity for the next half century, materially enhancing the development of aeronautics in America. The results of the NACA's research appeared in more than 16,000 research reports of one type or another, distributed widely for the benefit of all. Many of the reports documenting R&D conducted under NACA auspices are still being used today. Since the creation of NASA in 1958, the critical R&D function has continued but is not well known. This work documents the historical R&D program of the agency by focusing on flight research.

| 14. SUBJECT TERMS | | | 15. NUMBER OF PAGES |
|---|---|---|---|
| flight research, National Advisory Committee for Aeronautics (NACA), High Speed Research Facility, X-planes | | | 543 |
| | | | **16. PRICE CODE** |

| 17. SECURITY CLASSIFICATION OF REPORT | 18. SECURITY CLASSIFICATION OF THIS PAGE | 19. SECURITY CLASSIFICATION OF ABSTRACT | 20. LIMITATION OF ABSTRACT |
|---|---|---|---|
| UNCLASSIFIED | UNCLASSIFIED | UNCLASSIFIED | UL |

NSN 7540-01-280-5500

Standard Form 298 (Rev. 2-89)
Prescribed by ANSI Std. 239-1
298-1

www.ingramcontent.com/pod-product-compliance
Lightning Source LLC
Chambersburg PA
CBHW080232180526
45167CB00006B/2249